Coincidence Studies of Electron and Photon Impact Ionization

PHYSICS OF ATOMS AND MOLECULES

Recent volumes in this series:

ATOMS AND LIGHT: INTERACTIONS
John N. Dodd

COINCIDENCE STUDIES OF ELECTRON AND PHOTON IMPACT IONIZATION
Edited by Colm T. Whelan and H. R. J. Walters

DENSITY MATRIX THEORY AND APPLICATIONS, Second Edition
Karl Blum

ELECTRON COLLISIONS WITH MOLECULES, CLUSTERS, AND SURFACES
Edited by H. Ehrhardt and L. A. Morgan

INTRODUCTION TO THE THEORY OF LASER–ATOM INTERACTIONS, Second Edition
Marvin H. Mittleman

INTRODUCTION TO THE THEORY OF X-RAY AND ELECTRONIC SPECTRA OF FREE ATOMS
Romas Karazija

PHOTON AND ELECTRON COLLISIONS WITH ATOMS AND MOLECULES
Edited by Philip G. Burke and Charles J. Joachain

POLARIZATION BREMSSTRAHLUNG
Edited by V. N. Tsytovich and I. M. Ojringel

POLARIZED ELECTRON/POLARIZED PHOTON PHYSICS
Edited by H. Kleinpoppen and W. R. Newell

SELECTED TOPICS ON ELECTRON PHYSICS
Edited by D. Murray Campbell and Hans Kleinpoppen

THEORY OF ELECTRON–ATOM COLLISIONS, Part 1: Potential Scattering
Philip G. Burke and Charles J. Joachain

VUV AND SOFT X-RAY PHOTOIONIZATION
Edited by Uwe Becker and David A. Shirley

A Chronological Listing of Volumes in this series appears at the back of this volume.

Coincidence Studies of Electron and Photon Impact Ionization

Edited by

Colm T. Whelan

University of Cambridge
Cambridge, England

and

H. R. J. Walters

The Queen's University of Belfast
Belfast, Northern Ireland

Plenum Press • New York and London

Library of Congress Cataloging-in-Publication Data

Coincidence studies of electron and photon impact ionization / edited by Colm T. Whelan and H.R.J. Walters.
p. cm. -- (Physics of atoms and molecules)
Includes bibliographical references and index.
ISBN 0-306-45689-3
1. Electron impact ionization--Measurement. 2. Photoionization of gases--Measurement. 3. Photoionization cross sections--Measurement. 4. Helium--Measurement. I. Whelan, Colm T. II. Walters, H. R. J. III. Series.
QC702.7.E38C65 1997
530.4'44--dc21 97-29190
CIP

Proceedings of the European Conference on Coincidence Studies of Electron and Photon Impact Ionization, held September 5 – 7, 1996, in Belfast, Northern Ireland

ISBN 0-306-45689-3

A Division of Plenum Publishing Corporation
233 Spring Street, New York, N. Y. 10013

http://www.plenum.com

10 9 8 7 6 5 4 3 2 1

Printed in the United States of America

PREFACE

The great advantage of coincidence measurements is that by suitable choice of the kinematical and geometrical arrangement one may probe delicate physical effects which would be swamped in less differential experiments. The measurement of the triple differential and higher-order cross sections presents enormous technical difficulties, but refined experiments of this type provide an insight into the subtleties of the scattering process and offer a welcome, if severe, test of the available theoretical models. The last few years have been an exciting time to work in the field and much has been learned. Profound insights have been gleaned into the basic Coulomb few body problem in atomic physics: the experimental study of the fundamental (e,2e) processes on hydrogen and helium targets continues to add to our knowledge and indeed to challenge the best of our theoretical models; significant advances have been made in the understanding of the "double excitation problem," that is the study of ionization processes with two active target electrons: important measurements of (e,3e), (γ,2e), excitation-ionization and excitation autoionization have been reported and strides have been made in their theoretical description; the longstanding discrepancies between theory and experiment for relativistic (e,2e) processes were resolved, spin dependent effects predicted and observed and the first successful coincidence experiments on surfaces and thin films were announced. Theory and experiment have advanced in close consort. The papers presented here cover the whole gambit of research in the field. Much has been achieved but much remains to be done.

The European Union has been at the forefront in support of this area of research and in particular funded a series of three annual Euroconferences which took place in Colmar, Trieste and Belfast. It is the papers contributed to the Belfast meeting that constitute the basis of this book. While we attracted a large number of very senior scientists both from Europe and beyond to these meetings it has been the practice to place a heavy emphasis on the participation of younger scientists. It was felt very important that they should participate fully and with this in mind prizes were created "The Ehrhardt Prizes" to be awarded to two young scientists in the early stages of their research career who had made the most significant contribution to coincidence studies. There was no restriction on nationality and a prize could be awarded to either a theoretician or an experimentalist. The 1996 prizes were announced at the meeting; one went to Steve Jones of the University of Missouri-Rolla and the other to Jens Rasch of the University of Cambridge.

The Conference was primarily funded through a Euroconference grant (ERB 5050 PL934136) from the European Union and further financial support was given by the British Engineering and Physical Science Research Council (EPSRC) through the "Computational Collaborative Project No 2: Continuum States of Atoms and Molecules" program. We are also grateful to Belfast City Council, whose generosity helped to make the whole meeting an extremely pleasant event. Sincere thanks are also due to Professor Stefani, of the University of Rome, who had the onerous responsibility of being the overall coordinator of the series of Euroconferences, a task he accomplished with his usual exemplary efficiency and courtesy. A large number of people helped with the organization; among these special thanks are due to Robert Allan, Jens Rasch, and Shaun Lucey who, amongst other things, put us on the Web and Eoin Whelan who designed our logo.

Colm T. Whelan
University of Cambridge

H. R. J. Walters
The Queen's University of Belfast

CONTENTS

RECENT ABSOLUTE (e,2e) MEASUREMENTS ON ATOMIC HYDROGEN AND HELIUM AT LOW AND INTERMEDIATE ENERGIES

H. Ehrhardt and J. Röder

Fachbereich Physik
Universität Kaiserslautern
D-67663 Kaiserslautern
Germany

INTRODUCTION

Electron impact ionization of atoms is one of the most important processes in terrestrial and extraterrestrial plasmas. Especially atomic hydrogen and helium are of interest, not only since these targets are the most abandoned atoms in the universe, but also because they are simple and comparison to theory is quite valuable. In addition $H + e^- \rightarrow p + e^- + e^-$ is one of the few true three particle processes in nature in which the three particles have long range forces and are in the continuum. Other important three particle problems of this type are

$$He + h\nu \rightarrow \alpha + e^- + e^-$$

$$He + p \rightarrow p + p + e^-$$

and similar positron processes. But positron collisions are difficult to measure in absolute units and in details such as differential cross sections.

In the last few years quite several theories for cross sections of electron impact ionization of atomic hydrogen and helium at low and intermediate impact energies between 15.6 eV and 54.4 eV have been published[1-5]. Especially at intermediate impact energies and for more complex atoms the theories are extremely complicated since quite different processes occur and relevant amplitudes of a rather large number of partial waves interfere. Such processes are the capture of the colliding electron and the ejection of two atomic electrons, the exchange of momentum between the participating electrons and the polarisation of the ionic core by the two outgoing electrons. Of course the dominating long range interactions are the electron-electron and the electron-ion Coulomb interactions. The theoretical description of the electron impact ionisation of atomic hydrogen is much simpler since some of the interactions (e.g. particle exchange and core polarisation) are not present and the atomic wave function is known exactly.

The comparison of theory with experiments is especially valuable if triple differential cross sections

$$\frac{d^3\sigma}{d\Omega_1 d\Omega_2 dE}$$

in absolute units are compared. The measurement of cross sections in absolute units is of greatest importance since quite often theory is able to reproduce the shapes of the cross sections as functions of both angles of the outgoing electrons, but large discrepancies (as large as a factor 100 or even more) in absolute units are observed.

The easiest target atom for theory, namely the hydrogen, is the most difficult for the experimentalist. The molecular hydrogen has to be dissociated by a microwave discharge at a high rate, then the atomic hydrogen has to be conducted to the scattering center of the electron impact spectrometer, a beam of atomic hydrogen has to be formed without too many recombination processes into molecular hydrogen at the walls of the gas beam nozzle, the H-concentration in the gas beam has to be measured etc. There is only one machine in the world which can measure the coincidence cross sections of the process

$$H + e^- \rightarrow p + e^- + e^-$$

in the physically most interesting range of impact energies from threshold to about 100 eV. This machine was built and operated up to now at the University of Kaiserslautern in Germany, but we have transferd it a few weeks ago to the group of Professor Pochat and Professor Robaux in Brest (France). As we have heard, the machine is already working well and producing reliable absolute triple differential cross sections.

Nowadays we can measure triple differential cross sections for many collision parameters such as the energies of the incident and outgoing electrons, in and out-of-plane scattering with for instant constant angle between the outgoing electrons at a given pair of energies etc. This variety of parameter choices enables to test the effects of different interactions. Under certain symmetries the singlet cross sections are zero, for a difference angle of the two outgoing electrons of 180° the e^--e^--interaction is very small and the e^--ion-interactions are large. This advantage of selecting quite different scattering geometries is only valuable if closest cooperation between theorists and experimentalists is given and if a plan of procedure with clear purposes is agreed on in advance. Otherwise the variety of results is extremely large and interpretation is nearly impossible.

EXPERIMENTAL METHOD AND SET-UP

Figure 1 gives a schematic drawing of the apparatus. An electron beam of about 400 nA produced by an electron gun crosses a gas jet at right angle, thus defining the collision region. The two emerging electrons from the scattering center are detected in coincidence using two identical electrostatical 127°-tandem-analysers whose angular positions with respect to the electron beam can be varied independently from 30° to about 150°. Additionally one analyser is capable to move up to 90° out of the scattering plane so that three degrees of freedom in the angular motion of the spectrometer are being covered. The overall energy resolution of the apparatus is in the order of 180 meV for the detection of two coincidence electrons.

The atomic hydrogen is prepared by a microwave generator of 2.45 GHz and 150 Watt. Operation duration (before getting dirty and consequently strong recombination of atoms on the discharge walls) is approximately 100 hours. The length of the teflon tube

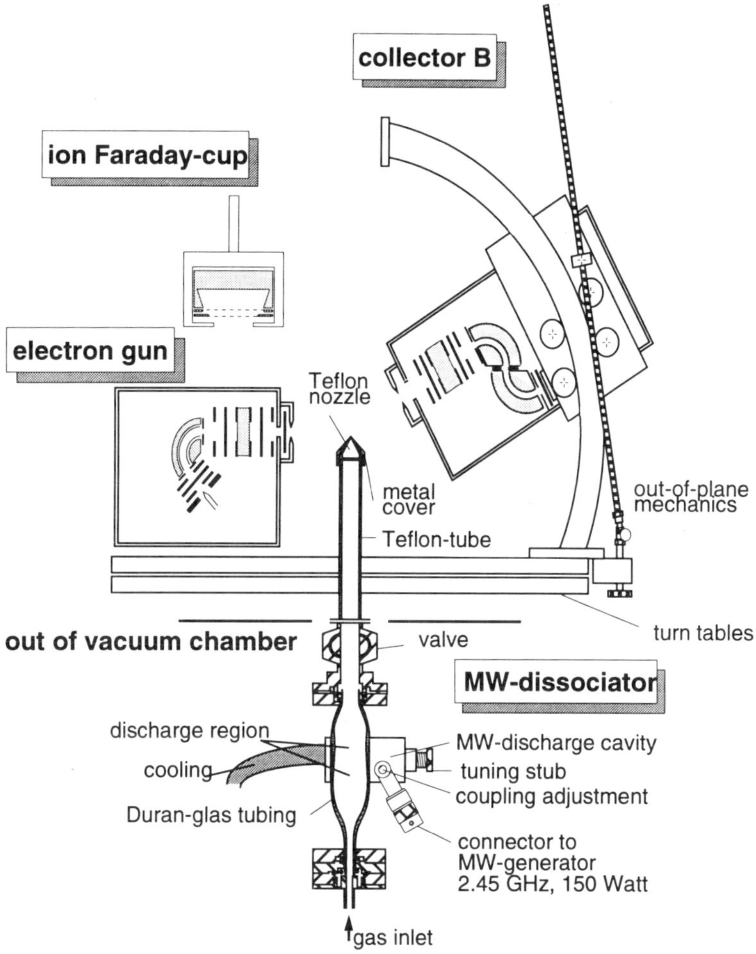

Figure 1. Schematical drawing of the apparatus giving a side view. The electron Faraday cup and the second detector are not shown in order to have a better overview.

between dissociator and gas nozzle is about 40 cm, but still about 20 to 40% of the H_2 is dissociated. The ion Faraday-cup can be moved above the gas nozzle in order to measure the ion current, which is proportional to the total ionisation cross section σ_{tot} (see below).

The method for the determination of absolute cross sections depends on the fact that the rate of electrons detected in coincidence is proportional to the triple differential cross section (TDCS), where the factor of proportionality can be subdivided into three terms describing the properties of the interaction region and both analysers as well as the coincidence energy resolution ΔE_{TDCS}:

$$N_{TDCS} = \frac{d^3\sigma}{d\Omega_1 d\Omega_2 dE_2} \cdot (n \cdot N_e \cdot l) \cdot (\Delta\Omega \cdot \varepsilon)_1 \cdot (\Delta\Omega \cdot \varepsilon)_2 \cdot \Delta E_{TDCS}$$

The factor $(n \cdot N_e \cdot l)$ with n being the target density, N_e the rate of electrons in the electron beam and l the scattering length can be derived from

$$N_{ion} = \sigma_{tot} \cdot (n \cdot N_e \cdot l)$$

where σ_{tot} is the total ionization cross section and N_{ion} the rate of ions measured by the Faraday-cup. The expression

$$N_{DCSD} = \frac{d^2\sigma}{d\Omega dE} \cdot (n \cdot N_e \cdot l) \cdot (\Delta\Omega \cdot \varepsilon) \cdot \Delta E_{DDCS}$$

for the double differential cross section can be used to determine the analyser properties $(\Delta\Omega \cdot \varepsilon)_{1,2}$. The double differential cross sections have been determined by us earlier[6,7]. The energy resolutions can be taken from the measurements of elastic peaks. The overall error for the normalization of the TDCS is estimated to be 25%[8], a rather low value if one keeps in mind the number of different determinations in the whole procedure.

Triple differential cross sections at higher impact energies (larger than ca. 150 eV) can be brought on absolute scale by extrapolation of the generalized oscillator strength to zero momentum transfer and comparison with the well-known cross sections for photoionization[9].

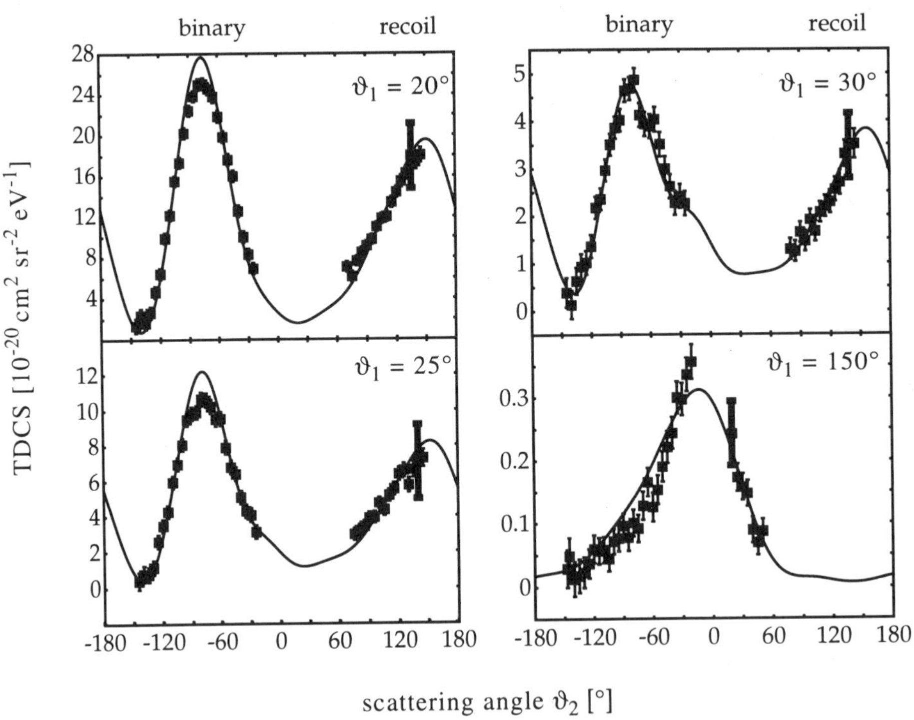

Figure 2. Triple differential cross section in coplanar asymmetric geometry for helium with primary energy $E_0 = 100$ eV, $E_1 = 71.5$ eV and $E_2 = 4$ eV. Large errorbars show the errors in the determination of the absolute values, full lines are the CCC-theory[10] results.

RESULTS AND DISCUSSION

Ionisation of Helium.

Figures 2, 3 and 4 show some results of triple differential cross sections of helium in absolute units. The energy of the impact electrons is 100 eV (Figure 2), 40 eV (Figure 3} and 32.6 eV (Figure 4), i.e. in an low to intermediate energy range where all the interactions are of importance which have been discussed above. The energy of one of the outgoing electrons was choosen to be 4 eV in all the cases, resulting in asymmetric conditions for the incident energy of 100 eV and equal energy sharing conditions for the impact energy of 32.6 eV. The full lines are results of the Convergent Close Coupling (CCC) theory of Bray et al.[10]. The angle ϑ_1 is the angle of the fast ("scattered") electron in the scattering plane with respect to the direction of the incoming electron, the angles ϑ_2 and φ_2 are the polar angles of the slow ("ejected") electron. The positions of the binary and the recoil peak are indicated. The most important results of the three figures are as follows:

Figure 2: The agreement between theory and experiment is perfect (within the error bars of the experiment). The theoretical problem is dominated by the high value of E_0 and E_1 compared to E_2. This means that some interactions have died out already due to high E_0, e.g. exchange, distortion in the incoming channel. The high value of E_1/E_2 means that electron 1 moves (for large distances) essentially field free since electron 2 and the ionic core resemble

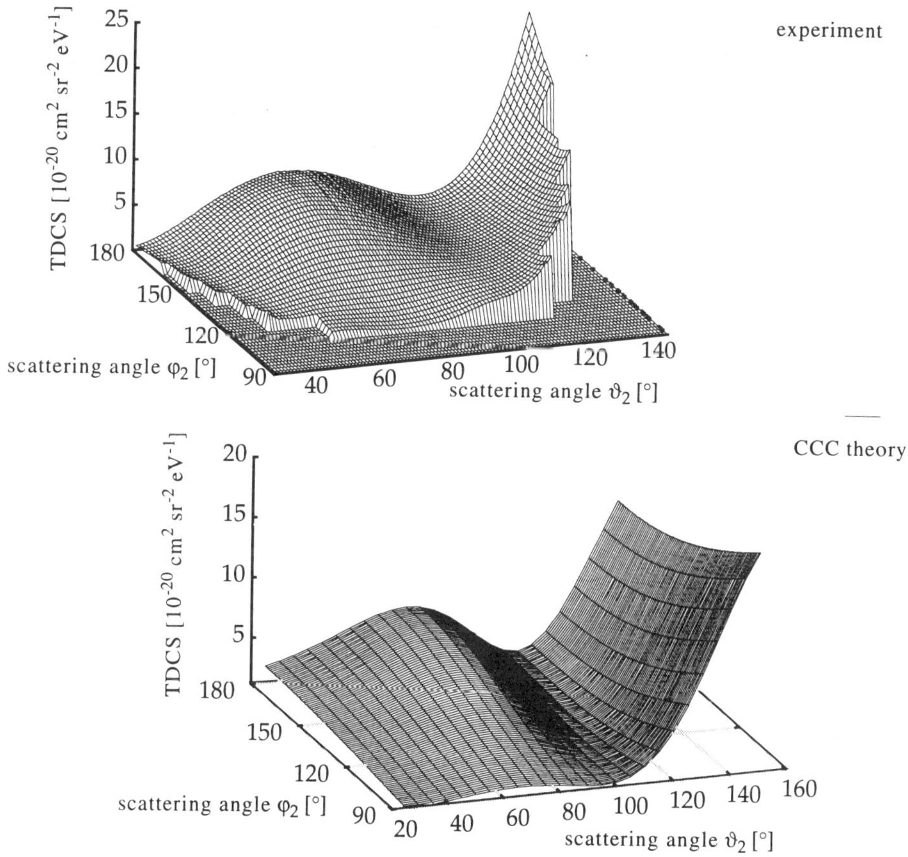

Figure 3. Triple differential cross section in out-of-plane geometry for helium with primary energy $E_0 = 40$ eV, $E_1 = 11.5$ eV and $E_2 = 4$ eV. The curve $\varphi_2 = 180°$ corresponds to in-plane scattering.

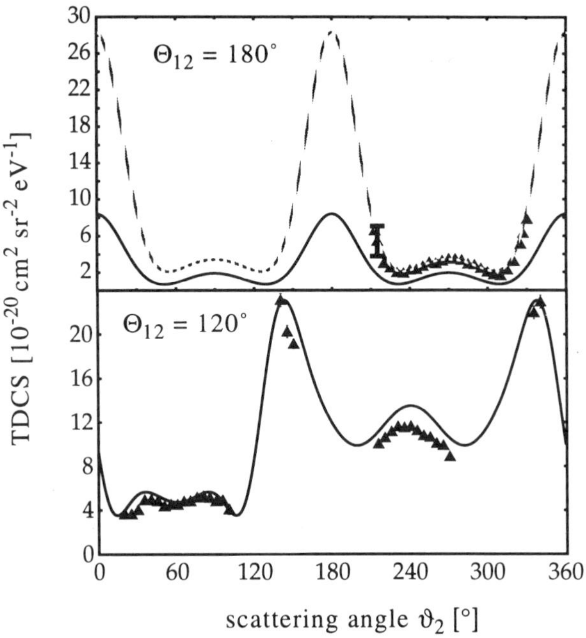

Figure 4. Triple differential cross section in coplanar fixed Θ_{12} geometry for helium with primary energy $E_0 = 32.6$ eV, $E_1 = E_2 = 4$ eV. The big errorbar shows the error in the determination of the absolute values, full line is the CCC-theory[10], dashed line is the DW theory of Pan and Starace[2,16]. In the case Θ_{12} = 120°the results of the CCC theory have been multiplied by about a factor of three.

a neutral atom, or somewhat different, the charge of the slow electron is shielded by the charge of the ion core. As a result of the screening the electron-electron interaction is weak. Of course this argument is only valid for large distances, for small distances the full problem has to be treated.

Figure 3: The TDCS in out of plane geometry determined by the experiment for $E_0 = 40$ eV and $E_2 = 4$ eV is compared with the results of the CCC theory.The situation is still similar to Figure 2, although the primary energy E_0 is reduced and also the ratio E_1/E_2. But still $E_1 \approx 3 \cdot E_2$ and electron 1 moves (for large distances) in a strongly shielded Coulomb field. The three-dimensional plots are indicating, that the backward scattering of the slow electron is also strong out of the scattering plane. Agreement between theory and experiment is very good, both with respect to absolute values as well as with respect to the angular dependencies.

Figure 4: The primary energy E_0 is 8 eV above threshold and the energies E_1 and E_2 are equal. The electron-electron interaction is practically not shielded and essentially of importance till infinitely long distances. The measurements made for fixed angles Θ_{12} of the outgoing electrons are compared with the results of the CCC theory and the Distorted Wave calculation of Pan and Starace[2,16]. The angular dependencies from theory and experiment are still very similar. For the absolute values the situation is different. The cross section calculated by Pan and Starace for Θ_{12} = 180° is in very good agreement with the experimental findings, whereas the results of the CCC theory are about a factor of three smaller. This behaviour is to be seen for all cross sections with Θ_{12} = 180° and energy sharing conditions from primary energies of about $E_0 = 100$ eV down to threshold. Whereas there is good agreement between the measurements and the theoretical data of Pan and Starace, the CCC results are always smaller and the difference factor growth with smaller primary energies. These

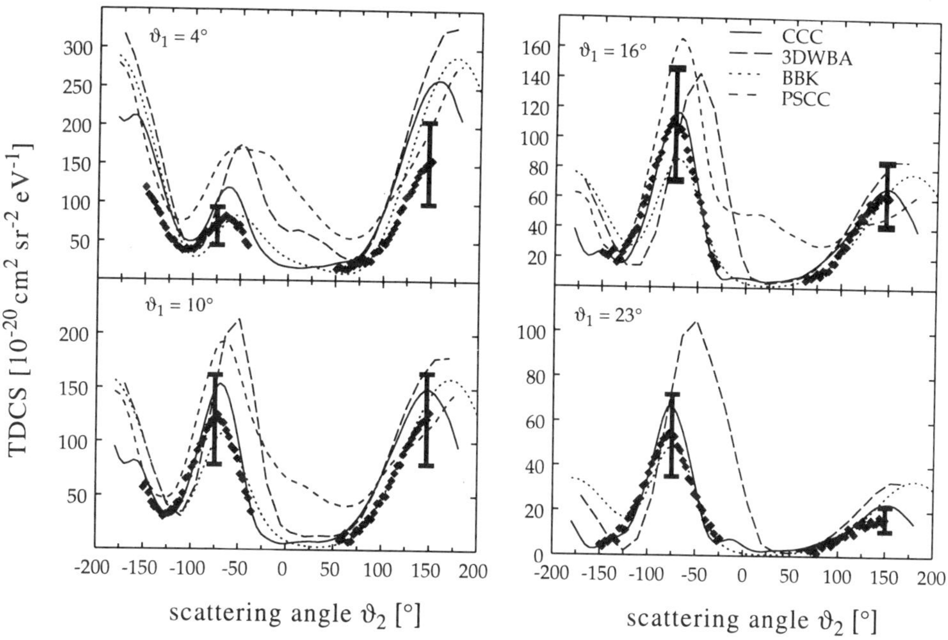

Figure 5. Triple differential cross section for the electron impact ionization of atomic hydrogen with primary energy E_0 = 54.4 eV, E_1 = 35.8 eV and E_2 = 5 eV. The big errorbar shows the error in the determination of the absolute values, full line is the CCC-theory of Bray et al.[10], long dashed line is the 3DWBA theory of Jones et al.[12], dotted line is the BBK theory of Brauner et al.[13], short dashed line is the PSCC theory of Curran et al.[14]

findings are consistent with the situation found for the determination of absolute single and double differential cross sections, where the CCC results deviate by about the same values from the experiment as for triple differential cross sections for the same kinematical conditions. Given the very good agreement in the angular dependencies found by the experiment and calculated by the CCC theory this shows the importance of absolute measurements, especially since the CCC results for the total ionisation cross section are in very good agreement with available experiments. Recent TDCS calculations of Berakdar et al.[4] for the primary energy of 32.6 eV, $\Theta_{12} = 120°$ and energy sharing conditions are in the angular range $0 \leq \vartheta_2 \leq 140°$ in very good agreement with the experimental data shown in Figure 4. Even if there are some differences in the angular dependencies for the larger angels, the further development of the BBK[13] theory by Berakdar and coworkers have overcome the difficulties of this theory in calculating much to small cross sections at low energies.

Ionisation of Hydrogen.

Figures 5 represent experimental results of triple differential cross sections in absolute units for atomic hydrogen. For Figures 5 the primary energy of 54.4 eV can be called intermediate since it is four times the ionization potential (IP). As in Figures 2 the ratio E_1/E_2 is very different. The agreement between experiment and the CCC-theory is also good in this case but also the results of the BBK theory are within the experimental errorbars. Since the values of the BBK theory tend to be smaller than the experimental data it would be interesting to see calculations of Berakdar et al. for this case, which can indicate in which energy regime the problems with absolute values start to arise for the BBK results. The other shown theoretical results in modified Distorted Wave Approximation[12] and with a Pseudostate

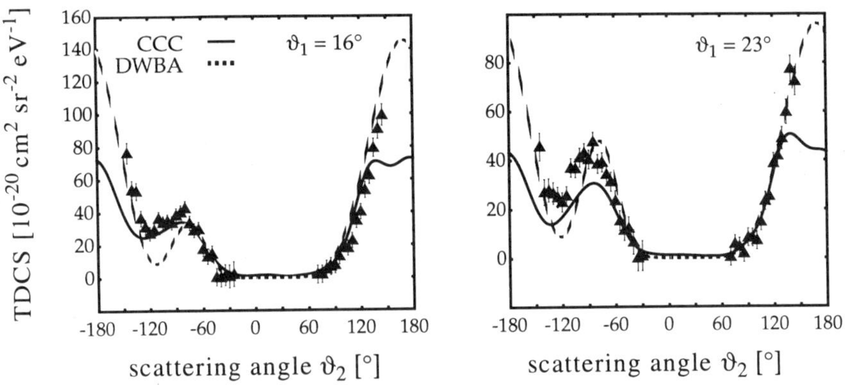

Figure 6. Triple differential cross section for the electron impact ionization of atomic hydrogen with primary energy $E_0 = 27.2$ eV, $E_1 = 9.6$ eV and $E_2 = 4$ eV. Full line is the CCC-theory of Bray et al.[10], dashed line is the DWBA theory of Rasch et al.[15]

Close Coupling formalism[14] are not able to reproduce the experimental data. Attention should be paid to the wings in the recoil peak calculated by the CCC theory at about $\vartheta_2 = -160°$.

In Figure 6 measurements of the triple differential cross section of atomic hydrogen at the primary energy of 27.2 eV with $E_2 = 4$ eV are shown. The relative results are as well as the DWBA calculations of Rasch et al.[14] normalised to the CCC theory. Different to the situation for helium as targetgas the CCC theory gives now results which are not in a good agreement with the experiments concerning the angular behaviour of the cross sections. This is due to the wings in the recoilpeak, which are not shown in the experimental data or the results of the DWBA, and which are now stronger than for the case shown in figure 5.

The situation with respect to comparison between theory and experiment is even more dramatic in Figure 7. Results are plotted for low primary energy electrons (E_0= 17.6 eV) on atomic hydrogen and $E_1 = E_2 = 2$ eV. The different theoretical results reproduce all very good the angular dependencies, if Θ_{12}, the angle difference (ϑ_1-ϑ_2) is close to 180°. This kinematics means that the two electrons move away from the remaining ion in opposite direction, the electron-electron interaction is very weak. Deviations between theory and experiment with respect to angular dependencies arise for Θ_{12} = 120° to 90° as to be seen for the case Θ_{12} = 90°. The absolute values deviate dramatically. In Figure 8 the theoretical values have been adjusted to experiment by normalizing the theory by a factor between 0.6 and 4. But also every theory has to be adjusted with different normalisation values for different angles Θ_{12}. This figure also shows the importance of absolute measurements and indicates, that the situation is worse in comparison to helium as target, since there the renormalisation of the theory is only depending of the energies of interest and not of the angular dependencies looked for these energies.

CONCLUSIONS

In the last few years enormous progress has been made in calculating triple differential cross section of electron impact ionization of helium and atomic hydrogen. Angular dependencies are in good agreement with new experiments for nearly all collision parameters. Large discrepancies still exist for low impact energies and kinematics in which Θ_{12} is small and the energies of the two outgoing electrons are similar, i.e. E_1/E_2 close to 1 or 2. It may be trivial to stress that the electron-electron interaction in the continuum is the most important reason

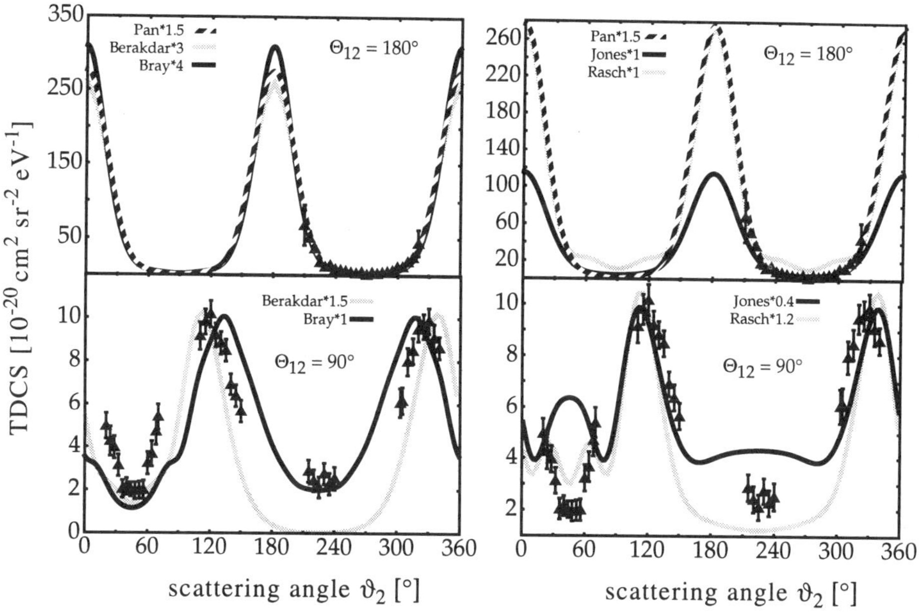

Figure 7. Triple differential cross section for the electron impact ionization of atomic hydrogen with primary energy $E_0 = 17.6$ eV and $E_1 = E_2 = 2$ eV. The theoretical data is renormalised to the experimental data with the factors given in the figure. Theories as indicated: Berakdar[17], Bray et al.[18], Jones et al.[1] ,Pan et al.[2] and Rasch et al.[15] .

for deviations between theoretical predictions and the real world.

ACKNOWLEDGEMENT

The authors acknowledge very much the continuous support of the Deutsche Forschungsgemeinschaft for this type of low energy collision physics.

REFERENCES

1 S. Jones, D.H. Madison, and M.K.Srivastava,
J. Phys. B: At. Mol. Opt. Phys. 25:1899 (1992)
2 C. Pan and A.F. Starace, Phys. Rev. 45: 4588 (1992)
3 J. Röder, J. Rasch, K. Jung, C.T. Whelan, H. Ehrhardt, R.J. Allan, and H.R.J. Walters, Phys. Rev. A 53: 225 (1996)
4 J. Berakdar, J. Röder, J.S. Briggs, and H. Ehrhardt,
J. Phys. B: At. Mol. Opt. Phys. (1996)
5 J. Röder, H. Ehrhardt, I. Bray, D. Fursa, and I.E. McCarthy,
J. Phys. B: At. Mol. Opt. Phys. (1996)
6 K. Jung, R. Müller-Fiedler, P. Schlemmer, H. Ehrhardt, and H. Klar,
J. Phys. B: At. Mol. Opt. Phys. 18: 2955 (1985)
7 R. Müller-Fiedler, K. Jung and H. Ehrhardt,
J. Phys. B: At. Mol. Opt. Phys. 19 (1986) 1211
8 T. Rösel, J. Röder, L. Frost, K. Jung and H. Ehrhardt,
J.Phys. B: At. Mol. Opt. Phys. 25 (1992) 3859

9 R. Müller-Fiedler, P. Schlemmer, K. Jung and H. Ehrhardt, Z. Phys. A-Atoms and Nuclei 320 (1985) 89
10 I. Bray and D. Fursa, Phys. Rev. Lett. 76: 2674 (1996)
11 I. Bray, D.A. Konovalow, and I.E. McCarthy, Phys. Rev. A 50: 2818 (1994)
12 S. Jones, D.H. Madison, A. Franz, and P.L. Altick, Phys. Rev. A 48: 22 (1993)
13 M. Brauner, J.S. Briggs, H. Klar, J.T. Broad, T. Rösel, K. Jung, and H. Ehrhardt, J. Phys. B: At. Mol. Opt. Phys. 24: 657 (1991)
14 E.P. Curran and H.R. Walters, J. Phys. B: At. Mol. Opt. Phys. 20: 337 (1987)
15 Rasch and Whelan (University of Cambridge, U.K.) (1995), private communication
16 C. Pan and A.F. Starace (1995), private communication
17 J. Berakdar (1995), private communication
18 I. Bray and D. Fursa (1995), private communication

SECOND BORN CALCULATIONS OF (e-2e) CROSS SECTIONS AT LOW ENERGY USING A PSEUDOSTATE SET

F. Rouet, G. Nguyen Vien, R. J. Tweed, O. Robaux, J. Langlois

Laboratoire des Collisions Electroniques et Atomiques
Faculté des Sciences et des Techniques
6, avenue Le Gorgeu, B.P. 809
29285 Brest cedex, France

INTRODUCTION

Here we present a novel method for the calculation of cross sections for electron impact ionization of atomic targets. Calculations are carried out in coplanar geometries for incident energies lower than 54.4eV and a hydrogen target. Many theories have been proposed over the years to solve the atomic ionization problem, most of them employing variations of the Born series. Among the first Distorted-Wave Born Approximation (DWBA) calculations for ionization of atomic hydrogen are those of Madison et al.[1] and Bransden et al.[2] where post-collision interaction (PCI) between the scattered and ejected electrons was included only via shielding of the nucleus by the slower one. This approximation is valid when the energies of the outgoing electrons are very different. In fact, the major problem for the final state is how to take account of the Coulomb repulsion between a pair of final state free electrons. An attempt to solve this problem was made by Brauner, Briggs and Klar[3] who proposed an approximation for the collision wavefunction appropriate to the final state of the system. This wavefunction satisfies the Schrödinger equation exactly in the asymptotic region but its use involves rather complicated and lengthy calculations. Some simplified treatments have been proposed. Dal Cappello and Joulakian[4] retained only the angular dependent normalisation factor of the BBK ansatz. This approach is equivalent to the treatment of Whelan et al.[5] where the dominant angular behaviour of the TDCS at low energies due to electron-electron interaction in final state is taken into account by the the Gamow factor N_{ee}.

Other work has concentrated on improvement of the initial state wavefunction so as to make it as full as possible a description of the collisional system as a whole, including all reaction channels. Curran and Walters[6] used a close-coupled wavefunction for the projectile-target system with a basis including continuum pseudostates. Bray and Fursa[7] made a much more elaborate calculation including a very large pseudostate basis. This has the effect of taking account of PCI through coupling between continuum based channels.

Coincidence Studies of Electron and Photon Impact Ionization
Edited by Whelan and Walters, Plenum Press, New York, 1997

In this work, we adopt a hybrid approach in which we combine a pseudostate distorted wave second Born treatment taking account of ante-collisional effects (elastic and inelastic scattering of the colliding electron prior to ionization of the target) and the N_{ee} correction compensating for neglect of the long-range part of the PCI. The second Born approximation and its variants (modified Glauber, Eikonal Born and unitarised Eikonal Born) have been shown by Byron et al.[8,9,10] to give a satisfactory treatment of ionization at 150 and 250 eV in coplanar asymmetric geometries. They demonstrate show the importance of the second Born terms and give results in good agreement with experiment in the case of the ionization of hydrogen[11,12] and of helium.[13] These models use the closure approximation in which all the intermediate target states are associated with the same energy denominator (corresponding to an average state energy of zero). The closure approximation is probably valid only in the case when the energy of the ejected electron is much smaller than that of the scattered electron and would be unadapted to other types of kinematics.

The method which we propose for calculation of the TDCS in coplanar symmetric and asymmetric geometries at low energy incorporates the second Born term by use of a pseudostate approximation : the infinite set of target intermediate states required by the development of the second order scattering amplitude is replaced by a small number of exact states and bound and continuum pseudostates. The pseudostate approximation respects the differences between the energy denominators but spans only a part of phase-space corresponding to the basis set used for its development, generally confined to a finite volume. It is complete in this subspace. In this paper, we obtain pseudostates by diagonalisation of the atomic hamiltonian on a basis including Slater-type orbitals, hydrogenic functions and harmonic oscillator-type functions. In the previous work of Rouet et al.,[14] we have shown that the target polarisation plays an important role in the low energy collisions. So the pseudostates are chosen to give the good polarisability of the exact states and to satisfy the Thomas-Reiche-Kuhn sum rule.

THEORY

Since we are interested in low energy collisions, we have to account for exchange. The triple differential cross section is then given by

$$\frac{d^3\sigma}{d\Omega_a d\Omega_b dE} = \frac{k_a k_b}{4k_0} \sum_S (2S+1)|f^S(\vec{k_a}, \vec{k_b})|^2 \tag{1}$$

The scattering amplitude for channels with total spin S is

$$f^S(\vec{k_a}, \vec{k_b}) \simeq f^S_{B1}(\vec{k_a}, \vec{k_b}) + f^S_{B2}(\vec{k_a}, \vec{k_b}) \tag{2}$$

where $f^S_{B1}(\vec{k_a}, \vec{k_b})$ and $f^S_{B2}(\vec{k_a}, \vec{k_b})$ are respectively the Born 1 and the Born 2 contributions. To each order we have a direct and an exchange amplitude

$$\begin{cases} f^S_{B1}(\vec{k_a}, \vec{k_b}) = f_1(\vec{k_a}, \vec{k_b}) + (-1)^S g_1(\vec{k_a}, \vec{k_b}) \\ f^S_{B2}(\vec{k_a}, \vec{k_b}) = f_2(\vec{k_a}, \vec{k_b}) + (-1)^S g_2(\vec{k_a}, \vec{k_b}) \end{cases} \tag{3}$$

The direct scattering amplitudes are given by

$$\begin{cases} f_1(\vec{k_a}, \vec{k_b}) = -4\pi^2 < \Psi^-_{f,\vec{k_a},\vec{k_b}} | V_f | \Psi^+_{i,\vec{k_0}} > \\ f_2(\vec{k_a}, \vec{k_b}) = -8\pi^2 < \Psi^-_{f,\vec{k_a},\vec{k_b}} | V_f G^+_i V_i | \Psi^+_{i,\vec{k_0}} > \end{cases} \tag{4}$$

The exchange amplitudes g_1 and g_2 are given by exchanging $\vec{k_a}$ and $\vec{k_b}$ in these expressions. The initial state wavefunction is approximated by an anti-symmetrized product of the target ground state wavefunction $\psi_{100}(\vec{r})$ and a continuum wavefunction $\chi(\vec{k_0}, \vec{r'})$, where $\vec{k_0}$ is the momentum of the incident electron

$$\Psi_{i,\vec{k_0}}(\vec{r}, \vec{r'}) = \mathcal{A}\psi_{100}(\vec{r})\chi(\vec{k_0}, \vec{r'}) \tag{5}$$

The partial wave expansion of the incident wavefunction is

$$\chi(\vec{k_0}, \vec{r'}) = \sqrt{\frac{2}{\pi}} \frac{1}{k_0 r'} \sum_{l'm'} i^{l'} e^{i\delta_{l'}} Y^*_{l'm'}(\hat{k_0}) Y_{l'm'}(\hat{r}') F_{l'}(k_0, r') \tag{6}$$

Substitution into the Schrödinger equation leads to the following integro-differential equation satisfied by the radial function associated with each angular momentum and spin

$$\left[\frac{\partial^2}{\partial r'^2} - \frac{l'(l'+1)}{r'^2} + k_0^2 + V_s(r')\right] F_{l'}(k_0, r') + \int W_{se}(r, r') F_{l'}(k_0, r) dr = 0 \tag{7}$$

where the static potential $V_s(r')$ and the static exchange potential operator $W_{se}(r, r')$ are as defined by McDowell et al.[15] The integro-differential equation (7) is solved using their single channel distorted wave polarised orbital method where the normalization and the phase-shifts are obtained by matching to a JWKB solution.

The final state wavefunction is an anti-symmetrized product of two Coulomb functions corresponding to the momenta $\vec{k_a}$ et $\vec{k_b}$ respectively

$$\Psi_{f,\vec{k_a},\vec{k_b}}(\vec{r}, \vec{r'}) = \mathcal{A}\phi_{\vec{k_a}}(\vec{r'})\phi_{\vec{k_b}}(\vec{r}) \tag{8}$$

The electrostatic repulsion between the two outgoing electrons is neglected in the calculation of the matrix elements but the cross sections are corrected afterwards through the so called Gamov factor

$$N_{ee} = \frac{\gamma}{e^\gamma - 1} \tag{9}$$

with

$$\gamma = \frac{2\pi}{|\vec{k_a} - \vec{k_b}|} \tag{10}$$

PSEUDOSTATES

The second order contribution to the scattering amplitude can be considered as arising from a transition via an intermediate state. The process can be pictured in the following way

$$e^-(\vec{k_0}) + H(1s) \longrightarrow e^-(\vec{k_\alpha}) + H(\alpha) \longrightarrow e^-(\vec{k_a}) + e^-(\vec{k_b}) + H^+ \tag{11}$$

with

$$k_\alpha^2 = k_0^2 + 2E_0 - 2E_\alpha \tag{12}$$

where E_α is the energy of the intermediate state of the target. The spectral decomposition of the Green's operator is given by

$$G_i^+ = \sum_\alpha \lim_{\epsilon \to 0^+} \int d\vec{k'} \frac{|\Psi_{\vec{k'},\alpha} >< \Psi_{\vec{k'},\alpha}|}{k_\alpha^2 - k'^2 + i\epsilon} = \sum_\alpha |\alpha >< \alpha| G_\alpha^+ \tag{13}$$

with

$$< \vec{r'}|G_\alpha^+|\vec{r}> = -\frac{k_\alpha}{4\pi}\sum_{l'=0}^{\infty}(2l'+1)P_{l'}(\hat{r},\hat{r'})F_{l'}(k\alpha,r_<)\{-G_{l'}(k\alpha,r_>)+iF_{l'}(k\alpha,r_>)\} \quad (14)$$

where $F_{l'}$ and $G_{l'}$ are respectively the regular and irregular solutions of a differential equation of the type

$$\left[\frac{\partial^2}{\partial r'^2} - \frac{l'(l'+1)}{r'^2} + k^2 + V_{\alpha l'}(r')\right] F_{l'}(k,r') = 0 \quad (15)$$

All the intermediate states, belonging to the discrete or to the continuous spectrum, are taken into account through the summation over α.

An approximate representation of the Green's operator may be obtained by replacing k'^2 by an average value. The calculation is then highly simplified by using the closure relation satisfied by the target states. We prefer to use a set of pseudostates to provide a restricted representation of the target. In the calculation presented here we use:

1. the exact states $1s$, $2s$, $2p$, $3s$, $3p$ and $3d$,
2. two pseudostates for each angular momentum, representating the rest of the discrete spectrum,
3. three continuum pseudostates for each angular momentum.

This is hereafter called basis 1.

The radial part of the wavefunctions associated with these pseudostates are developped on a basis including:

1. Slater-type orbitals (STO)

$$\Phi_{STO}(n_1,\xi_1,r) = \frac{(2\xi_1)^{n_1+\frac{3}{2}}}{\sqrt{(2n_1+2)!}} r^{n_1+1}e^{-\xi_1 r} \quad (16)$$

2. Slater-Laguerre functions (SLF)

$$\Phi_{SLF}(n_2,l,\xi_2,r) = \frac{2}{n_2^2}\sqrt{\frac{(n_2-l-1)!(\xi_2)^3}{[(n_2+l)!]^3}}\, r\left(\frac{2\xi_2 r}{n_2}\right)^l e^{-\frac{\xi_2 r}{n_2}} L_{n_2-l-1}^{2l+1}\left(\frac{2\xi_2 r}{n_2}\right) \quad (17)$$

3. and harmonic oscillator-type functions (HOF)

$$\Phi_{OSH}(n_3,l,\xi_3,r) = \left[\frac{2(\xi_3)^{\frac{3}{2}}n_3!}{\Gamma(n_3+l+\frac{3}{2})}\right]^{\frac{1}{2}} r(\xi_3 r^2)^{\frac{l}{2}} e^{-\frac{\xi_3 r^2}{2}} L_{n_3}^{l+\frac{1}{2}}(\xi_3 r^2) \quad (18)$$

For each trial set of n_1,n_2,l, the exponents ξ_1, ξ_2 et ξ_3 are considered as ajustable parameters whose values are optimized so as to make the pseudo-energies coincide with the abscissae corresponding to the Gauss-Laguerre quadrature. In fact more stringent requirements are imposed as discussed below.

The usual variational principle is used to establish matrix equations for the expansion coefficients of the wavefunction.[16] Thus if $\mathcal{C}_\alpha$ is the vector containing the coefficients for the states α of energies E_α, $\mathcal{H}_\alpha$ is the matrix of the atomic hamiltonian calculated between the basis functions φ_i and $\mathcal{U}_\alpha$ is the matrix overlap integrals of these functions, we have to solve the matrix equation

$$\mathcal{H}_\alpha \mathcal{C}_\alpha = E_\alpha \mathcal{U}_\alpha \mathcal{C}_\alpha \tag{19}$$

The target polarisation plays an important role in low energy collisions.[14] It is thus natural to seek for a representation of the target leading to a good static polarizability of the ground state and the first few excited states. The dipole polarisability of any bound state n is given (in atomic units) by

$$\alpha_n^{stat} = 2\left\{\sum_{n'} \frac{|<n'|Z|n>|^2}{E_{n'} - E_n} + \int d\alpha' \frac{|<\alpha'|Z|n>|^2}{E_{\alpha'} - E_n}\right\} \tag{20}$$

where the the dipolar matrix elements are given by

$$<\beta|Z|n> = <\psi_\beta|r\cos\theta|\psi_n> \tag{21}$$

In the case of hydrogen atom, it is necessary to include fine and hyperfine structure corrections to the energies to remove some degeneracies and avoid singular terms.

The calculation of the static polarizability provides a test of the quality of the whole set of pseudostates. The Thomas-Reiche-Kuhn rule for oscillator strengths has been used as a supplementary condition

$$\sum_{n'} f_{n'n} = 2\left\{\sum_{n'} (E_{n'} - E_n)|<n'|Z|n>|^2 + \int d\alpha' (E_{\alpha'} - E_n)|<\alpha'|Z|n>|^2\right\} = 1 \tag{22}$$

This rule is not so directly related to the collision problem but has the advantages that it does not require knowledge of polarisabilities or of any other properties. Additionally the weights of the transitions to the pseudostates are greater than those to exact excited states. Finally, it provides an excellent test of completeness properties of the pseudostates which are crucial to our use of them in a Born2 type calculation.

In table 1, we give the energies of the exact and pseudostates used in our calculations. The polarisability calculated for the ground state with this basis is 3.96 (the exact value is 4.5) and the sum of oscillator strengths is 0.86 instead of 1.

In addition we have made calculations with two other basis using the same exact states as ours. The basis 2 given by Callaway[17] has one bound and two continuum states for l=0 and l=1, and one continuum pseudostate for l=2. The basis 3 of Callaway and Unnikrishnan[18] has one bound pseudostate for l=0 and l=1, four continuum pseudostate for l=0, and two for l=1 and for l=2.

Table 1. Pseudostate energies of basis 1.

	$l=0$	$l=1$	$l=2$
Bound	-1.00000		
states	-0.25000	-0.25000	
	-0.11111	-0.11111	-0.11111
Bound	-0.03786	-0.03869	-0.04209
pseudostates	-0.00998	-0.00999	-0.00999
Continuum	0.41577	0.41577	0.41577
pseudostates	2.29428	2.29428	2.29428
	6.28995	6.28995	6.28995

RESULTS AND DISCUSSION

Firstly we consider the case of equal energy sharing between the ejected and scattered electrons in a coplanar geometry. The cross section is plotted as a function of the angle $(\theta_a + \theta_b)$ between the direction of the two electrons for a fixed angle θ_a of one of them. The experimental data reported by Brauner et al.[19] shows two clearly defined peaks for incident energies of 27.2 eV whereas at a lower energy of 15.6 eV one or other of the peaks becomes extremely small.

For our calculations at 27.2 eV shown in figure 1, we have compared the results of experiment and of a DWB1 calculation including the N_{ee} factor with results of our present calculations using different number of states in replacement of the sum over the infinite number of intermediate target states in the DWB2 approximation. The DWB1 calculation gives a completely false ratio between the intensities of the two peaks. Inclusion of the 1s target state only in the DWB2 calculation causes a marked decrease of the cross section accompanied by an improvement in the peak intensity ratio. As further bound intermediate states are added the cross section rises globally and the peak intensity ratio steadily improves. With all bound states and bound pseudostates present the cross section shape is satisfactory as compared to experiment, but the magnitude for the low angle peak is only half that predicted by the DWB1 approximation. Addition of continuum pseudostates leads to a marked rise in the magnitude of the cross section and to the appearance of a shoulder on the low angle peak. A denser representation of the intermediate continuum states might have the effect of bringing the position of the calculated peak into agreement with experiment.

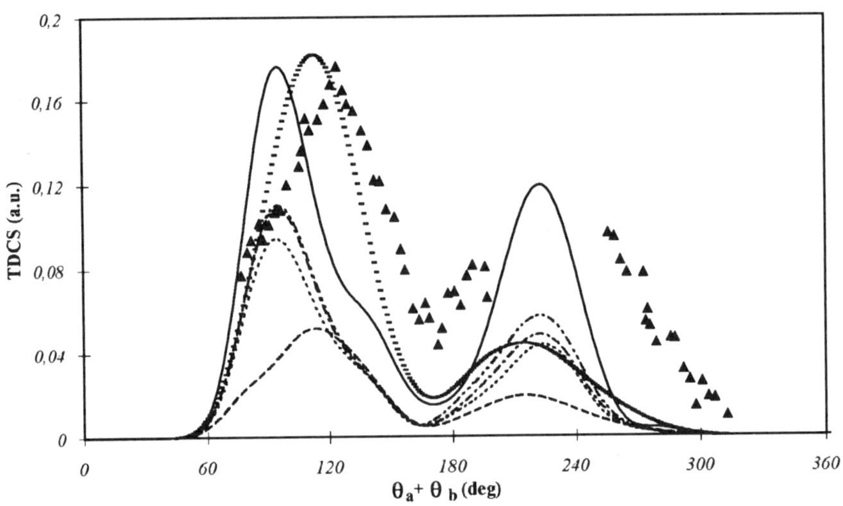

Figure 1. Coplanar asymmetric TDCS with $k_0^2 = 27.2eV$, $k_a^2 = k_b^2 = 6.8eV$ and $\theta_a = 45^o$. Distorted wave Born 2 calculations obtained by using in the basis 1 : the state $1s$ (_ _ _ _), the states $1s$, $2s$ et $2p$ (.....), all the bound states of basis (_ .. _ ._), the bound states and pseudostates (_ ... _ ..) and the basis "nearly" complete (______) ; distorted wave Born 1 calculations (-) ; ◆ experimental data of Brauner et al.[19]

Unfortunately we have some problems of numerical instability with our computer code when k_α^2 approches zero. For $k_\alpha^2 < 0$, F_α et G_α take imaginary arguments. So at

15.6 eV we have omitted all but the lowest lying continuum pseudostates. Our DWB2 results shown in figure 2 are in much better agreement with experiment than the DWB1 calculation. We consider two different angles θ_a and maintain the same relative normalisation of theory and experiment. Our DWB2 results have maxima at about the same angle as the measurements, unfortunatly in a region which is inaccessible to experiment for mechanical reasons. We find secondary maxima which seem to correspond to shoulders on the experimental data. Although Callaway's pseudostate bases were not designed for the study of ionization, it is interesting to see how well they perform in the context of our DWB2 model. A general improvement compared to the DWB1 approximation is found, similar to that obtained with our own pseudostates, but less pronouced.

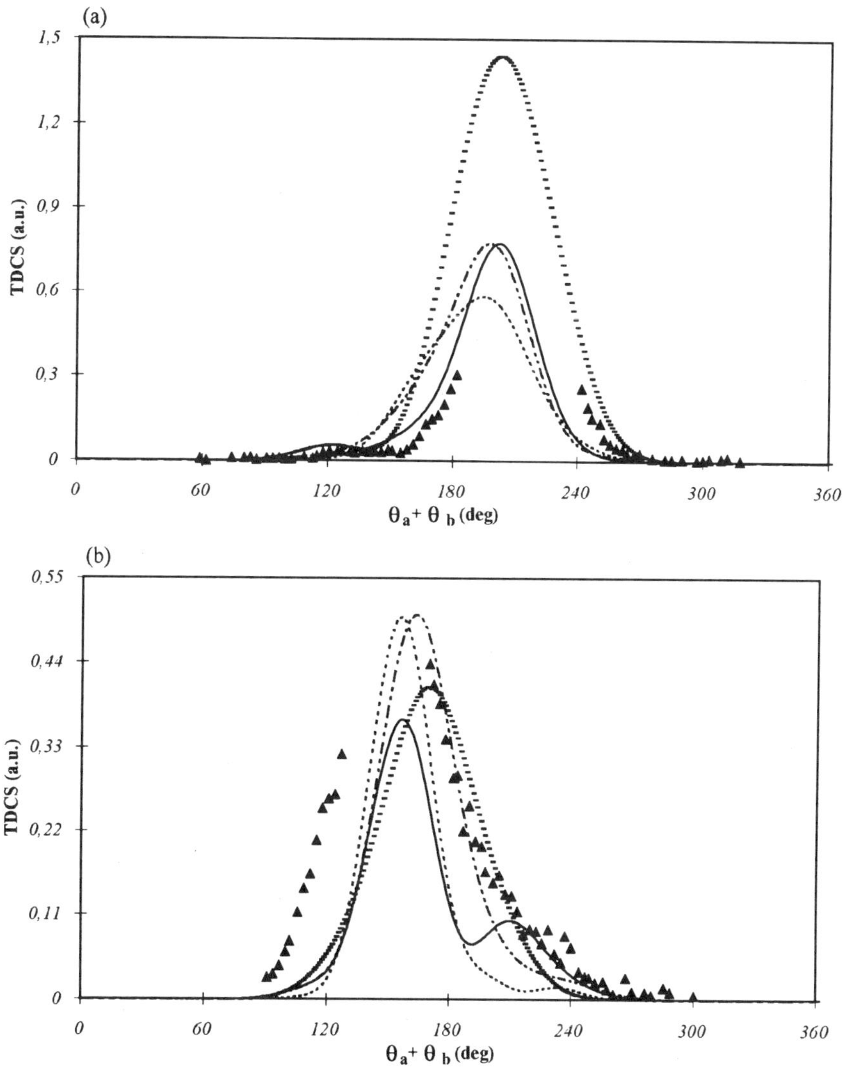

Figure 2. Coplanar asymmetric TDCS with $k_0^2 = 15.6eV$, $k_a^2 = k_b^2 = 1eV$ and $\theta_a = 30^o$ (a), 150° (b). Distorted wave Born 2 calculations obtained by using the basis 1 (_____), the basis 2 (.....) and the basis 3 (_ ... _ ..) ; distorted wave Born 1 calculations (-) ; ◆ experimental data of Brauner et al.[19]

Finally, in figure 3, we show results obtained with our pseudostates at 54.4 eV incident energy in the case of an asymmetric geometry and unequal energy sharing. The faster (scattered) electron is detected at fixed angle and the cross section is plotted as a function of the angle of the slower (ejected) electron. We give results for angles of 4° and 23° of the 35.8 eV fast electron.

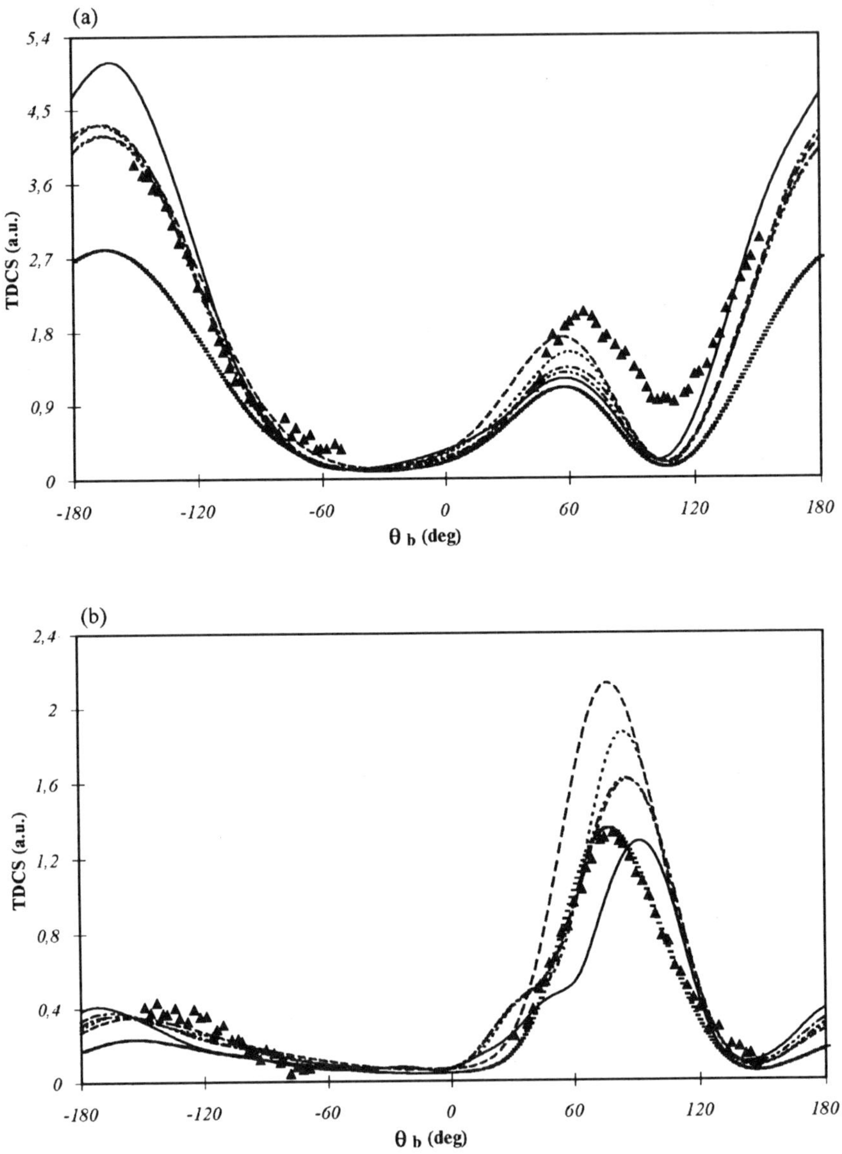

Figure 3. Same as figure 1 but with $k_0^2 = 54.4eV$, $k_a^2 = 35eV$, $k_b^2 = 6.8eV$ and $\theta_a = 4^o$ (a), 23° (b).

Our DWB2 calculations, like those of Curran and Walters[5] give a much stronger recoil peak (negative θ_b) than does DWB1, and so improve the agreement with experiment. It would appear that we get better results with an incomplete pseudostate set than when the continuum pseudostates are added. For 4° scattering angle these cause an exaggeration of the size of the recoil peak, whereas at 23°, although improving the intensity ratio of the peaks, they shift the binary peak (positive θ_b) towards higher angle as compared to experiment. However, calculations by Tweed et al.,[20] including PCI effects without improving the initial state, tend to improve the position and magnitude of this peak. The N_{ee} correction has little effect when the ejected and scattered electrons have very different energies. It should be noted that, since our pseudostates are obtained on a L^2 basis, they cannot describe long range Coulomb interactions between the ejected and scattered electrons. The methods developped by Nicolas et al.[21] do this rather well and will shortly be incorporated in our model.

CONCLUSION

The present DWB2 results are very encouraging. Our calculations demonstrate the importance of including the excitation of intermediate target bound and continuum states when calculating cross sections for ionization by low to intermediate energy electrons. Our model is relatively simple and can readily be extended to other targets than hydrogen. It is also capable of improvement, firstly by including a denser representation of intermediate continuum states, secondly by improving the treatment of post collision interaction in the final state.

REFERENCES

1. D.H. Madison, R.V. Calhoum and W.N. Shelton, Triple-differential cross sections for electron-impact ionization of helium, *Phys. Rev. A* 16:552 (1977).
2. B.H. Bransden, J.J. Smith and K.H. Winters, Distorted-wave approximation for the triple-differential cross section for ionization of helium by electron impact, *J. Phys. B: At. Mol. Phys.* 11:3095 (1978).
3. M. Brauner, J.S. Briggs and H. Klar, Triply-differential cross sections for ionization of hydrogen atoms by electrons and positrons, *J. Phys. B: At. Mol. Phys.* 22:2265 (1989).
4. C. Dal Cappello and B. Joulakian, *(e-2e) and Related Processes*, Kluwer, Dortrecht (1993).
5. C.T. Whelan, R.J. Allan, J. Rasch, H.R.J. Walters, X. Zhang, J. Röder, K. Jung and H. Ehrhardt, Coulomb three-body effects in (e,2e) collisions: the ionization of H in coplanar symmetric geometry, *Phys. Rev. A* 50:4394 (1994).
6. E.P. Curran and H.R.J. Walters, Triple differential cross sections for electron impact ionization of atomic hydrogen-a coupled pseudostate calculation, *J. Phys. B: At. Mol. Phys.* 20:337 (1987).
7. I. Bray and D.V. Fursa, Calculation of ionization within the close-coupling formalism, *Phys. Rev. A* 54:2991 (1996).
8. F.W. Byron, C.J. Joachain and B. Piraux, Triple differential cross sections for ionization of atomic hydrogen by fast electrons: a second Born treatment, *J. Phys. B: At. Mol. Phys.* 13:L673 (1980).
9. F.W. Byron, C.J. Joachain and B. Piraux, On the theory of (e, 2e) reactions in atomic hydrogen and helium, *Phys. Lett.* 106A:299 (1984).
10. F.W. Byron, C.J. Joachain and B. Piraux, Eikonal-Born theory of (e,2e) reaction in atomic hydrogen, *J. Phys. B: At. Mol. Phys.* 18:3203 (1985).
11. B. Lohmann, I.E. McCarthy, A.T. Stelbovics and E. Weigold, Electron impact ionization of atomic hydrogen: comparison of asymmetric (e,2e) measurments with theo-

ries, *Phys. Rev. A* 30:758 (1984).
12. H. Ehrhardt, G. Knoth, P. Schlemmer and K. Jung 1985, Absolute H(e, 2e)p cross section measurements: comparison with first and second order theory, *Phys. Lett.* 110A:92 (1985).
13. H. Ehrhardt, M. Fischer, K. Jung, F.W. Byron, C.J. Joachain and B. Piraux, Triple differential cross sections for the ionization of helium by fast electrons, *Phys. Rev. Lett.* 48:1807 (1982).
14. F. Rouet, R.J. Tweed and J. Langlois, The effect of target atom polarization and wavefunction distortion in (e-2e) ionization of hydrogen, *J. Phys. B: At. Mol. Phys.* 29:1767 (1996).
15. M.R.C. McDowell, L.A. Morgan and V.P. Myerscough, Electron impact excitation of H and He^+ I. 1s $\rightarrow$ ns transitions, *J. Phys. B: At. Mol. Phys.* 6:1435 (1973).
16. R.J. Tweed and J. Langlois, Pseudostate bases including an L^2 description of the continuum for use in e^--He collision calculations, *J. Phys. B: At. Mol. Phys.* 24:1779 (1991).
17. J. Callaway, The variational method in atomic scattering, *Phys. Rep.* 45:89 (1978).
18. J. Callaway and K. Unnikrishnan, Electron impact excitation of the n=2 and n=3 states of hydrogen atom at intermediate (14-100 eV) energies, *Phys. Rev. A* 48:4292 (1993).
19. M. Brauner, J.S. Briggs, H. Klar, J.T. Broad, T. Rösel, K. Jung and H. Ehrhardt, Triply differential cross sections for ionization of hydrogen atoms by electrons: the intermediate and threshold energy regions, *J. Phys. B: At. Mol. Phys.* 24:657 (1991).
20. R.J. Tweed, C. Tannous and P. Marchalant, Modelling of three-body effects in a double continuum, *Journal de Physique* 107:3 (1993).
21. M. Nicolas, R.J. Tweed and O. Robaux, Semiclassical approach to PCI in (e-2e) ionization of H: coplanar geometry energy sharing, *J. Phys. B: At. Mol. Phys.* 29:791 (1996).

EXCITATION - IONIZATION AND EXCITATION - AUTOIONIZATION OF HELIUM

Pascale J. Marchalant[1], Colm T. Whelan[2] and H.R.J. Walters[1]

[1] Department of Applied Mathematics and Theoretical Physics
The Queen's University of Belfast
Belfast BT7 1NN
United Kingdom

[2] Department of Applied Mathematics and Theoretical Physics
University of Cambridge
Silver Street
Cambridge CB3 9EW
United Kingdom

1. INTRODUCTION

In this article we survey some new theoretical results on (e, 2e) excitation - ionization and excitation - autoionization of ground state helium. A fuller account may be found in references [1-3]. Our interest lies in the high energy regime and in geometries for which the ejected electron is very much slower than the incident and scattered electrons. Under these conditions we may think of the collision as exciting a continuum state ψ_f of the isolated atom. In the case of excitation - ionization this state consists of the ejected electron, momentum $\boldsymbol{\kappa}$, moving in the field of an excited He^+ ion, for example, the system { He^+(2s or 2p) + $e^-(\boldsymbol{\kappa})$ }. In excitation - autoionization the helium atom makes a transition to an unstable doubly excited bound state He^{**}(nl, n′l′) (nl $\neq$ 1s, n′l′ $\neq$ 1s), ie, a resonant state of the atom, which then ejects an electron leaving a residual He^+ ion. The particular case we consider in this article is where this ion is left in its ground state He^+(1s) . The autoionization reaction then interferes with ordinary direct ionization to He^+(1s) since both processes lead to the same final state { He^+(1s) + $e^-(\boldsymbol{\kappa})$ }

In the high energy regime envisaged here it is natural to think in terms of a perturbative treatment. Accordingly, we have used both the first and second Born approximations to evaluate our (e, 2e) cross sections. For double excitation processes, such as excitation - ionization and excitation - autoionization, perturbation theory is somewhat more interesting than for single excitations. Thus, as we shall see below, the first Born amplitude is non - zero only by virtue of correlation effects in the atom states. In the absence of correlation the second Born term becomes the leading contribution to the Born series. This promotes the suspicion that, even with correlation effects "switched on", the second Born term may dominate the first Born amplitude up to very much higher impact energies than one would expect in single excitation processes. In addition, as we shall see in section 2, the second Born term brings in a two - step mechanism which has been a subject of much debate in this and other similar contexts[4,5]. In this two - step mechanism the double excitation process results from two consecutive virtual single excitations of the $1s^2$ electrons in the ground state of helium. With one exception[6], which will be discussed in section 3.1, it is the inclusion of the second Born term, with its implicit two - step process, which distinguishes the present results from previous theoretical investigations of excitation - ionization[7–13] and excitation - autoionization[14–18].

On the experimental front, absolute experimental data exist for coplanar (e, 2e) excitation - ionization of helium to the $He^+(2s)$ and $He^+(2p)$ states at impact energies of (approximately) 5.5 keV, 1.5 keV and 600 eV and for ejected electron energies ranging from 5 to 75 eV[10,19–21], kinematics which should be favourable to the form of approximation adopted here. Unfortunately, these experiments do not distinguish between the $He^+(2s)$ and $He^+(2p)$ final states so comparison can only be made with the summed cross section to $He^+(2s + 2p)$. The value to theory of having such **absolute** measurements cannot be over-emphasized. It is important, however, to understand how the experimental data have been put on an absolute scale. For the experiments quoted above this has been achieved by measuring the $He^+(2s + 2p)$ cross section relative to the ground state $He^+(1s)$ cross section and then normalising the latter. In comparing our calculations for excitation - ionization with the experimental data we should therefore also monitor how theory and experiment agree for the corresponding $He^+(1s)$ cross section. While we shall restrict our comparisons to the above - mentioned coplanar asymmetric (e, 2e) experiments it should be noted that there are other measurements of excitation - ionization of helium in the literature. There are experiments on (e, 2e) excitation - ionization in impulsive non - coplanar geometry[22–24] and in coplanar symmetric geometry[25,26], there are measurements of total excitation - ionization cross sections[27–32], and very recently[32] there has been a determination of a doubly differential cross section for excitation - ionization to $He^+(2p)$ at scattered electron energies of 60 and 130 eV.

For (e, 2e) excitation - autoionization there are measurements in the ejected energy range 32 to 37 eV by Weigold et al[33], Pochat et al[14], Lower and Weigold[34], McDonald and Crowe[35–37], and by Samardzic et al[18]. This ejected energy range includes the $(2s^2)^1S$, $(2s2p)^1P$, $(2s2p)^3P$ and $(2p^2)^1D$ resonant states of helium. A problem with such (e, 2e) measurements is the obtaining of sufficiently good energy resolution to distinguish clearly the resonant structures. Only in recent years have relatively good energy resolutions been achieved and for that reason we shall compare only with the most recent data of Lower and Weigold[34] and McDonald and Crowe[35–37]. Although the experiment of Samardzic et al[18] also has good energy resolution, it is at too low an impact energy for our perturbative treatment to be considered as reliable. Because we ignore electron exchange between the incident/scattered electron and the target electrons, the final helium state ψ_f in our theory must have the same total spin as the initial state ψ_0, which we assume to be the ground state, thus ψ_f must be a spin singlet continuum state. Consequently, we cannot obtain results for exciting the $(2s2p)^3P$ resonance. Consistent with our model, however, it is found that the $(2s2p)^3P$ resonance only makes a very weak contribution to the experimental data with which we make comparison. Finally, it should be remarked that the experimental measurements on (e, 2e) excitation - autoionization, unlike those on excitation - ionization, are not absolute.

We begin in section 2 by descibing the first and second Born approximations and how they have been implemented in the present work. In section 3 we make comparison between our calculations and the experimental data on excitation - ionization and excitation - autoionization. Our conclusions are presented in section 4.

2. THE THEORY

We consider an electron with momentum $\mathbf{k}_0$ incident upon the ground state ψ_0 of the helium atom. If this electron is scattered with momentum $\mathbf{k}_f$ and the atom is excited to the final (continuum) state ψ_f, the first Born amplitude is given by[§]

$$f^{Born1} = f^{B1} = -\frac{1}{2\pi}\left\langle e^{i\mathbf{k}_f\cdot\mathbf{r}_1}\psi_f(\mathbf{r}_2,\mathbf{r}_3)\left|\left(-\frac{2}{r_1}+\frac{1}{r_{12}}+\frac{1}{r_{13}}\right)\right|e^{i\mathbf{k}_0\cdot\mathbf{r}_1}\psi_0(\mathbf{r}_2,\mathbf{r}_3)\right\rangle \quad (1)$$

It is convenient to introduce the contracted notation

$$f^{Born1} = -\frac{1}{2\pi}\left\langle \mathbf{k}_f\psi_f\left|V\right|\mathbf{k}_0\psi_0\right\rangle \quad (2)$$

The integration over $\mathbf{r}_1$ in (1) is easily performed to yield the more useful expression

$$f^{Born1} = -\frac{4}{q^2}\left\langle\psi_f(\mathbf{r}_2,\mathbf{r}_3)\left|-1+e^{i\mathbf{q}\cdot\mathbf{r}_2}\right|\psi_0(\mathbf{r}_2,\mathbf{r}_3)\right\rangle \quad (3)$$

where

$$\mathbf{q} \equiv \mathbf{k}_0 - \mathbf{k}_f \quad (4)$$

is the momentum transfer in the collision and where we have used the symmetry

$$\psi_{0,f}(\mathbf{r}_2,\mathbf{r}_3) = \psi_{0,f}(\mathbf{r}_3,\mathbf{r}_2) \quad (5)$$

appropriate to spin singlet states ψ_0 and ψ_f.[¶] Formulae (1) and (3) give the first Born amplitude in the so - called "length" form. For exact wave functions ψ_0 and ψ_f there are other equivalent expressions that may be written down. The most important of these is the so - called "velocity" form

$$f^{Born1} = f^{B1} = \frac{2}{(\epsilon_f-\epsilon_0)q^2}\left\langle\psi_f(\mathbf{r}_2,\mathbf{r}_3)\left|\overleftarrow{\nabla}_2^2\, e^{i\mathbf{q}\cdot\mathbf{r}_2} - e^{i\mathbf{q}\cdot\mathbf{r}_2}\,\overrightarrow{\nabla}_2^2\right|\psi_0(\mathbf{r}_2,\mathbf{r}_3)\right\rangle \quad (6)$$

where $\epsilon_{0,f}$ is the energy of the state $\psi_{0,f}$ and the arrows indicate the direction in which ∇_2^2 operates. For inexact wave functions ψ_0 and ψ_f, (3) and (6) will no longer be equal in general. In this case close agreement between the computed values of (3) and (6) is a necessary, but not sufficient, condition for the calculated first Born term to be accurate.

The second Born amplitude is

$$f^{Born2} = f^{B1} + f^{B2} \quad (7)$$

where, in the notation (2), the second Born term f^{B2}, is given by

$$f^{B2} = -\frac{1}{8\pi^4}\lim_{\eta\to 0+}\sum_n\int d\mathbf{k}\frac{\left\langle\mathbf{k}_f\psi_f\left|V\right|\mathbf{k}\psi_n\right\rangle\left\langle\mathbf{k}\psi_n\left|V\right|\mathbf{k}_0\psi_0\right\rangle}{k_n^2-k^2+i\eta} \quad (8)$$

§ We use atomic units (au) in which $\hbar = m_e = e = 1$.

¶ In the first and second Born approximations, (1) and (7), only spin singlet states ψ_f can be excited.

$$k_n^2 = k_0^2 + 2(\epsilon_0 - \epsilon_n) \tag{9}$$

In (8) and (9) the sum on n is over all discrete and continuum states ψ_n of helium and ϵ_n is the energy of the state ψ_n. One of the simplest methods for performing the sum over n, and the one adopted here, is to use the closure approximation[38]. In this approximation ϵ_n in (9) is set to an average value $\bar{\epsilon}$ so that k_n^2 becomes independent of n, ie,

$$k_n^2 \to \overline{k}^2 = k_0^2 + 2(\epsilon_0 - \bar{\epsilon}) \tag{10}$$

The completeness of the atom states

$$\sum_n |\psi_n\rangle \langle\psi_n| = 1 \tag{11}$$

may then be used to trivially perform the sum over n giving

$$\overline{f}^{B2} = -\frac{2}{\pi^2} \lim_{\eta\to 0+} \int d\mathbf{k} \frac{\left\langle \psi_f \left| \left(e^{i(\mathbf{k}-\mathbf{k}_f)\cdot\mathbf{r}_2} + e^{i(\mathbf{k}-\mathbf{k}_f)\cdot\mathbf{r}_3} - 2\right) \left(e^{i(\mathbf{k}_0-\mathbf{k})\cdot\mathbf{r}_2} + e^{i(\mathbf{k}_0-\mathbf{k})\cdot\mathbf{r}_3} - 2\right) \right| \psi_0 \right\rangle}{|\mathbf{k}_0 - \mathbf{k}|^2 \, |\mathbf{k} - \mathbf{k}_f|^2 \left(\overline{k}^2 - k^2 + i\eta\right)} \tag{12}$$

where a bar over f^{B2} (f^{Born2}) indicates that the closure approximation has been employed. To calculate (12) we now need to know only the initial and final helium wave functions ψ_0 and ψ_f.

It is instructive to look at the first and second Born terms (1) and (8) in a model in which totally uncorrelated wave functions are used for the helium states ψ_n. Since only spin singlet states contribute to (1) and (8), we can restrict ourselves to singlet states. In this uncorrelated model we take

$$\psi_0(\mathbf{r}_2, \mathbf{r}_3) \to \psi_{00}(\mathbf{r}_2, \mathbf{r}_3) = \phi_0(\mathbf{r}_2)\phi_0(\mathbf{r}_3) \tag{13}$$

$$\psi_f(\mathbf{r}_2, \mathbf{r}_3) \to \psi_{fg}(\mathbf{r}_2, \mathbf{r}_3) = N_{fg}\left(\phi_f(\mathbf{r}_2)\phi_g(\mathbf{r}_3) + \phi_f(\mathbf{r}_3)\phi_g(\mathbf{r}_2)\right) \tag{14}$$

and, in general,

$$\psi_n(\mathbf{r}_2, \mathbf{r}_3) \to \psi_{pq}(\mathbf{r}_2, \mathbf{r}_3) = N_{pq}\left(\phi_p(\mathbf{r}_2)\phi_q(\mathbf{r}_3) + \phi_p(\mathbf{r}_3)\phi_q(\mathbf{r}_2)\right) \tag{15}$$

where

$$\langle \phi_p | \phi_q \rangle = \delta_{pq} \tag{16}$$

for any p and q. In (13) to (15) the atom states are built up from a set of orthonormal¶ orbitals ϕ, N_{pq} is a normalisation constant ($N_{pq} = 1/\sqrt{2}$ if $p \neq q$; $N_{pq} = 1/2$ if $p = q$), and it is convenient to introduce a double index notation, ie, pq, to label a single helium state.

Substituting (13) and (14) into (1) gives for the first Born amplitude

$$f^{Born1} = -\frac{N_{fg}}{\pi}\left(\langle \mathbf{k}_f\phi_f|V_1|\mathbf{k}_0\phi_0\rangle\langle\phi_g|\phi_0\rangle + \langle \mathbf{k}_f\phi_g|V_1|\mathbf{k}_0\phi_0\rangle\langle\phi_f|\phi_0\rangle\right) \tag{17}$$

where

$$\langle \mathbf{k}_p\phi_p|V_1|\mathbf{k}_q\phi_q\rangle \equiv \left\langle e^{i\mathbf{k}_p\cdot\mathbf{r}_1}\phi_p(\mathbf{r}_2) \left| -\frac{1}{r_1} + \frac{1}{r_{12}} \right| e^{i\mathbf{k}_q\cdot\mathbf{r}_1}\phi_q(\mathbf{r}_2) \right\rangle \tag{18}$$

If $\psi_f \equiv \psi_{fg}$ is a doubly excited state then

$$\phi_f \neq \phi_0 \qquad \phi_g \neq \phi_0 \tag{19}$$

¶ We present our formalism as if all the orbitals ϕ were discrete, it being understood that the orthonormality condition (16) is to be suitably generalised for continuum orbitals.

and we see from (17) that the first Born amplitude must be zero. We learn from this observation that correlation in the initial and final helium wave functions ψ_0 and ψ_f should be very important in double excitation processes. With correlated wave functions $f^{Born1} \neq 0$.

Now substituting (13) to (15) into the second Born term (8), and assuming (19), we get

$$f^{B2} = A + B \tag{20}$$

where

$$A \equiv -\frac{N_{fg}}{4\pi^4} \lim_{\eta\to 0+} \int d\mathbf{k} \frac{\langle \mathbf{k}_f \phi_f | V_1 | \mathbf{k}\phi_0 \rangle \langle \mathbf{k}\phi_g | V_1 | \mathbf{k}_0\phi_0 \rangle}{k_{g0}^2 - k^2 + i\eta} \tag{21}$$

$$B \equiv -\frac{N_{fg}}{4\pi^4} \lim_{\eta\to 0+} \int d\mathbf{k} \frac{\langle \mathbf{k}_f \phi_g | V_1 | \mathbf{k}\phi_0 \rangle \langle \mathbf{k}\phi_f | V_1 | \mathbf{k}_0\phi_0 \rangle}{k_{f0}^2 - k^2 + i\eta} \tag{22}$$

$$k_{pq}^2 = k_0^2 + 2(\epsilon_0 - \epsilon_{pq}) \tag{23}$$

and ϵ_{pq} is the energy of the helium state ψ_{pq}. The first observation on (20) is that, unlike the first Born term (17), the second Born term (20) is non - zero for doubly excited final states, it is then the leading term in the Born series for the uncorrelated model. This suggests that, even with more realistic wave functions for ψ_0 and ψ_f, the first Born amplitude may not dominate the second Born term until very high impact energies. The second observation is that the result (20) may be interpreted as a virtual two - step mechanism in which the incident electron first collides with one of the electrons in the orbital ϕ_0, exciting it to the orbital ϕ_g or ϕ_f, and then in a second virtual collision with the incident electron, the electron remaining in the orbital ϕ_0 is excited to the orbital ϕ_f or ϕ_g; the contributions, A and B, from the two possible routes to the final doubly excited state are added coherently, see (20). For correlated wave functions ψ_0 and ψ_f the second Born term does not reduce to such a simple form as (20). However, the result (20) suggests that it may still be dominated by the two - step mechanism which proceeds through the singly excited intermediate states ψ_{f0} and ψ_{g0}, or rather their equivalent correlated counterparts. Thus, for example, in exciting the $(2s2p)^1P$ resonance we would expect the $(1s2s)^1S$ and $(1s2p)^1P$ states of helium, acting as intermediate states in the second Born term, to carry the bulk of what we think of as the two - step mechanism in the uncorrelated model.

For ionization leading to an ejected electron of momentum $\boldsymbol{\kappa}$ and a He^+ ion in the state $He^+(n_f l_f m_f)$, ψ_f must correspond to the wave function for an electron with momentum $\boldsymbol{\kappa}$ incident upon $He^+(n_f l_f m_f)$ but with ingoing scattered wave boundary conditions. For ψ_f we here use a close - coupling approximation in which ψ_f is expanded in He^+ states ψ_{nlm}^+, ie¶,

$$\psi_f(\mathbf{r}_2, \mathbf{r}_3) = \sum_{nlm} \left(F_{nlm}(\mathbf{r}_2)\psi_{nlm}^+(\mathbf{r}_3) + F_{nlm}(\mathbf{r}_3)\psi_{nlm}^+(\mathbf{r}_2) \right) \tag{24}$$

In (24) the ψ_{nlm}^+ may be either eigenstates or pseudostates of He^+. The required boundary conditions on ψ_f are obtained by demanding that for open channels

$$\begin{aligned} F_{nlm}(\mathbf{r}) \quad \overset{r\to\infty}{\longrightarrow} \quad & \left[1 - \frac{\alpha_{n_f l_f}^2}{i\left(k_{n_f l_f} r + \mathbf{k}_{n_f l_f} \cdot \mathbf{r}\right)} \right] \\ & \times \exp\left[i\left(\mathbf{k}_{n_f l_f} \cdot \mathbf{r} - \alpha_{n_f l_f} \ln\left(k_{n_f l_f} r + \mathbf{k}_{n_f l_f} \cdot \mathbf{r} \right) \right) \right] \delta_{nn_f}\delta_{ll_f}\delta_{mm_f} \\ & + g_{nlm,n_f l_f m_f}(\hat{\mathbf{r}}) \frac{\exp\left[-i\left(k_{nl} r - \alpha_{nl} \ln 2k_{nl} r \right) \right]}{r} \end{aligned} \tag{25}$$

¶ Recall that we need only spin singlet states of helium here.

where

$$\mathbf{k}_{n_f l_f} \equiv \boldsymbol{\kappa} \tag{26}$$

$$k_{nl}^2 = \kappa^2 + 2\left(\epsilon^+_{n_f l_f} - \epsilon^+_{nl}\right) \tag{27}$$

$$\alpha_{nl} = -1/k_{nl} \tag{28}$$

ϵ^+_{nl} is the energy of the state $He^+(nlm)$ and a "hat" denotes a unit vector. For closed channels $F_{nlm}(\mathbf{r}) \to 0$ as $r \to \infty$. With the normalisation (25), and assuming that ψ_0 and the ψ^+_{nlm} are normalised to unity, the triple differential cross section (TDCS) for ionization leading to $He^+(n_f l_f m_f)$, with the ejected (scattered) electron appearing in the solid angle $d\Omega_\kappa$ ($d\Omega_f$) about the direction of $\boldsymbol{\kappa}$ ($\mathbf{k}_f$) and having energy in the interval dE, is

$$\frac{d^3\sigma_{n_f l_f m_f}}{dE\, d\Omega_\kappa\, d\Omega_f} = \frac{1}{(2\pi)^3} \frac{\kappa k_f}{k_0} |f| \tag{29}$$

where $f = f^{Born1}(f^{Born2})$ for the first (second) Born approximation.

In our calculations for direct ionization to $He^+(1s)$ and for excitation - ionization to $He^+(2s$ or $2p)$ we have taken (24) to be a 1s - 2s - 2p close - coupling eigenstate expansion. For the excitation - ionization we consider this to be the simplest realistic approximation that can be used for ψ_f, for two reasons. Firstly, since the 2s and 2p states of He^+ are degenerate in energy, the coupling between them is very important and so should be treated exactly. Secondly, since ionization to $He^+(1s)$ is so much larger than to $He^+(n=2)$ (typically by two orders of magnitude), the coupling through the 1s state can be important. Approximations that neglect the couplings between 1s, 2s and 2p in ψ_f must therefore be viewed with caution. For the excitation - autoionization calculations ψ_f corresponds to an electron of momentum $\boldsymbol{\kappa}$ incident upon the ground state $He^+(1s)$. Here we have used a 1s - 2s - 2p - 3s - 3p - 3d eigenstate approximation for ψ_f. This approximation contains the $(2s^2)^1S$, $(2s2p)^1P$ and $(2p^2)^1D$ resonances coherently embedded in the $e^-(\boldsymbol{\kappa}) + He^+(1s)$ continuum and therefore represents both the direct and resonant ionization to $He^+(1s)$ and the interference between them.

In calculating the first Born terms (3) and (6) we have taken ψ_0 to be the multiconfiguration variational wave function of Tweed and Langlois[39]. This wave function gives a ground state energy of -2.90125 au, which is close to the exact value -2.9037 au[40]. However, for the calculation of the second Born term we have employed the simpler Hartree - Fock wave function of Byron and Joachain[41]. The use of this simpler wave function was dictated by the greater complexity of the second Born calculation. We justify this simplification by the non - zero result for the second Born term in the uncorrelated model, a result suggesting that the second Born term may not be as sensitive to correlation as the first Born amplitude, which is only non - zero by virtue of correlation. Furthermore, if the two - step mechanism, which does not require correlation in ψ_0 or ψ_f, dominates the second Born term, then correlation should be even less important.

In the calculations reported here, the second Born term has been evaluated in the closure approximation, taking the average energy $\bar{\epsilon}$ to correspond to the ionization threshold for helium, ie,

$$(\bar{\epsilon} - \epsilon_0) = 0.904 \text{ au} = 24.59 \text{ eV} \tag{30}$$

This was a choice originally used successfully by Byron and Joachain and co - workers[42,43] in studying direct ionization of helium to $He^+(1s)$ in coplanar asymmetric geometry. We have adopted the same value of $\bar{\epsilon}$ in all of our calculations, ie, dirct ionization to $He^+(1s)$, excitation - autoionization, and excitation - ionization to $He^+(n=2)$. In the last two cases this choice was based upon a consideration of the two - step mehanism. For excitation - autoionization of the $(2s^2)^1S$, $(2s2p)^1P$ and $(2p^2)^1D$ states, this mechanism proceeds through the intermediate states $(1s2s)^1S$ and $(1s2p)^1P$. These states have excitation energies of 20.6 and 21.2 eV respectively, which are close to the value (30). Because of the interference with direct ionization to $He^+(1s)$, we need also to get a good representation of the direct process. In the asymmetric situations considered here (30) is known to be a good choice for the direct process, (30), therefore, is a good compromise both for the two - step mechanism and for the direct ionization. For excitation - ionization to $He^+(n = 2)$ the two - step mechanism not only involves the states $(1s2s)^1S$ and $(1s2p)^1P$, discussed above, but also the intermediary $(1s\ \boldsymbol{\kappa})$. This latter has an energy of 24.59 eV + ejected electron energy. If the energy of the ejected electron is not too large, then (30) is again a good compromise average energy.

Finally, we need to specify the quantization axis for the final ion state $He^+(n_f l_f m_f)$. This, of course, only matters if $He^+(n_f l_f m_f)$ is not an s - state. In the results reported here, we have taken the quantization axis to be in the direction of the outgoing ejected electron, ie, in the direction of $\boldsymbol{\kappa}$, consequently the quantization axis varies with the direction of $\boldsymbol{\kappa}$. With this convention it can be shown[1,2] that the TDCS in the first Born approximation, when viewed as a function of $\boldsymbol{\kappa}$ for given fixed $\mathbf{k}_0$ and $\mathbf{k}_f$, must be symmetric about the direction of the momentum transfer $\mathbf{q} = \mathbf{k}_0 - \mathbf{k}_f$, irrespective of the final state of the He^+ ion. No such symmetry is implied by the second and higher Born approximations and so depertures from this symmetry indicate a breakdown of the first Born approximation.

3. RESULTS

In this section we present a representative sample of our results. All of the cases we look at correspond to coplanar geometry. We adopt the convention that angles are measured in the range 0^0 to 360^0 relative to the incident direction, ie, $\mathbf{k}_0$. The angle of the fast scattered electron, θ_f, is assumed to be described in an anticlockwise direction with respect to $\mathbf{k}_0$, the angle of the slower ejected electron, θ_κ, in a clockwise direction. We denote by E_0, E_κ and E_f the energies of the incident, ejected and scattered electrons respectively.

3.1 Excitation - Ionization of Helium to $He^+(n = 2)$

We have calculated the TDCS for ionization to $He^+(1s)$ and to $He^+(n = 2)$ using both the length and velocity forms (3) and (6). For the cases studied, which correspond to the experiments of Dupré et al[10] and Stefani et al[19–21], we find very good agreement between the length and velocity results for ionization to the ground state $He^+(1s)$ and also good accord for the n=2 triple differential cross sections; in the latter case this agreement is noticeably better than that obtained in the analogous first Born calculations of Kheifets et al[12] where a 1s - 2s - 2p - 3s - 3p - 3d close - coupled wave function was used for ψ_f. Unfortunately, we cannot conclude from these comparisons that our first Born results are highly reliable since, as stated following (6), agreement between length and velocity forms is only a necessary but not a sufficient condition for a caculation to give accurate first Born numbers. Our first Born calculations for ionization to $He^+(1s)$ agree well with the results of Kheifets et al[12] and, although differing in some details from their work, our first Born cross sections for $n = 2$ excitation - ionization are comparable.

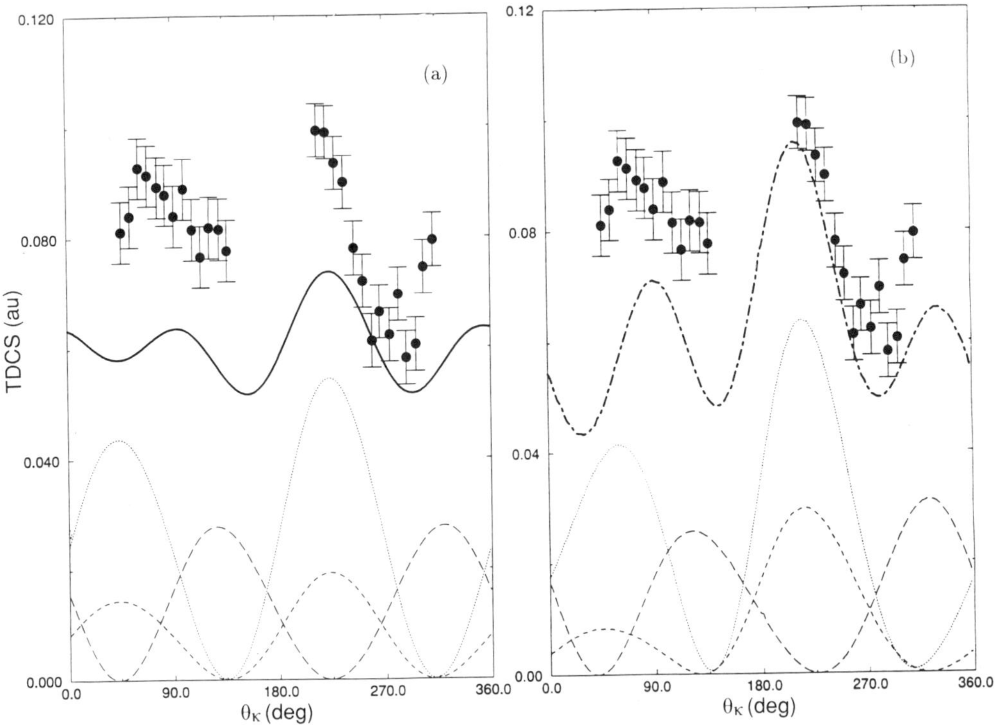

Figure 1. TDCS (in au) for $n = 2$ excitation - ionization of helium at $E_f = 5500$ eV, $E_\kappa = 5$ eV, $\theta_f = 0.35^0$ in (a) the first Born approximation, (b) the second Born approximation. Total $n = 2$ cross section, solid curve in (a) and dash - dot curve in (b); 2s cross section, dotted curve; $2p_0$ cross section, short - dashed curve; $2p_1$ cross section, long - dashed curve. Solid circles are experimental data from Dupré et al[10].

We now make comparison of our first and second Born calculations with the experimental data of Dupré et al[10] and Stefani et al[19–21]. These experiments do not resolve the final 2s and 2p states of He^+ and so the experimental data correspond to the summed cross section $He^+(2s + 2p)$. It is important to note that these experiments are **absolute** and so provide a very stringent test of the theory. In both the first and second Born approximations shown in figures 1 to 6, the length form (3) of the first Born term has been used.

In figure 1 we compare our first and second Born calculations for excitation - ionization to $He^+(n = 2)$ with the experimental data of Dupré et al[10] for the case $E_f =$ 5500 eV, $E_\kappa = 5$ eV and $\theta_f = 0.35^0$. We show not only the total $n = 2$ cross section, ie, the cross section to $He^+(2s + 2p)$, but also its individual components to $He^+(2s)$, $He^+(2p_0)$ and $He^+(2p_1)$ where, see section 2, the axis of quantization is in the direction of $\boldsymbol{\kappa}$ (the momentum of the ejected electron). With this choice of quantization axis the cross section to $He^+(2p_{-1})$ is identically equal to that to $He^+(2p_1)$.

To begin with, let us look at the first Born results, figure 1(a). Although not quite in agreement with the measurements, the first Born approximation does predict the correct magnitude of the TDCS, this is something of an achievement in itself. Examining the components of the first Born approximation, we see that the $He^+(2s)$ and $He^+(2p_0)$ cross sections have the classic shape, familiar from ionization to the ground state $He^+(1s)$ (see figure 2), consisting of binary and recoil peaks in the directions of $\mathbf{q}$ ($\theta_\kappa = 44^0$)

and $-\mathbf{q}$ ($\theta_\kappa = 224^0$) respectively. The He^+ ($2p_1$) cross section shows a similar structure but shifted to larger θ_κ by 90^0¶. This difference in the $He^+(2p_1)$ cross section arises from another dynamical symmetry of the first Born approximation which requires the $He^+(2p_1)$ cross section to be identically zero in the directions of $\mathbf{q}$ and $-\mathbf{q}$[1,2]. Figure 1(a) shows that the binary peaks of the $He^+(2s)$ and $He^+(2p_0)$ cross sections are smaller than their recoil peaks, while the reverse is true for $He^+(2p_1)$. Since the $He^+(2p_1)$ cross section is substantial, it fills in the minima between the binary and recoil peaks of the $He^+(2s)$ and $He^+(2p_0)$ curves with the result that the total $n = 2$ cross section does not display the classic $n = 1$ shape (see figure 2). Figure 1(b) shows what happens in the second Born approximation. Here we see improved agreement between theory and experiment . The agreement in the recoil peak region is particularly good, although a significant discrepancy is still present in the binary region. Comparing the components of the cross sections shown in figures 1(a) and 1(b), we see that, for $He^+(2s)$ and $He^+(2p_0)$, the second Born approximation has reduced the binary peak and increased the recoil peak, while rotating the former to slightly larger angles θ_κ and the latter to slightly smaller θ_κ, exactly as happens in ionization to the ground state $He^+(1s)$ (see figure 5). Although not obvious from figure 1, the second Born approximation has also reduced and increased the binary and recoil peaks of the $He^+(2p_1)$ cross section, now rotating these peaks to (very slightly) smaller and larger θ_κ respectively.

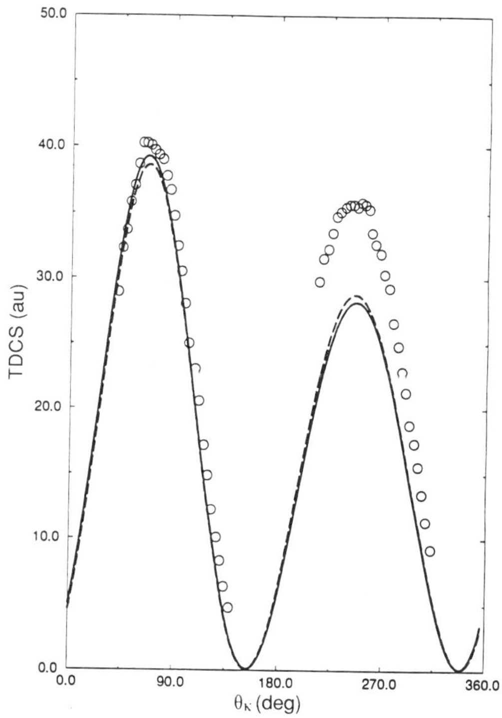

Figure 2. TDCS for ionization to $He^+(1s)$ at $E_f = 5500$ eV, $E_\kappa = 5$ eV and $\theta_f = 0.35^0$: solid curve, first Born approximation; dashed curve, second Born approximation; open circles, experimental data of Dupré et al[10].

¶ For $He^+(2p_1)$ we shall refer to the peaks near $\theta_q + 90^0$ and $\theta_q + 270^0$ as being the "binary" and "recoil" peaks respectively, where θ_q is the direction of the momentum transfer $\mathbf{q}$ measured, like θ_κ, clockwise from the incident direction. In all other cases the terms "binary" and "recoil" will refer to the directions near $+\mathbf{q}$ and $-\mathbf{q}$ respectively. In figure 1, $\theta_q = 44^0$.

Figure 2 compares our first and second Born results for ionization to the ground state $He^+(1s)$ with the experimental data of Dupré et al[10] for the same kinematics as figure 1, ie, $E_f = 5500$ eV, $E_\kappa = 5$ eV, $\theta_f = 0.35^0$. The significance of figure 2 lies in the fact that the $n = 2$ measurements of figure 1 have been normalised to the $He^+(1s)$ data shown in figure 2. From figure 2 we see that there is very little difference between the first and second Born curves, implying that the first Born approximation should be valid in this case. However, while the experimental data agree well with the binary peak of the theoretical curve, there is a pronounced discrepancy in the recoil peak. We have repeated the first Born calculations using different wave functions for ψ_0 and $\psi_f^{1,2}$, including a 1s - 2s - $\overline{3s}$ - $\overline{4s}$ - 2p - $\overline{3p}$ - $\overline{4p}$ - $\overline{3d}$ - $\overline{4d}$ close - coupling pseudostate¶ approximation for ψ_f, and have found neglegible change on the results presented in figure 2. We see no alternative but to conclude that the theory is reliable and that it is experiment which is most likely at fault, with consequent implications for the $n = 2$ measurements shown in figure 1. A more detailed discussion of this point, including the experiment, is given in references [1-2]. Figure 2 shows that the $n = 1$ cross section is about 500 times larger than the $n = 2$ cross section of figure 1, highlighting how much more difficult the $n = 2$ measurements are.

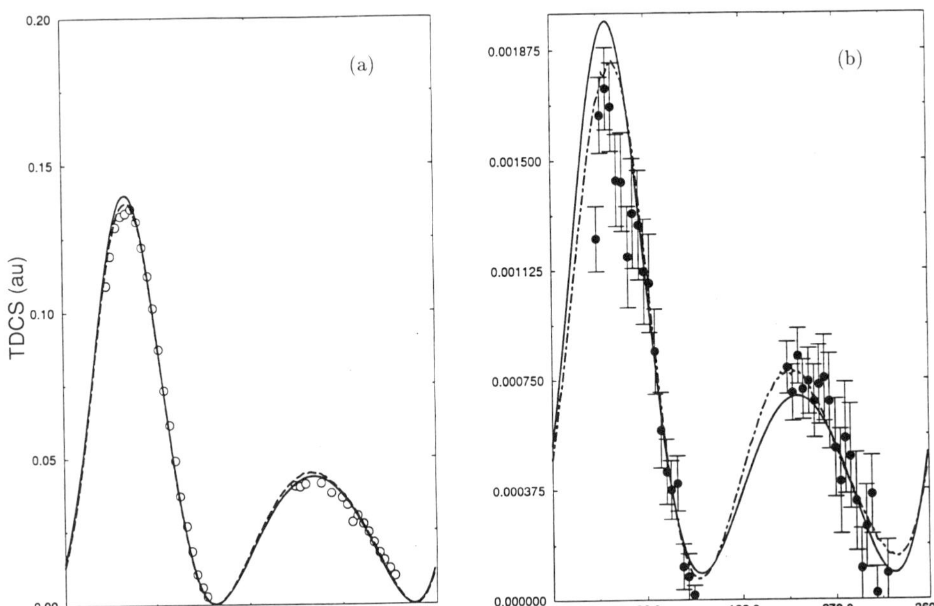

Figure 3. TDCS (in au) for (a) ionization to $He^+(1s)$ and (b) excitation - ionization to $He^+(n = 2)$ for $E_f = 5500$ eV, $E_\kappa = 75$ eV and $\theta_f = 1^0$: solid curve, first Born approximation; dashed curve, second Born approximation; circles, experimental data of Dupré et al[10] normalised to the second Born cross section for $n = 1$ ionization.

Figure 3 exhibits the data of Dupré et al[10] for a very much higher ejected energy, ie, $E_f = 5500$ eV, $E_\kappa = 75$ eV, $\theta_f = 1^0$. Figure 3(a) shows very little difference between the first and second Born cross sections for $n = 1$ and excellent agreement with the

¶ Pseudostates are denoted by a bar

measurements, which in this case have been normalised to the second Born theory¶. Figure 3(b) also shows little difference between the first and second Born results for $n = 2$, although the experimental data are clearly in better agreement with the second Born cross section, with which they are in excellent accord. The $n = 2$ cross section now has the same shape as that for $n = 1$, with the binary maximum being much larger than the recoil peak. Analysis of the components of the $n = 2$ curve[1,2] reveals that the $He^+(2s)$ and $He^+(2p_0)$ contributions now dominate, the $He^+(2p_1)$ cross section being almost neglegible.

Figures 4 to 6 compare our calculations with the experimental data of Stefani et al[19−21] at a very much lower impact energy ($E_f = 570$ eV). Figure 4 shows first and second Born cross sections for $n = 2$ excitation - ionization at $E_f = 570$ eV, $E_\kappa = 10$ eV, $\theta_f = 4^0$. From figure 4(a) we see that the first Born approximation is in agreement with the measurements both in shape and magnitude. Figure 4(b) shows that the second Born calculations describe well the recoil peak but are not so satisfactory in the binary region. However, to get this agreement with the second Born approximation the experimental data have been normalised upwards by a factor of 2.75 ! Could the

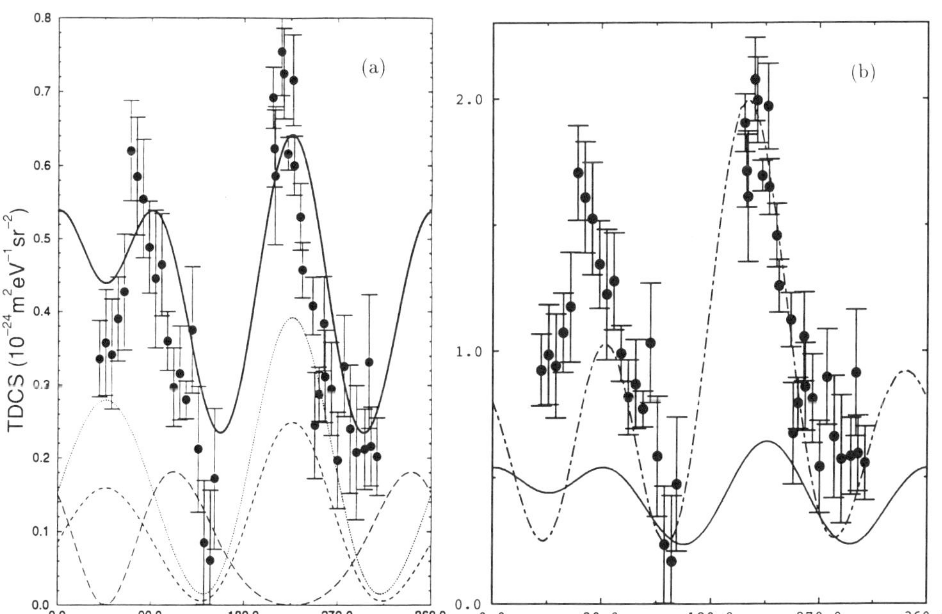

Figure 4. TDCS (in units of $10^{-24}m^2eV^{-1}sr^{-2}$) for $n = 2$ excitation - ionization of helium at $E_f = 570$ eV, $E_\kappa = 10$ eV and $\theta_f = 4^0$. Figure (a) shows cross sections in the first Born approximation : solid curve, total $n = 2$ cross section; dotted curve, 2s cross section; short - dashed curve, $2p_0$ cross section; long - dashed curve, $2p_1$ cross section; solid circles, experimental data of Stefani et al[19−21]. Figure (b) compares the first and second Born cross sections for total $n = 2$ excitation - ionization : solid curve, first Born approximation; dashed curve, second Born approximation; solid circles, experimental data of Stefani et al[19−21] normalised upwards by a factor of 2.75.

¶ For ejected energies other than $E_\kappa = 5$ eV, Dupré et al have to normalise their $n = 1$ data to a suitable theory, the normalisation of the $n = 2$ cross section then folloes automatically.

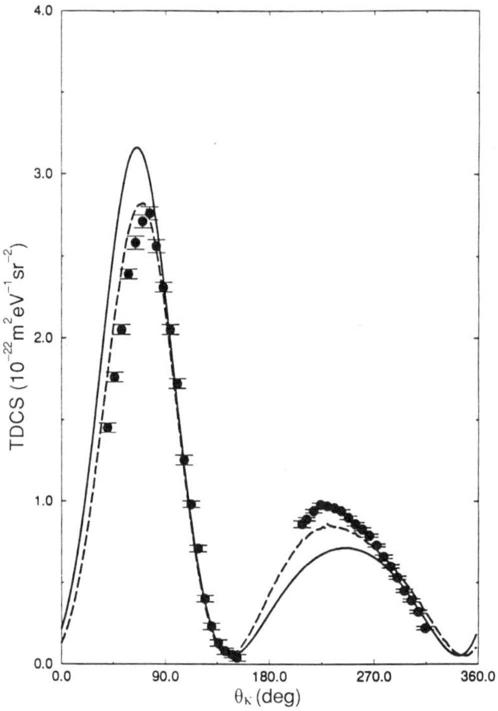

Figure 5. TDCS (in units of $10^{-22}\mathrm{m}^2\mathrm{eV}^{-1}\mathrm{sr}^{-2}$) for ionization to $\mathrm{He}^+(1\mathrm{s})$ at E_f = 570 eV, E_κ = 10 eV and $\theta_f = 4^0$: solid curve, first Born approximation; dashed curve, second Born approximation; solid circles, experimental data of Stefani et al[19−21].

experimental normalisation be in error ? In figure 5 we show our calculations for $n = 1$ ionization in the same kinematics. We see very good agreement between the measurements and the second Born curve which accurately describes the reduction (increase) of the binary (recoil) peak and its rotation to larger (smaller) θ_κ compared with the first Born approximation. Since the $n = 2$ measurements of figure 4 are normalised to the $n = 1$ data of figure 5, there is no reason to doubt the normalisation of the $n = 2$ measurements on this score. Also, since the second - order correction in figure 4(b) is so large we should perhaps ask about the size of third - and higher - order Born terms. Further theoretical work and experimental checks would be in order.

Figure 6 exhibits first and second Born calculations at a higher ejected energy, ie, $E_f = 570$ eV, $E_\kappa = 40$ eV, $\theta_f = 4^0$. Figure 6(a) shows the $n = 2$ excitation - ionization cross section. Here we see a marked improvement between theory and experiment on going from the first to the second Born approximation. The second Born approximation correctly predicts the movement of the binary peak to larger θ_κ and the pronounced enhancement of the recoil peak which is seen in the experimental data. Figure 6(b) for $n = 1$ ionization does not show quite as good agreement between theory and experiment as figure 5. However, if the $n = 1$ measurements were normalised downwards by 13% the agreement would be significantly enhanced. A corresponding renormalisation of the $n = 2$ measurements of figure 6(a) would also improve agreement between theory and experiment.

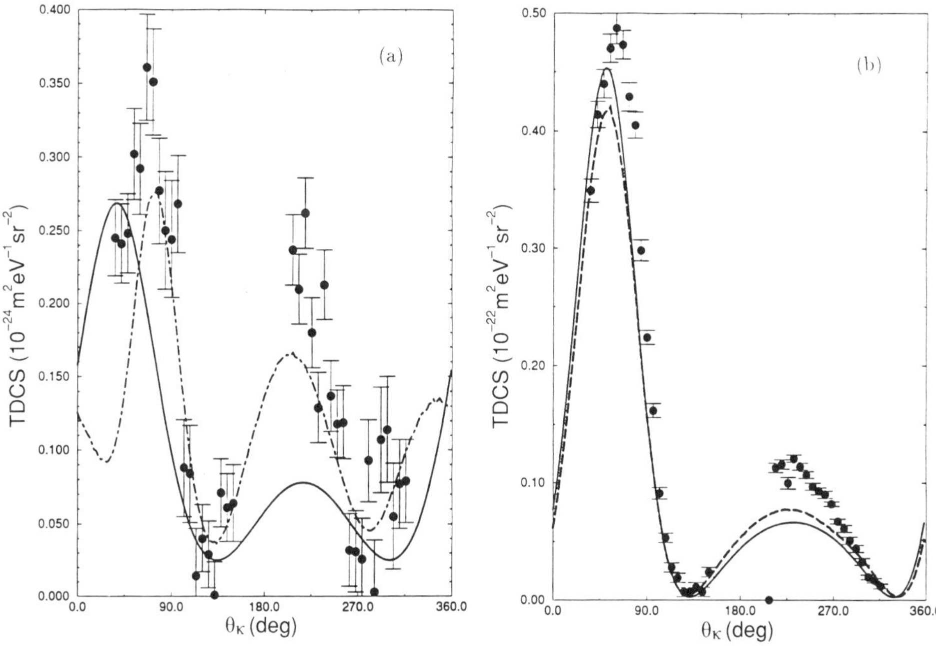

Figure 6. TDCS for (a) excitation - ionization to $He^+(n=2)$ (in units of $10^{-24}m^2eV^{-1}sr^{-2}$), and (b) ionization to $He^+(1s)$ (in units of $10^{-22}m^2eV^{-1}sr^{-2}$), at $E_f = 570$ eV, $E_\kappa = 40$ eV and $\theta_f = 4^0$: solid curve, first Born approximation; dashed curve, second Born approximation; solid circles, experimental data of Stefani et al[19–21].

In summary, for $n = 2$ excitation - ionization our calculations show the following :
(i) at $E_f = 5500$ eV, second - order effects are about 20 to 30% for $E_\kappa = 5$ eV and $\theta_f = 0.35^0$, decreasing with increasing E_κ;
(ii) at $E_f = 570$ eV, $E_\kappa = 10$ eV and $\theta_f = 4^0$ second - order effects increase the first Born cross section by a factor of about 2.75 on average, again second - order effects decrease with increasing E_κ;
(iii) generally speaking, our second Born approximation leads to improvement in the recoil peak region but not in the binary peak.

Finally, we comment upon the only other second Born calculation of excitation - ionization of helium, that of Franz and Altick[6] who have made calculations at $E_0 = 1$ keV, 645.4 eV and 350 eV. Like the present work, they consider the closure approximation (12)¶, but, unlike the present work, they employ only a single channel approximation to ψ_f, ie, they take no account of the important coupling between the 1s, 2s and 2p states of He^+ in descibing the motion of the slow ejected electron. Both their first and second Born results at $E_0 = 645.4$ eV, $E_f = 570$ eV, $E_\kappa = 10$ eV and $\theta_f = 4^0$ are smaller than our $He^+(2p)$ cross sections of figure 4, which is in the same kinematics, by about an order of magnitude ! The agreement with the absolute experimental data seen in figures 1 to 6, and with the work of Kheifets et al[12], gives credibility to the size of our computed numbers. Like figure 4, however, their second Born approximation predicts a large enhancement of the first Born recoil peak, by a factor of 2.

¶ They also consider a form of the second Born approximation which takes direct account of the interaction between the scattered electron and the 1s, 2s , 2p and 3s states of the final He^+ ion.

3.2 Excitation - Autoionization of Helium

In this section we show some first and second Born results for excitation - autoionization of the $(2s^2)^1S$, $(2s2p)^1P$ and $(2p^2)^1D$ resonances of helium which lie in the energy range 32 to 37 eV. The calculated positions and widths of these resonances in the 1s - 2s - 2p - 3s - 3p - 3d close - coupling approximation that we use for ψ_f are given in table 1 where they are compared with the recommended values of van den Brink et al[44]. The agreement between the two sets of numbers is quite good. In figures 7 to 11

Table 1. Energies and widths (in eV) of the $(2s^2)^1S$, $(2s2p)^3P$, $(2p^2)^1D$ and $(2s2p)^1P$ resonances in $e^- + He^+$ scattering.

	Present Work		Results of van den Brink et al[44]	
State	Energy	Width	Energy	Width
^1S	33.26	0.138	33.24 ± 0.04	0.138 ± 0.015
^3P			33.72 ± 0.02	0.08
^1D	35.36	0.069	35.32 ± 0.02	0.072 ± 0.018
^1P	35.60	0.04	35.555	0.038 ± 0.020

we compare our calculations with the experimental measurements of Lower and Weigold[34] and McDonald and Crowe[37]. These measurements give the TDCS as a function of the ejected electron energy E_κ for fixed incident energy E_0 and fixed angles θ_f and θ_κ. Unlike the data for excitation - ionization discussed in section 3.1, none of the experimental data quoted here are absolute. In figures 7 to 11 we have therefore normalised the data to the theory. A further complication is the energy resolution¶ of the experiments. The experiment of Lower and Weigold has a resolution of 150 meV, while that of McDonald and Crowe shown in figure 11 has a resolution of 80 meV. Since these resolutions are comparable to the resonance widths given in table 1, it is necessary to take them into account when comparing theory with experiment. Accordingly, the theoretical results shown in figures 7 to 11 have been convoluted over E_κ with the Gaussian

$$G(E_\kappa, \overline{E}_\kappa) = \frac{1}{\sqrt{2\pi}\sigma} \exp\left[-\frac{\left(E_\kappa - \overline{E}_\kappa\right)^2}{2\sigma^2}\right] \tag{31}$$

where E_κ ($\overline{E}_\kappa$) is the theoretical (experimental) ejected electron energy and the energy resolution of the experiment is

$$2(2\ln 2)^{\frac{1}{2}}\sigma \tag{32}$$

Figure 7 compares our first and second Born results, suitably convoluted with (31), with the data of Lower and Weigold at the highest measured incident energy, $E_0 = 400$ eV, for $\theta_f = 3^0$ and $\theta_\kappa = 40^0$. This corresponds to the binary region, see figure 6(b) for example, where the direct process $e^- + He \to e^- + e^- + He^+(1s)$ is largest and so interference between the direct and resonance contributions is also greatest. Distructive interference between the direct and resonance processes is evident in the minima which are seen near the resonance positions in figure 7. Overall, the second Born approximation appears to get the dip and peak near the ^1D and ^1P resonances more accurately than the first Born calculation. It should also be noticed that the average (background) value of the second Born cross section in figure 7(b) is lower than that of the first Born approximation, figure 7(a), as we would expect for direct ionization in the binary region, see figure 6(b).

¶ Quoted as full width at half maximum.

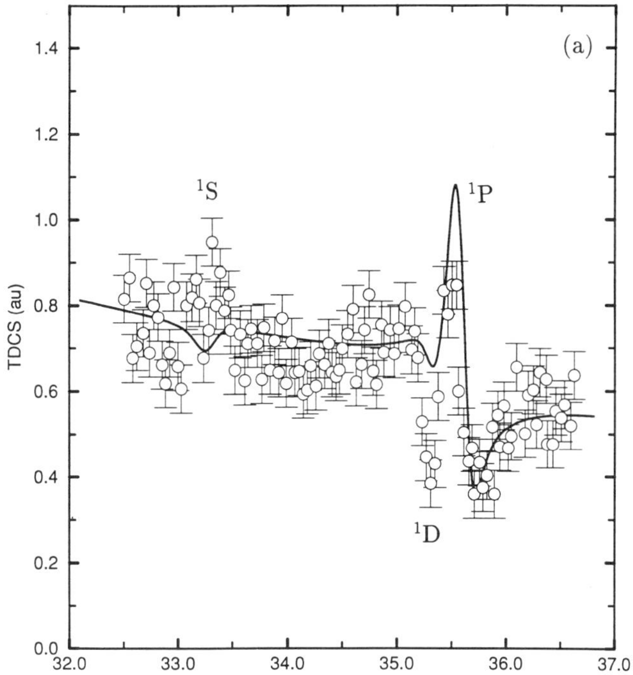

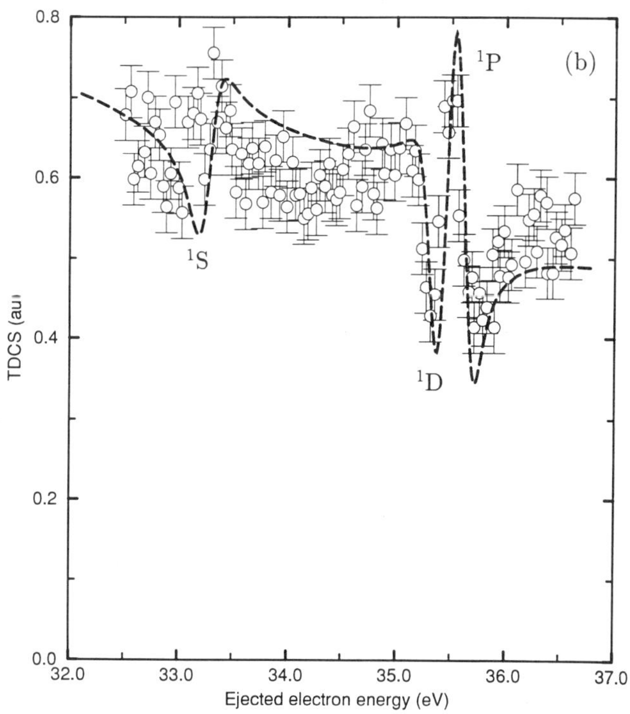

Figure 7. TDCS (in au) for ionization to $He^{+}(1s)$ at $E_0 = 400$ eV, $\theta_f = 3^0$ and $\theta_\kappa = 40^0$. Curves are theoretical results convoluted with experimental energy resolution of 150 meV : in (a) first Born approximation; in (b) second Born approximation. Open circles are experimental data of Lower and Weigold[34] normalised to give the best visual fit to the theoretical cross section in each case.

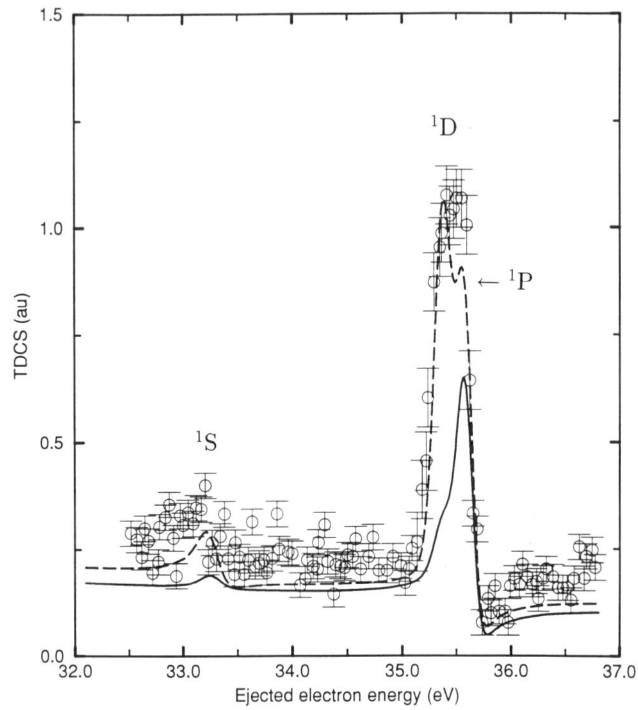

Figure 8. TDCS (in au) for ionization to $He^+(1s)$ at E_0 = 400 eV, θ_f = 3^0 and θ_κ = 220^0. Curves are theoretical results convoluted with experimental energy resolution of 150 meV : solid curve, first Born approximation; dashed curve, second Born approximation. Open circles are the experimental data of Lower and Weigold[34] normalised to give the best visual fit to the second Born cross section.

The situation shown in figure 8, where E_0 = 400 eV, θ_f = 3^0 and θ_κ = 220^0, lies in the recoil region where, see figure 6(b), the direct cross section is relatively small. Consequently, interference between the direct and resonance processes is also small and the resonance contributions essentially sit as peaks on top of the much smaller smooth direct ionization background. Here we see an interesting difference between the first and second Born approximations in which, after convolution, the relative heights of the ^{1}D and ^{1}P resonance contributions have been switched. In the first Born approximation the ^{1}D resonance appears as a shoulder on the low energy side of the ^{1}P peak, while in the second Born calculation two distinct peaks are seen with the ^{1}D peak being the larger. The agreement between the second Born appproximation and the experimental data, which have been normalised to the second Born curve in figure 8, is very encouraging. A similar normalisation of the measurements to the first Born curve would not produce such a good fit. Again, it should be noticed that the background value of the second Born cross section of figure 8 is larger than that in the first Born approximation, as we would expect in the recoil region, see figure 6(b).

Figure 9 shows results at the lower incident energy E_0 = 200 eV for θ_f = 13^0 and θ_κ = 50^0. Like figure 7 this corresponds to the binary region. Figure 9(a) compares the first and second Born results (after convolution with the experimental energy resolution), we see quite a significant difference in the structure of the two curves. Again, as expected in the binary region, the second Born cross section averages at a

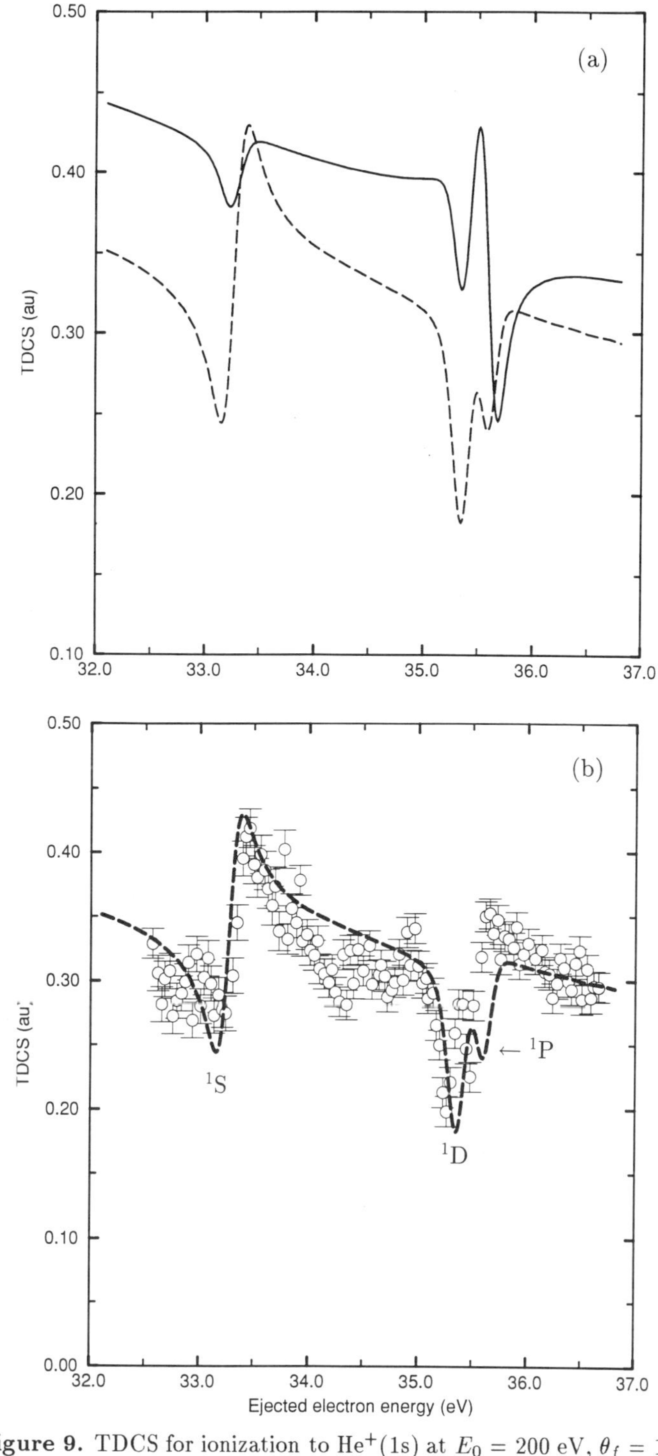

Figure 9. TDCS for ionization to $He^+(1s)$ at $E_0 = 200$ eV, $\theta_f = 13^0$ and $\theta_\kappa = 50^0$. Curves are theoretical results convoluted with experimental energy resolution of 150 meV. In (a) : solid curve, first Born approximation; dashed curve, second Born approximation. Figure (b) shows the experimental data of Lower and Weigold[34] normalised to give the best visual fit to the second Born curve.

lower value than the first Born numbers. Figure 9(b) compares the experimental data of Lower and Weigold with the second Born calculation, the agreement is very good. It is clear from figures 9(a) and 9(b) that the experiment would not be in such good accord with the first Born cross section, thus indicating the need for the second Born treatment.

Figure 10 illustrates results in the recoil region, $\theta_\kappa = 230^0$, at $E_0 = 200$ eV and $\theta_f = 13^0$. The picture is similar to figure 8 at 400 eV, except that the switch in the relative contributions of the ^{1}D and ^{1}P resonances, on going from the first to the second Born approximation, is even more pronounced and the experimental data of Lower and Weigold are even more definite in their preference for the second Born calculation.

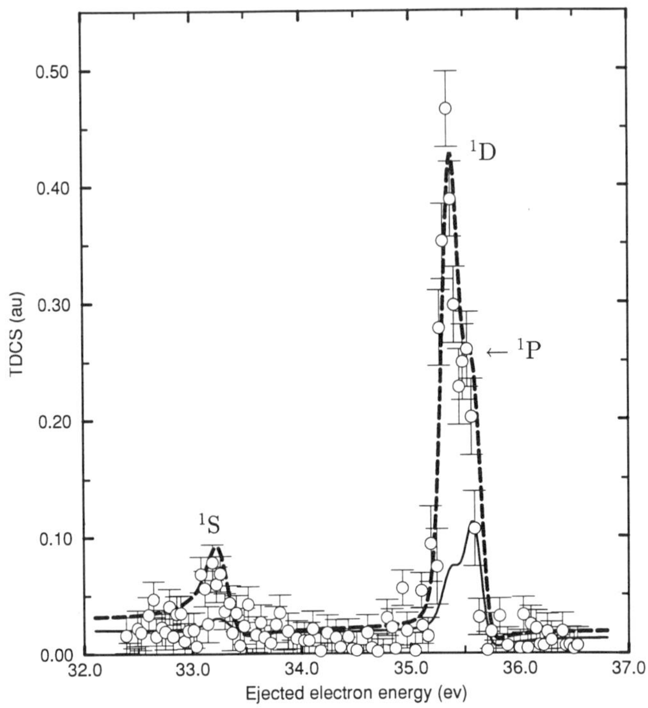

Figure 10. TDCS (in au) for ionization to $He^+(1s)$ at $E_0 = 200$ eV, $\theta_f = 13^0$ and $\theta_\kappa = 230^0$. Curves are theoretical results convoluted with experimental energy resolution of 150 meV : solid curve, first Born approximation; dashed curve, second Born approximation. Open circles are experimental data of Lower and Weigold[34] normalised to give best visual fit to the second Born curve.

Figure 11 shows the data of McDonald and Crowe[37] at $E_0 = 200$ eV, $\theta_f = 12^0$ and $\theta_\kappa = 240^0$, almost the same kinematics as the Lower and Weigold experiment of figure 10. The difference is that McDonald and Crowe have a finer energy resolution than Lower and Weigold, 80 meV compared with 150 meV, and so are better able to separate the ^{1}D and ^{1}P resonance contributions. From the measurements we see very clearly the failure of the first Born approximation which gives a ^{1}D peak significantly smaller than the ^{1}P peak. By contrast, the second Born calculation is in good agreement with the measurements, except that the second Born peaks are seen to be shifted to slightly higher ejected energies. This can be explained from the calculated positions of our ^{1}D and ^1P resonances, shown in table 1, which are slightly higher, by about 0.05 eV, than the true positions as estimated by van den Brink et al[44].

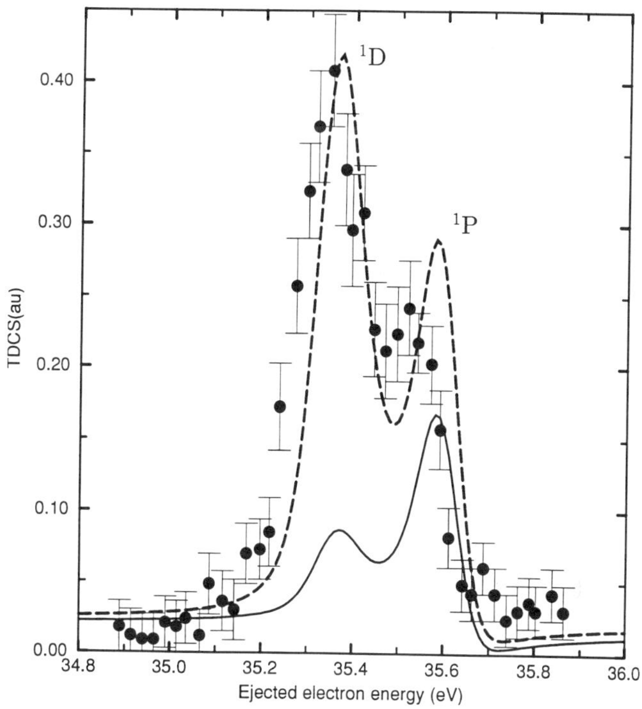

Figure 11. TDCS (in au) for ionization to $He^+(1s)$ at $E_0 = 200$ eV, $\theta_f = 12^0$ and $\theta_\kappa = 240^0$. Curves are theoretical results convoluted with experimental energy resolution of 80 meV : solid curve, first Born approximation; dashed curve, second Born approximation. Solid circles are experimental data of McDonald and Crowe[37] normalised to give best visual fit to the second Born curve.

It has been customary to reduce the results of experiment to the so - called Shore - Balashov parameters[45–47]. In the vicinity of autoionizing states the Shore - Balashov formula writes the TDCS in the form

$$\frac{d^3\sigma}{dE d\Omega_\kappa d\Omega_f} = f(\mathbf{k}_f, \boldsymbol{\kappa}) + \sum_r \frac{\left(a_r(\hat{\mathbf{k}}_f, \hat{\boldsymbol{\kappa}}, E_0)\epsilon_r + b_r(\hat{\mathbf{k}}_f, \hat{\boldsymbol{\kappa}}, E_0)\right)}{(1+\epsilon_r^2)} \tag{33}$$

where

$$\epsilon_r = \frac{2\,(E_\kappa - E_r)}{\Gamma_r} \tag{34}$$

In (33) individual resonance contributions are labelled by the index r while $f(\mathbf{k}_f, \boldsymbol{\kappa})$ represents the smooth direct background cross section. Formula (33) assumes that individual resonances with the same $L_r S_r$, where L_r (S_r) stands for the total orbital (spin) angular momentum of the r th resonant state, do not overlap and that there is no appreciable interaction between the scattered and autoionized electrons. These conditions are satisfied in figures 7 to 11. A problem in extracting these parameters from experiment is the need to deconvolve the energy resolution out of the data.

Table 2. Shore - Balashov parameters a_r, b_r and f_r, in au, for the $(2s^2)^1S$, $(2p^2)^1D$ and $(2s2p)^1P$ resonances at $E_0 = 200$ eV and $\theta_f = 13^0$. Theory : second Born approximation. Experiment : Lower and Weigold[34]. The experimental value of $f_{^1S}$ has been normalised to theory at $\theta_\kappa = 50^0$.

(a) $\theta_\kappa = 50^0$.

	a_r		b_r		f_r	
State	Theory	Experiment	Theory	Experiment	Theory	Experiment
^1S	0.250	0.204 ± 0.019	−0.024	0.104 ± 0.010	0.345	0.345 ± 0.017
^1D	0.075	0.157 ± 0.020	−0.278	-0.149 ± 0.021	0.319	0.327 ± 0.016
^1P	0.038	0.272 ± 0.045	−0.247	-0.215 ± 0.038	0.314	0.325 ± 0.017

(b) $\theta_\kappa = 230^0$.

	a_r		b_r		f_r	
State	Theory	Experiment	Theory	Experiment	Theory	Experiment
^1S	−0.063	-0.010 ± 0.010	0.084	0.019 ± 0.005	0.027	0.012 ± 0.002
^1D	0.013	0.052 ± 0.016	0.801	0.213 ± 0.016	0.021	0.012 ± 0.002
^1P	−0.286	-0.077 ± 0.029	0.522	0.115 ± 0.021	0.021	0.012 ± 0.002

In table 2 we compare our second Born results for the Shore - Balashov parameters with the experimental values obtained by Lower and Weigold[34] for the cases of figures 9 and 10, ie, $E_0 = 200$ eV, $\theta_f = 13^0$, $\theta_\kappa = 50^0$ and $\theta_\kappa = 230^0$. The theoretical parameters have been obtained by fitting the bare, ie, unconvoluted, theoretical cross section to the Shore - Balashov formula. In each case an excellent fit to this formula was obtained, justifying the validity of the form of the expression (33). Since the experiments are not absolute, the experimental parameters have been normalised to the theory by setting $f(\mathbf{k}_f, \boldsymbol{\kappa})$ equal to the theoretical value at the position of the ^1S resonance for $\theta_\kappa = 50^0$. In table 2 we denote by f_r the value of $f(\mathbf{k}_f, \boldsymbol{\kappa})$ at the position of the r th resonance. Although the experiment is not absolute, the relative normalisation at different θ_κ is supposedly determined for given fixed incident energy E_0 and scattering angle θ_f. Table 2 shows poor agreement between theory and experiment despite the good accord seen in figures 9 and 10, this is strange. At $\theta_\kappa = 230^0$, table 2, the theoretical a_r and b_r parameters are about four times larger than experiment, while the theoretical f_r parameter is almost twice as large. A similar situation, but not such a pronounced deviation, occurs at 400 eV[1,3]. This implies a substantial normalisation discrepancy between theory and experiment, by a factor of 4 in the resonance contributions and by a factor of 2 in the background cross section. In particular, it suggests that the theoretical binary to recoil ratio for the direct ionization process is twice as large as seen in experiment. Usually, when first and second Born theories fail for direct ionization, this discrepancy is the other way around, see figure 6(b). Could there be a problem with extracting the Shore - Balashov parameters from the experimental data and/or with the relative normalisation of the measurements ?

In summary, for excitation - autoionization, the agreement between the second Born calculations and experiment is very encouraging and there is clear improvement upon the first Born approximation.

4. CONCLUSIONS

There is clear evidence of the importance of the second Born term in the excitation - ionization and excitation - autoionization processes that we have studied here. Yet discrepancies still exist between theory and experiment. The present second Born calculations are, however, only an initial step. Even within the framework of the plane wave second Born approximation, significant improvements, on what has been done here, are possible. One priority is to make a more accurate evaluation of the second - order term using a pseudostate basis rather than the relatively crude closure approximation. A second priority is to explicitly extract the two - step mechanism to see exactly how important it really is. A third improvement would be to use a fully correlated wave function for ψ_0 in the evaluation of the second - order term. In addition, there is also scope to improve both the first and second Born terms by using a better approximation for ψ_f, one that will give a definitive converged answer. Again, the use of pseudostates, this time in the close - coupling expansion (24), would be the obvious route to achieve this goal. These developments are in progress.

ACKNOWLEDGEMENTS

We gratefully acknowledge support from the British Council under the "Alliance" scheme. We are indebted to G. Stefani, L. Avaldi and R. Camilloni for generously supplying us with their measurements, and to J. Lower and E. Weigold for help with the data on excitation - autoionization. We would also like to thank C.P. Campbell for his assistance with the preparation of this article. One of us (PJM) thanks the European Social Fund for a postgraduate studentship.

REFERENCES

1. P.J. Marchalant, *(e, 2e) Excitation - Ionization of Helium*, PhD Thesis, The Queen's University of Belfast, (1996).
2. P.J. Marchalant, C.T. Whelan and H.R.J. Walters, Second - order effects in (e, 2e) excitation - ionization of helium to $He^+(n = 2)$, *J. Phys. B*, to be published.
3. P.J. Marchalant, C.T. Whelan and H.R.J. Walters, Second - order effects in (e, 2e) excitation - autoionization of helium, *J. Phys. B*, to be published.
4. J.H. McGuire, Multiple - electron excitation, ionization, and transfer in high - velocity atomic and molecular collisions, *Adv. At. Mol. Phys.* 29:217 (1991).
5. R.J. Tweed, Double processes in e^- – He collisions, *Z. Phys. D* 23:309 (1992).
6. A. Franz and P.L. Altick, Electron - impact ionization - excitation cross sections of helium, *J. Phys. B* 28:4639 (1995).
7. F. Mota - Furtado and P.F. O'Mahony, (e, 2e) reactions leaving the residual ion in an excited state, in : *Invited Papers and Progress Reports of the 2nd European Conference on (e, 2e) Collisions and Related Problems*, H. Ehrhardt, ed., Universität Kaiserslautern, Germany, p7 (1989).
8. F. Mota - Furtado and P.F. O'Mahony, Theoretical triple differential cross sections for the simultaneous ionization and excitation of helium, in : *Invited Papers and Progress Reports of the 3rd European Conference on (e, 2e) Collisions and Related Problems*, G. Stefani, ed., Area della Ricerca di Roma, Italy, p1 (1990).

9. O. Robaux, R.J. Tweed and J. Langlois, Wavefunction effects in theoretical (e, 2e) cross sections for ionization of helium with excitation of the residual ion, *J. Phys. B* 24:4567 (1991).
10. C. Dupré, A. Lahmam - Bennani, A. Duguet, F. Mota - Furtado, P.F. O'Mahony and C. Dal Cappello, (e, 2e) triple differential cross sections for the simultaneous ionization and excitation of helium, *J. Phys. B* 25:259 (1992).
11. A. Franz and P.L. Altick, Initial - and final - state correlation in ionization - excitation of helium by electron impact, *J. Phys. B* 25:L257 (1992).
12. A.S. Kheifets, I. Bray, I.E. McCarthy and B. Shang, Theoretical triple differential cross section of the helium atom ionization with excitation to the $n = 2$ ion state, *Phys. Rev. A* 50:4700 (1994).
13. B. Nath, R. Biswas and C. Sinha, Electron - impact triple differential cross sections of the helium atom for simultaneous ionization and excitation, *J. Phys. B* 29:5909 (1996).
14. A. Pochat, R.J. Tweed, M. Doritch and J. Peresse, Double and triple differential cross sections for autoionisation of helium by electron impact, *J. Phys. B* 15:2269 (1982).
15. R. J. Tweed and J. Langlois, Theoretical triple differential cross sections for the autoionisation of helium by electron impact : limits on the validity of first - order models, *J. Phys. B* 19:3583 (1986).
16. A. S. Kheifets, Triple differential cross section calculation for the helium autoionization by electron impact, *J. Phys. B* 26:2053 (1993).
17. I.E. McCarthy and B. Shang, Calculation of the resonant ionization of helium, *Phys. Rev. A* 47:4807 (1993).
18. O. Samardzic, A.S. Kheifets, E. Weigold, B. Shang and M.J. Brunger, An (e, 2e) study of helium autoionization experiment and theory, *J. Phys. B* 28:725 (1995).
19. G. Stefani, L. Avaldi and R. Camilloni, On the relevance of the final ionic state to the (e, 2e) TDCS, *J. Phys. B* 23:L227 (1990).
20. G. Stefani, L. Avaldi, R. Camilloni and E. Fainelli, Resonant and direct ionization studied by (e, 2e) experiments, in : *Correlations and Polarization in Electronic and Atomic Collisions and (e, 2e) Reactions*, P.J.O. Teubner and E. Weigold, eds., Institute of Physics, Bristol, England, p109 (1992).
21. G. Stefani, L. Avaldi and R. Camilloni, private communication (1995).
22. A.J. Dixon, I.E. McCarthy and E. Weigold, Excitation of the $n = 2$ states of He^+ in the ionization of helium, *J. Phys. B* 9:L195 (1976).
23. J.P.D. Cook, I.E. McCarthy, A.T. Stelbovics and E. Weigold, Non - coplanar symmetric (e, 2e) momentum profile measurements for helium : an accurate test of helium wavefunctions, *J. Phys. B* 17:2339 (1984).
24. A.D. Smith, M.A. Coplan, D.J. Chornay, J.H. Moore, J.A. Tossel, J. Mrozek, V.H. Smith Jr. and N.S. Chant, Distortion effects in the (e, 2e) spectroscopy of helium at high momentum, *J. Phys. B* 19:969 (1986).
25. I.E. McCarthy, A. Ugbade, E. Weigold and P.J.O. Teubner, (e, 2e) reaction as a probe for details of the helium wave function, *Phys. Rev. Lett.* 33:459 (1974).
26. A.J. Murray and F.H. Read, Coplanar doubly - symmetric helium (e, 2e) measurements with excitation of the residual ion, *J. Phys. B* 25:L579 (1992).
27. J.L. Forand, K. Becker and J.W. McConkey, Absolute electron impact emission cross section of the He^+ $2^2P \rightarrow 1^2S$ line at 304 Å produced by simultaneous ionisation - excitation, *J. Phys. B* 18:1409 (1985).
28. J.O.P. Pedersen and F. Folkmann, Simultaneous excitation and ionisation of helium by fast projectiles, *J. Phys. B* 23:441 (1990).
29. K-H. Schartner, B. Lommel and D. Detleffsen, Production of $He^+(2p)$ in collisions of He with energetic H^+, H_2^+, H_3^+ ions and electrons, *J. Phys. B* 24:L13 (1991).

30. S. Fuelling, R. Bruch, E.A. Rauscher, E. Träbert and P.H. Heckmann, Ionization plus excitation of helium in fast H^+ + He and e^- + He collisions, *Nucl. Instrum. Methods B* 56/57:275 (1991).
31. M. Bailey, R. Bruch, E.Rauscher and S. Bliman, Cross sections for the ionization - excitation of helium by fast electrons and H^+, H_2^+ and H_3^+ ions : $(np)^2P^o$ levels, $n = 2 - 5$, *J. Phys. B* 28:2655 (1995).
32. P.A. Hayes and J.F. Williams, Simultaneous Ionization and Excitation to the He^+ 2^2P state, *Phys. Rev. Lett.* 77:3098 (1996).
33. E. Weigold, A. Ugbade and P.J.O. Teubner, Observation of autoionizing transitions in helium using the (e, 2e) technique, *Phys. Rev. Lett.* 35:209 (1975).
34. J. Lower and E. Weigold, Correlations in the autoionising region of He measured by the (e, 2e) technique, *J. Phys. B* 23:2819 (1990).
35. D.G. McDonald and A. Crowe, Triple differential cross sections in the vicinity of the $(2s^2)^1S$ state of helium, *Z. Phys. D* 23:371 (1992).
36. D.G. McDonald and A. Crowe, An (e, 2e) study of double excitation - autoionization mechanisms in helium, *J. Phys. B* 26:2887 (1993).
37. A. Crowe and D.G. McDonald, Processes other than simple outer shell ionization, in : *(e, 2e) and Related Processes*, C.T. Whelan, H.R.J. Walters, A. Lahmam - Bennani and H. Ehrhardt, eds., Kluwer, Dordrecht, p383 (1993).
38. H.R.J. Walters, Perturbative methods in electron - and positron - atom scattering, *Phys. Repts.* 116:1 (1984).
39. R.J. Tweed and J. Langlois, Wavefunction effects in the calculation of cross sections for electron impact ionisation of helium at high energy, *J. Phys. B* 20:5213 (1987).
40. K. Fankowski and C.L. Pekeris, Logarithmic terms in the wave functions of the ground state of two - electron atoms, *Phys. Rev.* 146:46 (1966).
41. F.W. Byron Jr. and C.J. Joachain, Correlation effects in atoms. I. helium, *Phys. Rev.* 146:1 (1966).
42. H. Ehrhardt, M. Fischer, K. Jung, F.W. Byron Jr., C.J. Joachain and B. Piraux, Triple differential cross sections for the ionization of helium by fast electrons, *Phys. Rev. Lett.* 48:1807 (1982).
43. F.W. Byron Jr. and C.J. Joachain, Theory of (e, 2e) reactions, *Phys. Repts.* 179:211 (1989).
44. J.P. van den Brink, G. Nienhuis, J. van Eck and H.G.M. Heideman, Coherences between autoionising states of different excitation energies, *J. Phys. B* 22:3501 (1989).
45. B.W. Shore, Analysis of absorption profiles of autoionizing lines, *J. Opt. Soc. Am.* 57:881 (1967).
46. V.V. Balashov, S.S. Lipovetskiĭ and V.S. Senashenko, A unified desription of the resonance line profile in the energy spectra of scattered and emitted electrons, *Sov. Phys. - JETP* 36:858 (1973).
47. R.J. Tweed, Autoionization of helium by electron impact : a theory incorporating exchange, *J. Phys. B* 9:1725 (1976).

AN ANALYTICAL APPROACH TO RESONANT AND DIRECT FRAGMENTATION OF MANY-BODY COULOMB SYSTEMS

J. Berakdar

Atomic and Molecular Physics Laboratories,
Research School of Physical Sciences and Engineering,
Institute of Advanced Studies, Australian National University,
Canberra, ACT 0200, Australia

INTRODUCTION

The description of the correlated motion of few Coulomb interacting particles is one of the fundamental unsolved problems in theoretical physics. At energies above the total fragmentation threshold the infinite-range of Coulomb forces leads to phase distortions of the plane-wave motion of the escaping particles [1, 2, 3, 4]. The determination of this correlated phase modifications leads to asymptotic scattering states of the many-body Coulomb systems [1, 2, 3, 4]. On the other hand evaluation of scattering amplitudes requires the knowledge of the many-body scattering state in the whole Hilbert space. In fact, the reaction zones most important for such amplitudes are often confined to a small region around the origin, the so-called condensation region, where all particles are close together. Therefore, a correct propagation of asymptotic states to finite ranges (in phase space) is crucial for the correct description of scattering reactions. Thus, we first discuss asymptotic scattering states of N interacting charged particles. Based on that we develop then a theoretical model for calculating scattering amplitudes of reactions leading to four-body Coulomb continuum systems. Subsequently, we introduce a method to incorporate the reaction dynamics at finite distances of three interacting charged particles whilst maintaining the requirement of correct asymptotic behaviour of scattering states.

N-BODY ASYMPTOTIC SCATTERING STATES AND THEIR FINITE-DISTANCE PROPAGATIONS

For N particles of masses m moving in the N-fold continuum of a residual ion of mass M with $m/M \ll 1$, the non-relativistic time-independent Schrödinger equation at the total energy E can be formulated in the relative-coordinate representation as

$$(H_0 + V - E)\ \Psi(\mathbf{r}_1, \cdots, \mathbf{r}_N) = 0 \tag{1}$$

In Eq. (1) H_0 is the total kinetic-energy operator. The total potential operator V is given by

$$V = \sum_{j=1}^{N} \frac{ZZ_j}{r_j} + \sum_{\substack{i,j \\ j>i=1}}^{N} \frac{Z_i Z_j}{r_{ij}} \tag{2}$$

where $\mathbf{r}_j$ is the position of particle j with respect to the residual charge Z and $\mathbf{r}_{ij} := \mathbf{r}_i - \mathbf{r}_j$ denotes the relative coordinate between particles i and j . Assuming the continuum particles to escape with relative asymptotic momenta $\mathbf{k}_j$ (with respect to the charge Z) it has been suggested [1], due to unpublished work by Redmond, that for large interparticle distances the wave function $\Psi(\mathbf{r}_1, \cdots, \mathbf{r}_N)$ takes on the form

$$\lim_{\substack{r_{lm}\to\infty \\ r_n\to\infty}} \Psi(\mathbf{r}_1, \cdots, \mathbf{r}_N) \rightarrow (2\pi)^{-3N/2} \prod_{s=1}^{N} \xi_s(\mathbf{r}_s)\psi_s(\mathbf{r}_s) \prod_{\substack{i,j=1 \\ j>i}}^{N} \psi_{ij}(\mathbf{r}_{ij}) \,, \;\; \forall\; l,m,n \in [1,N]; m > l. \tag{3}$$

where the functions $\xi_j(\mathbf{r}_j)$, $\psi_j(\mathbf{r}_j)$, $\psi_{ij}(\mathbf{r}_{ij})$ are defined as

$$\xi_j(\mathbf{r}_j) := \exp(i\mathbf{k}_j \cdot \mathbf{r}_j) \tag{4}$$

$$\psi_j(\mathbf{r}_j) := \exp\Big[\mp i\alpha_j \ln(k_j\, r_j \pm \mathbf{k}_j \cdot \mathbf{r}_j)\Big] \,, \tag{5}$$

$$\psi_{ij}(\mathbf{r}_{ij}) := \exp\Big[\mp i\alpha_{ij} \ln(k_{ij}\, r_{ij} \pm \mathbf{k}_{ij} \cdot \mathbf{r}_{ij})\Big] \,. \tag{6}$$

The $+$ and $-$ signs refer to outgoing and incoming boundary conditions, respectively, and $\mathbf{k}_{ij}$ is the momentum conjugate to $\mathbf{r}_{ij}$, i. e. $\mathbf{k}_{ij} := (\mathbf{k}_i - \mathbf{k}_j)/2$. The Sommerfeld-parameters α_j, α_{ij} are given by

$$\alpha_{ij} = \frac{Z_i Z_j}{v_{ij}}, \qquad \alpha_j = \frac{ZZ_j}{v_j}. \tag{7}$$

In Eq. (7) v_j denotes the velocity of particle j relative to the residual charge whereas $\mathbf{v}_{ij} := \mathbf{v}_\mathrm{i} - \mathbf{v}_\mathrm{j}$. Recently, using a multiple-scattering scheme, an analytical approximate expression for $\Psi(\mathbf{r}_1, \cdots, \mathbf{r}_N)$ has been derived [5]

$$\Psi(\mathbf{r}_1, \cdots, \mathbf{r}_N) \approx \mathcal{N} \prod_{m>l,j=1}^{N} \xi_j(\mathbf{r}_j)\varphi_j(\mathbf{r}_j)\varphi_{lm}(\mathbf{r}_{lm}) \tag{8}$$

where

$$\varphi_j(\mathbf{r}_j) = {}_1F_1[\alpha_j, 1, -i(k_j r_j + \mathbf{k}_j \cdot \mathbf{r}_j)] \tag{9}$$

and $\mathcal{N}$ is a normalisation constant. It has been shown [5] that the part of the total Hamiltonian which has been neglected to arrive at the eigenstate (8) diminishes faster than the Coulomb potential in the asymptotic regime of large interparticle distances. It is straightforward to show that, for large interparticle separations, the function (8) tends to (3) which confirms the assumption of the N-body asymptotic states in the form (3).

APPLICATION TO THE FOUR-BODY PROBLEM

Apart from the Rutherford scattering, the simplest application using the wavefunction (8) is the description of three-body Coulomb continuum states which are, for example, achieved as final states of electron-impact ionization and photo double ionization of atomic and molecular systems. In this case the wave function (8) simplifies to the three-body wave function proposed in Refs. [2, 6] and extensively used by various authors. The second step in complexity is the description of the four-body scattering states. These states arise in the final channel of various reactions such as the double ionization of atomic systems by electron and positron impact as well as the photo triple ionization. Experimentally there has been an increased interest in such reactions due to recent advances in coincidence-detection techniques [7, 8, 9, 10, 11, 12, 13]. The first measurement of fully differential cross sections of reactions leading to four-body Coulomb continuum states have been reported in Refs. [7, 8]. In this study argon and krypton targets have been double ionized by a fast projectile electron under conditions where little momentum is being transferred to the target atoms. To describe the final state, the standard theoretical treatment [14, 15, 16, 17, 18, 19, 20, 21] has been to reduce the four-body Coulomb continuum problem to a three-body problem by employing a first-order Born-type approximation (FBA) in the projectile-target potential which is justified for the conditions under which the experiment has been performed. Recently, the first measurements for electron-impact double ionization have been performed at intermediate incident energies using the cold-target recoil-ion momentum spectroscopy [9, 11]. In this case the FBA model is inappropriate. The last statement was also inferred from recent measurements [13] of the cross section for the double ionization of magnesium by electron impact at moderate incident energy. In view of these recent experiments it appears timely to consider the applicability of the wavefunction (8) for the case of the four-body problem.

Despite recent major advances in computational power the calculations of scattering-reaction amplitudes involving more than three particles by directly employing the wavefunction (8) is still out of reach. This is basically due to the spatial coupling between the continuum particles. Thus, further simplifications are needed. An approach well investigated in the treatment of electron-impact single ionization of atomic systems is that of effective charges where correlations between the continuum particles are completely subsumed into velocity-dependent electron-ion interactions [22, 23, 24]. Here we adopt this philosophy and write the total potential of the four-body system in the form

$$\sum_{j=1}^{3} \overline{V}_j := \frac{\overline{Z}_j}{r_j} = V. \tag{10}$$

Obviously, the effective charges $\overline{Z}_j$ have to be position dependent in order to satisfy Eq. (10). To incorporate some features of the reaction dynamics in the functions $\overline{Z}_j$ we introduce for the effective two-body potential $\overline{V}_j = \overline{Z}_j / r_j$ the linear transformation

$$\overline{V}_j = a_{j1}V_1 + a_{j2}V_2 + a_{j3}V_3 + a_{j4}V_{12} + a_{j5}V_{13} + a_{j6}V_{23}, \; j \in [1, 3] \tag{11}$$

where $V_i = ZZ_i/r_i$, $V_{ij} = Z_iZ_j/r_{ij}$ are the physical two-body potentials. Equivalent transformations of the charges $\overline{Z}_j = \overline{V}_j r_j$ immediately derive from Eqs. (11). In Ref. [5] the functional forms of the expansion coefficients a_{ij} have been derived by imposing the following requirements: **(a)** all three continuum particles must be treated on equal footing. **(b)** When two particles i, j approach each other ($r_{ij} \to 0, r_{ij}/r_k \to 0$) their mutual interaction increases as Z_iZ_j/r_{ij} and dominates the other interactions appearing in Eq. (1). Hence, the interactions of particle i and particle j with the nucleus must change in a way which simulates their mutual Coulomb interaction regardless the direction in which particle i is approached by particle j. **(c)** If one particle is moving in the vicinity of the nucleus it experiences the full nuclear charge. Assuming the remaining two continuum particles to be far away from the nucleus, their interaction with the nucleus must remain finite. **(d)** The Schrödinger equation (1) must be invariant under the transformation (11). **(e)** If one particle, say particle ℓ, is far away from the remaining three particles this particle (ℓ) "sees" a net charge of $Z + Z_i + Z_j$. In addition, if in the three-body system, formed by the nucleus, particle i and particle j, particle i approaches the nucleus, then particle j experiences a net nuclear charge of $Z + Z_i$, as

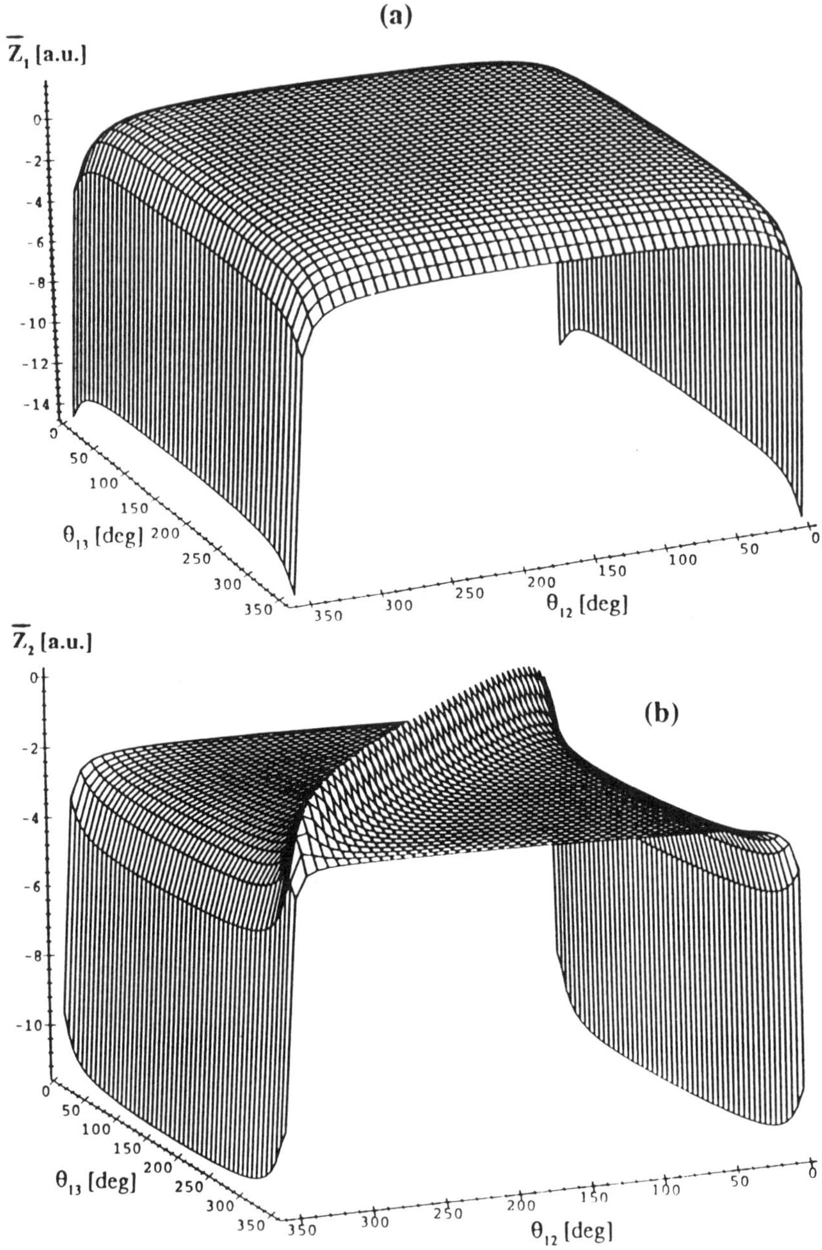

Fig. 1. For a four-body Coulomb system consisting of two electrons and one positron in the field of a residual charge $Z = 2$ the effective charges, given by Eqs. (12,13), are depicted for the case where all particles escape in the same plane with velocities $v_1 = v_2 = 1\ a.u.$ and $v_3 = 1.2\ a.u.$. The positron is taken to be particle 1. All angles are measured with respect to the direction $\hat{\mathbf{v}}_1$. Fig. (a) shows the angular dependence of the positron-nucleus effective product charge $\overline{Z}_1$ whereas in Fig. (b) the product charge of electron 2 with the nucleus ($\overline{Z}_2$) is investigated.

immediately concluded from Eq. (1). **(f)** The four-body Wannier ionization mode [25, 26] has to be satisfied. The above listed conditions resulted in the spatially dependent product charges [5]:

$$\overline{Z}_1 = ZZ_1 + \left[\frac{Z_1 Z_2}{(r_1+r_2)r_{12}} + \frac{Z_1 Z_3}{(r_1+r_3)r_{13}}\right] r_1^2 \tag{12}$$

$$\overline{Z}_2 = ZZ_2 + \left[\frac{Z_1 Z_2}{(r_1+r_2)r_{12}} + \frac{Z_2 Z_3}{(r_2+r_3)r_{23}}\right] r_2^2 \tag{13}$$

$$\overline{Z}_3 = ZZ_3 + \left[\frac{Z_1 Z_3}{(r_1+r_3)r_{13}} + \frac{Z_2 Z_3}{(r_2+r_3)r_{23}}\right] r_3^2. \tag{14}$$

For practical applications, the position dependence of $\overline{Z}_j$ has to be converted into velocity dependence by assuming, as done in Refs. [22, 23, 24], that ratios of relative distances are proportional to corresponding ratios of velocities. For the case of positron-impact double ionization of helium the positron-nucleus product charge $\overline{Z}_1$ is depicted in Fig. (1a) whereas Fig. (1b) shows the electron-nucleus product charge in a typical scattering geometry where all momenta of the outgoing particles lie in the same plane. From Fig. 1(a,b) the above listed properties of the effective charges are easily verified. Using this method pilot calculations of the fully differential cross sections for the electron and positron-impact double ionization of $He(^1S^e)$ have been performed [27].

THREE-BODY SYSTEMS IN THE CONDENSATION REGION

Now we focus on the three-body problem at finite distances. In what follows we consider three particles of arbitrary charges nad masses. Making use of the basic ideas of the Wannier theory for threshold ionization [25, 28], we split the Hilbert space into an "inner" and a "far zone" depending on whether the total potential is larger or smaller than the kinetic energy. The scattering state in the inner zone is derived with special regard to the ionization dynamics (in the sense specified below) rather than asymptotic considerations. Subsequently, this state is mapped onto the asymptotic solution at the boundary between the "inner" and the "far zone" to arrive at an asymptotically correct behaviour. To this end we follow Ref. [4] and reformulate the non-relativistic time-independent three-particle Schrödinger equation in the curvilinear coordinate system

$$\{\xi_k = r_{ij} + \hat{\mathbf{k}}_{ij}\cdot\mathbf{r}_{ij};\ \xi_m = r_{ij}\},\quad \epsilon_{ijk}\neq 0;\ j>i, k\in[1,3];\ m\in[4,6] \tag{15}$$

where $\hat{\mathbf{k}}_{ij}$ denote the directions of the momenta, $\mathbf{k}_{ij}$, conjugate to the interparticle distances r_{ij}. To decouple kinematical from dynamical properties the eigenfunction Ψ of the total Hamiltonian $\mathcal{H}$, at the total energy E, is written in the form

$$\Psi(\mathbf{r}_{ij},\mathbf{R}_k) = N\exp(i\,\mathbf{r}_{ij}\cdot\mathbf{k}_{ij} + i\,\mathbf{R}_k\cdot\mathbf{K}_k)\,\overline{\Psi}(\mathbf{r}_{ij},\mathbf{R}_k)\,. \tag{16}$$

In Eq.(16) the coordinate $\mathbf{R}_k$ refers to the particle k with respect to the centre of mass of the pair ij while $\mathbf{K}_k$ designates the momentum conjugate to this coordinate. The normalisation factor is denoted by N. The plane-wave part of Eq. (16) is an eigenfunction of the kinetic-energy operator with eigenvalue E. Thus, $\overline{\Psi}(\mathbf{r}_{ij},\mathbf{R}_k)$ is solely due to the presence of the total potential and is determined as an eigenfunction of a differential operator H, given in Refs. [4]. It is established [4] that the three-body operator H, expressed in the coordinate system (15), can be decomposed into two *parametrically* coupled differential operators; an operator $H_{\rm par}$ which is differential in the coordinates $\xi_{1\cdots3}$ only (and therefore the dependence of $H_{\rm par}$ on $\xi_{4\cdots6}$ can be treated parametrically) while the second operator is differential in internal degrees of freedom $r_{ij};\ j > i \in \{1,2,3\}$. An additional mixing term arises from the off-diagonal elements of the metric tensor. Furthermore, the differential operator $H_{\rm par}$ is exactly separable in the coordinates $\xi_{1\cdots3}$ since it factorizes as

$$H_{\rm par} = \sum_{j=1}^{3} H_{\xi_j},\ [H_{\xi_j}, H_{\xi_i}] = 0;\ \forall\, i,j \in \{1,2,3\}, \tag{17}$$

where (atomic units are used throughout)

$$H_{\xi_j} = \frac{2}{\mu_{lm} r_{lm}} \left[\partial_{\xi_j}\, \xi_j\, \partial_{\xi_j} + i k_{lm}\, \xi_j\, \partial_{\xi_j} - \mu_{lm}\, Z_{lm}\right];\ \epsilon_{jlm} \neq 0,\ j \in \{1,2,3\}. \tag{18}$$

In Eq. (18) the reduced mass of the pair ij is denoted by μ_{ij} and their product charge is referred to by Z_{lm}. Considering that the relative plane-wave motion has been factored out [Eq. (16)], then Eq. (18) is the Schrödinger equation for two-body Rutherford scattering expressed in parabolic coordinates [29]. Hence, within the approximation $H \approx H_{\rm par}$, the three-body system is considered as the sum of three spatially decoupled two-body Coulomb systems on the *two-body energy shell.* At a given total energy E, the exact regular eigenfunction of the total Hamiltonian $\mathcal{H}$, within the approximation $H \approx H_{\rm par}$, has the explicit form [4]:

$$\begin{aligned}\Psi_{\rm par}(\xi_{1\cdots6}; \mathbf{k}_{ij}) \ &= \ N \exp(i\, \mathbf{r}_{ij} \cdot \mathbf{k}_{ij} + i\, \mathbf{R}_k \cdot \mathbf{K}_k)\ {}_1F_1\,(i\beta_{23}, 1,\ -ik_{23}\,\xi_1\,) \\ &\quad {}_1F_1\,(i\beta_{13},\ 1,\ -ik_{13}\,\xi_2\,)\ {}_1F_1\,(i\beta_{12},\ 1,\ -ik_{12}\,\xi_3\,)\,. \end{aligned} \tag{19}$$

The Sommerfeld parameters β_{ij} are given by $\beta_{ij} = Z_{ij}\mu_{ij}/k_{ij}$. The range of validity of the approximation $H \approx H_{\rm par}$ has been discussed elsewhere [4]. The approximate (asymptotic) separability expressed in Eqs. (17,19) as well as the parametric dependence of H_{ξ_j} [Eq. (18)] on internal degrees of freedom have been exploited to introduce three-body coupling between each two-body subsystem to the third particle [4] and to incorporate in the wavefunction some rotationally invariant features of the total potential surface. This results in local Sommerfeld-parameters dependent on body-fixed coordinates. Although this approach is physically appealing and has proved quite successful in calculating scattering amplitudes [31, 32] its applicability is hampered by the complicated functional dependence of the resulting wavefunctions and by the fact that properties of the total potential are inherent to the particular three-body system under investigation. Hence, the dynamically coupled many-body wavefunction has to be constructed for each particular three-body system which is quite an involved task. Therefore, we adopt here a different approach to take advantage of Eqs. (17,18,19) which is motivated by the measurement process and the Wannier analysis of threshold ionization [25].

In a scattering experiment the measurable quantities (observables) are the *asymptotic* momenta $\mathbf{k}_{ij}$ of the emerging reaction fragments (spin and spatial degrees of freedom are considered to be decoupled). In the "reaction zone" these quantum numbers are undetermined. To quantify this picture we define, following the treatment of Wannier [25, 28], an inner *Coulomb zone* and an outer *far zone* depending on whether the total potential or the kinetic energy is the dominant quantity. In the context of this work the Coulomb and far zone are more appropriately termed *momentum-exchange* and *asymptotic zones*, respectively. As is well known [25, 28], the boundary between these regimes is the Wannier radius R_w which is a scalar quantity. In the Coulomb zone a two-body subsystem ij can assume any two-body quantum state defined by a particular $\mathbf{k}'_{ij}$, i. e. each two-body subsystem propagates *off the two-body energy shell*, as defined by the (asymptotic) measurement process. The description of this is well facilitated by Eqs. (17,18,19) since the momenta $\mathbf{k}_{ij}$ enter in Eq. (18) as dummy parameters and are determined only by the asymptotic boundary conditions and the total energy conservation. However, to ensure the invariance of the Schrödinger equation under the introduction of intermediate momenta $\mathbf{k}'_{ij}$, we must operate under the constraint that the total energy E is conserved. This means that the two-body subsystems exchange an indefinite amount of momentum in the momentum-exchange zone and virtually occupy all (two-body) continuum states available in the energy band $[0, E]$. As the system evolves towards the Wannier boundary R_w the reaction fragments take on the (asymptotic) momenta measured in a scattering reaction. To put the preceding in a mathematical language we note that Eqs. (18) contain the magnitudes of the momenta k_{ij} parametrically. Within the present model, however, the directions $\hat{\mathbf{k}}_{ij}$ are fixed by the (asymptotic) boundary conditions [compare Eqs. (15)]. As the exact normalized eigenfunctions of H_{par} [compare Eq. (17)] are known and characterized by k'_{ij}, or equivalently by $k'^2_{ij}/2\mu_{ij} = E'_{ij} \in [0, E]$, we write the general solution in the momentum-exchange zone as a linear superposition of $\Psi_{\mathrm{par}}(\xi_{1\cdots6}; k'_{ij})$:

$$\Psi^{\mathrm{in}}(\xi_{1\cdots6}) = \mathcal{N} \int d^3\mathbf{k}'_{ij} d^3\mathbf{K}'_k \, A_{\mathbf{k}'_{ij}}(\xi_{4\cdots6}) \Psi_{\mathrm{par}}(\xi_{1\cdots6}; \mathbf{k}'_{ij}) \delta(E - E') \delta^2(\hat{\mathbf{k}}_{ij} - \hat{\mathbf{k}}'_{ij}) \delta^2(\hat{\mathbf{K}}_k - \hat{\mathbf{K}}'_k) \tag{20}$$

where E' is the intermediate total energy determined by k'_{ij}. It is readily verified that the function Eq. (20) is an eigenfunction of $\mathcal{H}$ under the approximation $H \approx H_{\mathrm{par}}$. To account for the neglected part $H - H_{\mathrm{par}}$ the expansion coefficients $A_{\mathbf{k}'_{ij}}(\xi_{4\cdots6})$ have to be deduced, e.g. variationally. For many-body continuum states, however, this procedure is intractable. In the context of this work $A_{\mathbf{k}'_{ij}}(\xi_{4\cdots6})$ derives, however, from the observation that the wavefunction (20) is defined only in the momentum-exchange zone whose boundary is fixed by R_w. The Wannier radius R_w, however, scales inversely with E [28]. Hence, the determination of $A_{\mathbf{k}'_{ij}}(\xi_{4\cdots6})$, and thus of Eq. (20), is needed only at lower excess energies E ($\lim_{E\gg1} R_w \to 0$). Assuming $H \approx H_{\mathrm{par}}$, we remark further that the expansion coefficients $A_{\mathbf{k}'_{ij}}(\xi_{4\cdots6})$ indicate the occupation probabilities for the intermediate states characterized by $E'_{ij} \in [0, E]$. According to the Wannier threshold analysis [25], the correlated motion in the interaction region is ergodic, which implies $A_{\mathbf{k}'_{ij}}(\xi_{4\cdots6}) = 1$. This means a flat probability distribution of occupying the intermediate virtual states in the momentum-exchange zone.

The wavefunction (20) is asymptotically incorrect [30]. Hence, a mapping onto the asymptotic wavefunction (19) at the boundary R_w is necessary. Generally, this can be done analogously to the R-matrix approach [33]. Here we adopt a different procedure by introducing the exponential matching factor $f := \exp(-R/R_w)$ where the extent of the three-body system is measured by $R := r_{12} + r_{13}$. The wavefunction in the entire Hilbert space is then given by

$$\Psi_{\mathrm{ex}}(\xi_{1\cdots6}; E) = f\ \Psi^{\mathrm{in}} + (1 - f)\Psi_{\mathrm{par}}(\xi_{1\cdots6}; \mathbf{k}_{ij}). \tag{21}$$

First we remark that since R_w and R are scalar quantities, i. e. they depend on $\xi_{4\cdots6}$ only, the wavefunction $\Psi_{\mathrm{ex}}(\xi_{1\cdots6}; E)$ is an eigensolution of the total Hamiltonian within the approximation $H \approx H_{\mathrm{par}} = \sum_{j=1}^{3} H_{\xi_j}$. As the wavefunction (19) is asymptotically correct for large interparticle separations [2, 4] and satisfy the Kato [34] cusp conditions [4] (R is always large in this case) it follows that these (asymptotic) properties are directly reflected into the wavefunction Ψ_{ex} (since $\lim_{R\gg1} f \to 0$). For distances R larger than R_w ($f \to 0$) we fall back to Eq. (19) which means that in high-energy scattering ($R_w \to 0$) the escaping particles directly assume their experimentally measured momenta. For $R < R_w$ the two-body subsystems exchange an indefinite amount of energy. Of particular interest is the region of low excess energies where R_w extends to very large distances R, i. e. $f \to 1$. In this case the three particles keep exchange energies up to infinity for $E \to 0$ and the transformation of the total wavefunction from Ψ^{in} to Ψ_{par} occurs at very large distances. This has important implications in so far as, according to Eq. (21), properties of scattering amplitudes which are derived from asymptotic wavefunctions are smeared out at threshold. A second important conclusion at threshold is that if we extend the integral in Eq.(21) to run over bound states

[30] as well, the coupling to highly excited Rydberg states provides the major contribution to the wavefunction, given by Eq. (21), in agreement with the Wannier picture of double escape [25, 28].

To view the physical meaning of the this procedure in terms of transition amplitudes we introduce hyper-spherical momenta $\kappa = E'$; $\tan\alpha = (K'_k/k'_{ij})\sqrt{\mu_{ij}/\mu_k}$ where $\mu_k = m_k(m_i + m_j)/(m_i + m_j + m_k)$. The integrals in Eq. (21) simplify then to

$$\Psi_{\rm ex}(\xi_{1\cdots 6}; E) = f\, C \int_0^{\pi/2} d\alpha\, \sin^2 2\alpha\, \Psi_{\rm par}(\xi_{1\cdots 6}; \alpha) \,+\, (1-f)\Psi_{\rm par}(\xi_{1\cdots 6}; k_{ij}). \tag{22}$$

where $C = \mathcal{N} \mu_{ij}^{3/2} \mu_k^{3/2} E^2$. In a reaction leading to a three-body continuum state $\Psi_{\rm ex}$, scattering amplitudes are determined by transition matrices of the form (prior form)

$$T = \langle \Psi_{\rm ex}|W|i\rangle \tag{23}$$

where $|i\rangle$ is the initial state of the three-body system and W is the perturbation operator due to which the final state $\Psi_{\rm ex}$ is achieved. Upon substitution of Eq. (22) into Eq. (23) and making use of the continuous convergence of the integrals involved in (22,23), the total transition amplitude T can be written in the form

$$T = T^{\rm asy} + C \int_0^{\pi/2} d\alpha\, \sin^2 2\alpha\, T^{\alpha} \tag{24}$$

where

$$T^{\alpha} = \langle\, f\, \Psi_{\rm par}(\xi_{1\cdots 6}; \alpha) \mid W \mid i\rangle,\; T^{\rm asy} = \langle\, (1-f)\, \Psi_{\rm par}(\xi_{1\cdots 6}; k_{ij})|W|i\rangle. \tag{25}$$

In case the integral in (21) runs over bound-state energies of the system the integral in Eq. (24) includes a sum over these states. From Eq. (24) it is obvious that a transition to an asymptotic state defined by the measured momenta $\mathbf{k}_{ij}$ occurs via an infinite number of intermediate virtual transitions given by T^{α} which may interfere to result in the measured cross section. This picture is closely related to the interpretations of resonant states made in Ref. [35]. It should be stressed that this procedure is applicable to an arbitrary three-body Coulomb system regardless of the mass and charge states of these particles. Here we investigate the one-photon double ionization of $He(^1S^e)$ (DPI) and the electron-impact ionization of atomic hydrogen (e,2e) and include in the integral (21) continuum states only, i. e. the momenta k'_{ij} (energies E'_{ij}) are real (positive real) quantities. According to Eq. (21) momentum-exchange effects dominate at shorter distances ($R < R_w$). Therefore, for DPI processes the velocity form of the dipole operator is employed (first-order perturbation theory for the radiation field is assumed). Neglecting energy exchange between the two-body subsystems results in the wavefunction (19) which has been employed for the calculations of the relative angular distributions of electrons following DPI with remarkable success [36, 37]. The comparison between results obtained by representing the final state by (19) or (21) provides an insight into the amount of energy-exchange effects.

From Figs. 2(a-c) it is evident that energy-exchange effects are strongly dependent on the configuration in which the two electrons are emitted. The most drastic influence is observed when the two electrons escape with low and asymmetric energies [Fig. 2(b)] in which case the symmetry of the initial-state ($^1S^e$) imposes less severe restrictions on the angular distribution than in the case of equal-energy electrons [37]. With increasing photon energy, the radius R_w declines and the electrons swiftly assume their asymptotic momenta [Fig. 2(c)]. Hence, energy-exchange effects play a less prominent role in this regime. For the case of (e,2e) [Fig. (3)] we remark that the position of the recoil and binary peaks are well predicted. However, there is some discrepancy between theory and experiment in the binary region. This

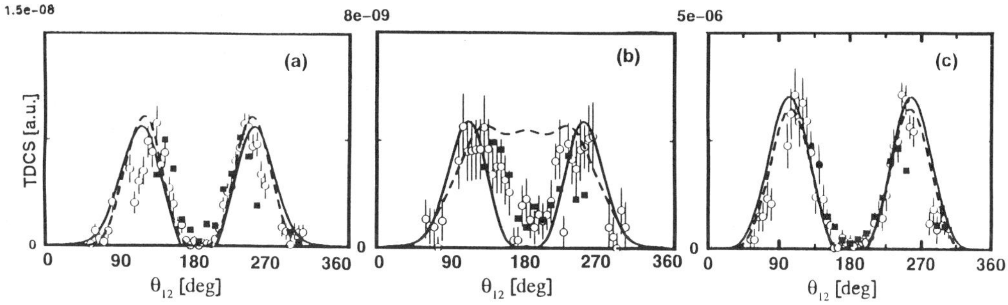

Fig. 2. (a) The Triply Differential Cross Section (TDCS) for the double ionization of $He(^1S^e)$ by linearly polarized photons. One electron is detected along the direction of the polarization vector (ϵ) while the other electron is detected, in coincidence with the first one, under an angle θ_{12} with respect to ϵ. The two equal-energy electrons escape with a total excess energy of $4\ eV$. The relative experimental data are due to [39]. Representing the final-state by Eq. (21) [or Eq. (19)] results in the solid [dashed curve]. Results of Eq.(19) are scaled down by a factor of 1.8. The initial state is represented by a Hyllerass wavefunction which contains radial and angular correlations [37]. In (b) same as (a) but the electron fixed to ϵ is detected with an energy of $3.3\ eV$ whereas in (c) the same geometry as in (a) is chosen but the excess energy is increased to $18.6\ eV$. The TDCS evaluated using Eq. (19) has been scaled down by a factor of 1.4.

discrepancy is due to the neglected part when employing the approximation $H = H_{\rm par}$ to arrive at Eq. (21) which basically means that the three-body system is considered as *spatially decoupled* three two-body systems. It is clear that an introduction of coupling (through $\xi_{4\cdots6}$) between these subsystems is feasible, however, cumbersome.

In conclusion, correlated N-body Coulomb scattering states and their asymptotic properties have been considered. For the four-body scattering problem an effective-charge appraoch has been presented which renders possible the calculations of scattering amplitudes. In the case of three-body Coulomb continuum problem, energy-exchange and off-shell effects have been included into analytical correlated three-body wavefunctions with correct asymptotic behaviour. In constructing these states the basic ideas of the Wannier theory of threshold double escape have been incorporated. To elucidate the extent of the energy-exchange effects, scattering amplitudes for photo-double ionization and electron-impact ionization leading to three-body continuum final state have been calculated.

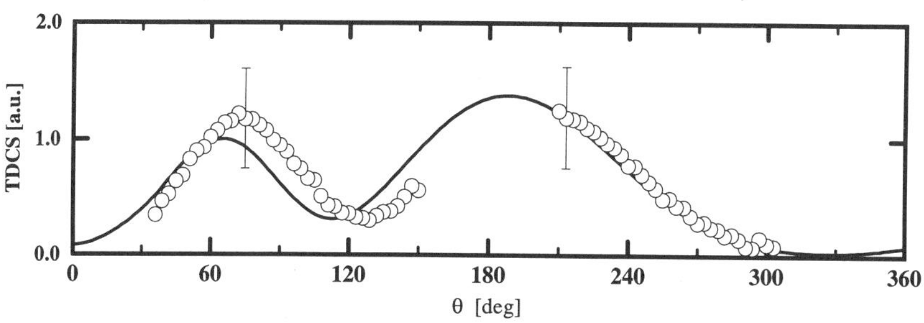

Fig. 3. The TDCS for the co-planar electron-impact ionization of atomic hydrogen calculated using the wavefunction (21). The incident energy is $54.4\ eV$. One electron is detected under an angle of $350°$ with respect to the incident direction $\mathbf{k}_i$ whereas the other one is detected under an angle θ with respect to $\mathbf{k}_i$ and with an energy of $5\ eV$. Experimental data are due to Ref. [38]. The error bars indicate the uncertainty in the absolute value of the TDCS.

ACKNOWLEDGEMENTS

I would like to thank Steve Buckman, Erich Weigold and Stephane Mazevet for helpful suggestions and comments and Alain Huetz for communicating his data. This work was supported by the Alexander von Humboldt Foundation and the Australian National University.

References

[1] L. Rosenberg, Phys. Rev. D **8**, 1833 (1973).

[2] M. Brauner, J. S. Briggs and H. Klar, J. Phys. B **22**, 2265 (1989).

[3] E. O. Alt and A. M. Mukhamedzhanov, Phys. Rev. A **47**, 2004 (1993).

[4] J. Berakdar Phys. Rev. A **53**, 2314 (1996); *ibid* **54**, 1480 (1996).

[5] J. Berakdar Phys. Rev. A (1997) in press

[6] G. Garibotti and J. E. Miraglia, Phys. Rev. A **21**, 572 (1980).

[7] A. Lahmam-Bennani, C. Dupŕe and A. Duguet, Phys. Rev. Lett. **63**, 1582 (1989).

[8] A. Lahmam-Bennani, A. Duguet, A. M. Grisogono and M. Lecas, J. Phys. B **25**, 2873 (1992).

[9] R. Dörner, V. Mergel, L. Spielberger, O. Jagutzki, S. Nuttgens, M. Unverzagt, H. Schmidt-Böcking, J. Ullrich, R. E. Olson, K. Tokesi, W. E. Meyerhof, W. Wu and C. L. Cocke, Nuclear Instr. & Meth. B **99**, 111 (1995).

[10] B. El Marji, A. Duguet, A. Lahmam-Bennani, M. Lecas and H. F. Wellenstein, J. Phys. B **26**, L733 (1995).

[11] O. Jagutzki, L. Spielberger, R. Dörner, S. Nüttgens, V. Mergel, H. Schmidt-Böcking, J. Ullrich, R. E. Olson and U. Buck, Z. Phys. D **36**, 5 (1996).

[12] L. Avaldi, P. Belotti, P. Bolognesi, R. Camilloni and G. Stefani, Phys. Rev. Lett **75**, 1915 (1995).

[13] M. J. Ford, J. H. Moore, M. A. Coplan, J. W. Cooper and J. P. Doering, Phys. Rev. Lett. **77**, 2650 (1996).

[14] B. Joulakian, C. Dal Cappello and M. Brauner, J. Phys. B **25**, 2863 (1992).

[15] B. Joulakian and C. Dal Cappello, Phys. Rev. A **47**, 3788 (1993).

[16] C. Dal Cappello and H. Le Rouzo, Phys. Rev. A **43**, 1395 (1991).

[17] R. J. Tweed, J. Phys. B **6**, 270 (1973).

[18] Yu. F. Smirnov, A. V. Pavlitchenkov, V. G. Levin and G. Neudatchin, J. Phys. B **11**, 20 (1978).

[19] J. Berakdar and H. Klar, J. Phys. B **26**, 4219 (1993).

[20] J. Berakdar, J. Phys. (France) IV **3**, C6-135 (1993).

[21] J. Berakdar, Phys. Rev. A **53**, 2281 (1996).

[22] R. K. Peterkop, *Theory of Ionization of Atoms by the Electron Impact* (Colorado Associated University Press, Boulder, 1977).

[23] S. Jetzke and F. H. M. Faisal, J. Phys. B **25**, 1543 (1992).

[24] M.R.H.Rudge and M.J.Seaton Proc.Roy.Soc. A,*283*, 262 (1965).

[25] G. Wannier, Phys. Rev. **90**, 817 (1953).

[26] H. Klar and W. Schlecht, J. Phys. B **9**, 1699 (1976).

[27] J. Berakdar Phys. Lett. A **220** , 237 (1996).

[28] A. R. P. Rau, Phys. Rev. A **4**, 207 (1971); Phys. Rep. **110**, 369 (1984).

[29] H. A. Bethe and E. E. Salpeter, *Quantum mechanics of one- and two-electron atoms* (Springer, Berlin, 1957).

[30] J. Berakdar, unpublished.

[31] J. Berakdar and J.S. Briggs, Phys. Rev. Lett. **72**, 3799 (1994); J. Phys. B. **27**, 4271 (1994); *ibid* **29**, 2289 (1996).

[32] J. Berakdar, J. Röder, J. S. Briggs and H. Ehrhardt J. Phys. B. *in press* (1996).

[33] K. A. Berrington, P. G. Burke, K. Butler, M. J. Seaton, P. J. Storey, K. T. Taylor and Yu Yan, J. Phys. B. **20**, 6379 (1987).

[34] T. Kato, Commun. Pure Appl. Math. **10**, 151 (1957).

[35] J. W. Cooper, U. Fano and F. Prats Phys. Rev. Lett. **10**, 538 (1963).

[36] F. Maulbetsch, M. Pont, J. S. Briggs and R. Shakeshaft, J. Phys. B. **28**, L341 (1995).

[37] F. Maulbetsch, Ph.D. thesis, Freiburg University (1995).

[38] J. Röder *private communication*, see also M. Brauner, J. S. Briggs, H. Klar, J. T. Broad, T. Rösel, K. Jung and H. Ehrhardt, J. Phys. B **24**, 657 (1991).

[39] A. Huetz, L. Andric, A. Jean, P. lablanquie, P. Selles and J. Mazeau, *private communication* and *Proceedings of XIX ICPEAC, Whistler*, edited by L. Dube' *et al* (AIP Press, New York, 1995).

IONIZATION DYNAMICS AND EXCHANGE EFFECTS IN PURE THREE-BODY COULOMB SCATTERING

J. Berakdar[1] and J.S. Briggs[2]

[1] Atomic and Molecular Physics Laboratories,
Research School of Physical Sciences and Engineering,
Institute of Advanced Studies, Australian National University,
Canberra, ACT 0200, Australia

[2] Fakultät für Physik, Albert-Ludwig-Universität
Hermann-Herder-Straße 3, 79104 Freiburg, Germany

INTRODUCTION

The measurement of two electrons in coincidence following ionization by electron impact has been performed in a variety of collision geometries [1, 2, 3, 4, 5, 6, 7, 8, 9, 10, 11, 12]. The different cases yield information on different dynamical aspects of the ionization process. Theoretically, upon analysis of ionization mechanisms certain patterns are deduced for the distributions of the vector momenta of reaction fragments. The major obstacle in a theoretical description of such scattering reactions is the inherent non-separability of many-body Coulomb interacting systems. Thus, these systems have to be represented by approximate Hamiltonians and different approximations may well lead to different interpretations of the scattering dynamics. Here we investigate pure three-body Coulombic systems above the complete break-up threshold. A cluster model is adopted in which the three-body system is approximated by three, spatially decoupled, two-body (Kepler) subsystems. This results in the well known 3C-approximation [13, 14]. Coupling between these, in the configuration space, non-interacting two-body (Rutherford) scattering systems can be introduced via position-dependent Sommerfeld parameters [15]. Analyses of the results yielded by these techniques show that dominant structures in the (theoretical) spectra of the outgoing particles can be assigned to the following underlying mechanisms:

- Single and double-binary collisions
- Coulomb density-of-state factors
- interplay between collisional ionization mechanisms and exchange effects
- Wannier ionization mode

- Quantum interference between contributing scattering amplitudes and three-body effects.

Rather than going into mathematical details (these can be found elsewhere [15, 16, 17]), a few examples for the electron-impact ionization of atomic hydrogen and helium are given to illustrate the above conclusions.

IONIZATION DYNAMICS

An extensive analysis of single and double-binary collisions as well as the Coulomb density-of-states factors have already been given [18, 16].

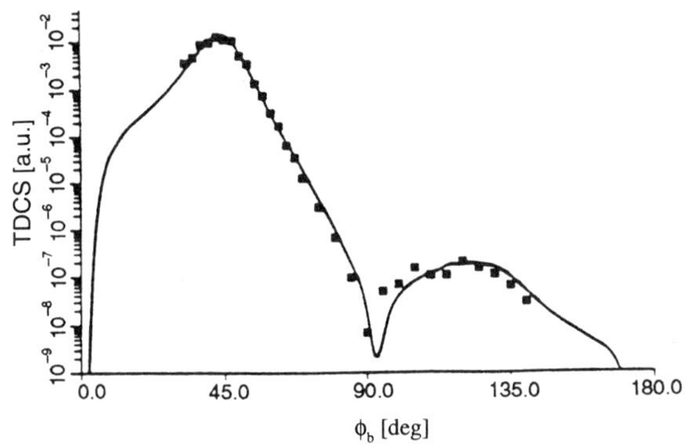

Figure 1: The Triply Differential Cross Section (TDCS) for the electron-impact single ionization of helium. The incident energy is $E_i = 500\ eV$. The equal energy electrons are detected under angles $\phi_a = \phi_b$ where the incident directions bisects the electronic relative angle. Calculations have been performed within the DS3C model. Experimental data (full squares) are due to Ref. [7]

Figure 1 summarizes some of those effects in the case of two electrons escaping with the same energy. If the two electrons escape with equal velocity vectors, their mutual density of states decreases exponentially and the cross section vanishes. A single-binary collision between the projectile electron and the bound one (the nucleus is a spectator) leads to a dominant binary peak at $\phi_b = 45^o$ where the two electrons escape perpendicular to each other. In the case where the projectile first recoils off the nucleus and subsequently scatters from the target electron, a corresponding peak appears when the two electrons emerge in the backward direction and perpendicular to each other. Again a vanishing electron-electron density of states leads to a zero cross section when these electrons move backwards with vanishingly small relative velocity.

At lower energies, exchange effects due to the indistinguishability of the escaping electrons become more important. Thus, the interplay between collisional ionization mechanisms is best studied at fairly low energies. The collision geometry suitable for such purposes is where two equal-energy electrons escape with relative, fixed angle of $\pi/2$ and the angle ϕ_b of one of the outgoing electrons is scanned with respect to the incident direction $\hat{\mathbf{k}}_i$ (Fig. 2). From simple symmetry considerations it can be inferred that the cross section is cylindrically symmetric around an axis rotated by an azimuthal angle of $\Phi = 45^o$ with respect to the incident direction $\hat{\mathbf{k}}_i$. Hence triplet cross sections must vanish at $\phi_b = 45^o$ and $\phi_b = 135^o$. The measured and calculated cross section are shown in Fig.(2a-d) for a variety of impact energies ranging from $E_i = 20\ eV$ to $63.6\ eV$. One notes a very strong change in the angular distributions as the energy of the two continuum electrons is increased from near threshold. At the lowest energies the patterns consist of two lobes and two minima. As the energy increases the angle between the two lobes diminishes until at the highest energy they have essentially merged into a single lobe. Here the underlying processes leading to the metamorphosis of the peak at $\phi_b = 45°$ is discussed. This location corresponds to two electrons escaping perpendicular to each other and symmetrically with respect to $\hat{\mathbf{k}}_i$ which is the configuration of a single-binary projectile-electron collision. Hence, at higher incident energies (Fig. 2d) the origin of this peak is an "isolated" projectile-electron encounter. As noted above, the triplet cross section vanishes at $\phi_b = 45°$. Therefore, as indicated by Fig.2e (see also [19]) the direct process dominates and only a minor influence of vanishing triplet states is observed.

Figure 2: The TDCS section in the coplanar, equal-energy, fixed-relative angle geometry. The two escaping electrons are detected perpendicular to each other with one detector being rotated around the incident direction $\hat{\mathbf{k}}_i$. The incident energy is $E_i = 20\ eV$ (a), $E_i = 25\ eV$ (b), $E_i = 30\ eV$ (c), $E_i = 64.6\ eV$ (d). Calculations have been performed within the DS3C theory. Experimental data (squares) are due to Ref.[20]. In (e) the singlet (dot-dashed curve) and triplet scattering (dotted curve) cross section are depicted along with their statistical average (solid curve). The collision geometry is the same as in (b). The triplet cross section is enlarged by a factor of 1.8.

As the energy is decreased, exchange effects become more important and the vanishing triplet cross section carves out a hole in the spin-unresolved cross section at $\phi_b = 45^o$ to leave a "butterfly" pattern for the angular distribution. These two competing processes are readily seen in Fig. 2e where the contributions of singlet and triplet are shown separately.

At much lower energies the Wannier ionization mode [21] where the two electrons travel outward on opposite sides of, and equal distances from the nucleus, becomes the dominant process. Figure 3a elucidates this geometry. However, the memory of the incident direction, which is irrelevant in the "half-collision" Wannier treatment, has a pronounced effect on the electrons' spectra (see Fig3b). This effect can be linked to the occupation probability of certain $LS\pi$ states. A more extensive study of this issue will be published elsewhere.

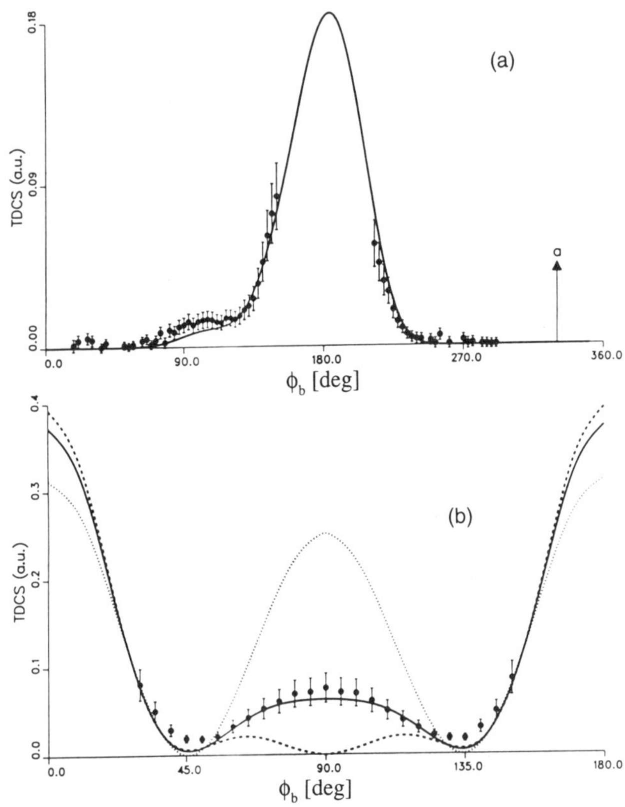

Figure 3a,b. In (b) the TDCS is depicted for co-linear configuration of two 1-eV escaping electrons, as a function of the angle ϕ_b of the interelectronic axis with respect to the beam direction. Continuous line: calculations using the DS3C wavefunction; singlet (dotted line) and triplet (dashed line) cross sections are also shown. The experimental data are taken from Ref.[22]. In (a) the same as in (b) but one electron is fixed at angle $\phi_a = 30^o$ with respect to $\hat{\mathbf{k}}_i$ (which is designated by the arrow) whereas ϕ_b is variable.

In a previous work we showed that interference between different scattering amplitudes can lead to striking structures in the observed spectra [12, 16, 17]. Here we look at a somewhat different geometry where the outgoing electrons emerge coplanar with $\hat{\mathbf{k}}_i$ but with substantially different energies. At higher impact energies the dominant features of the cross sections are identified [4, 23] as a *binary peak* originating from a single-binary electron-

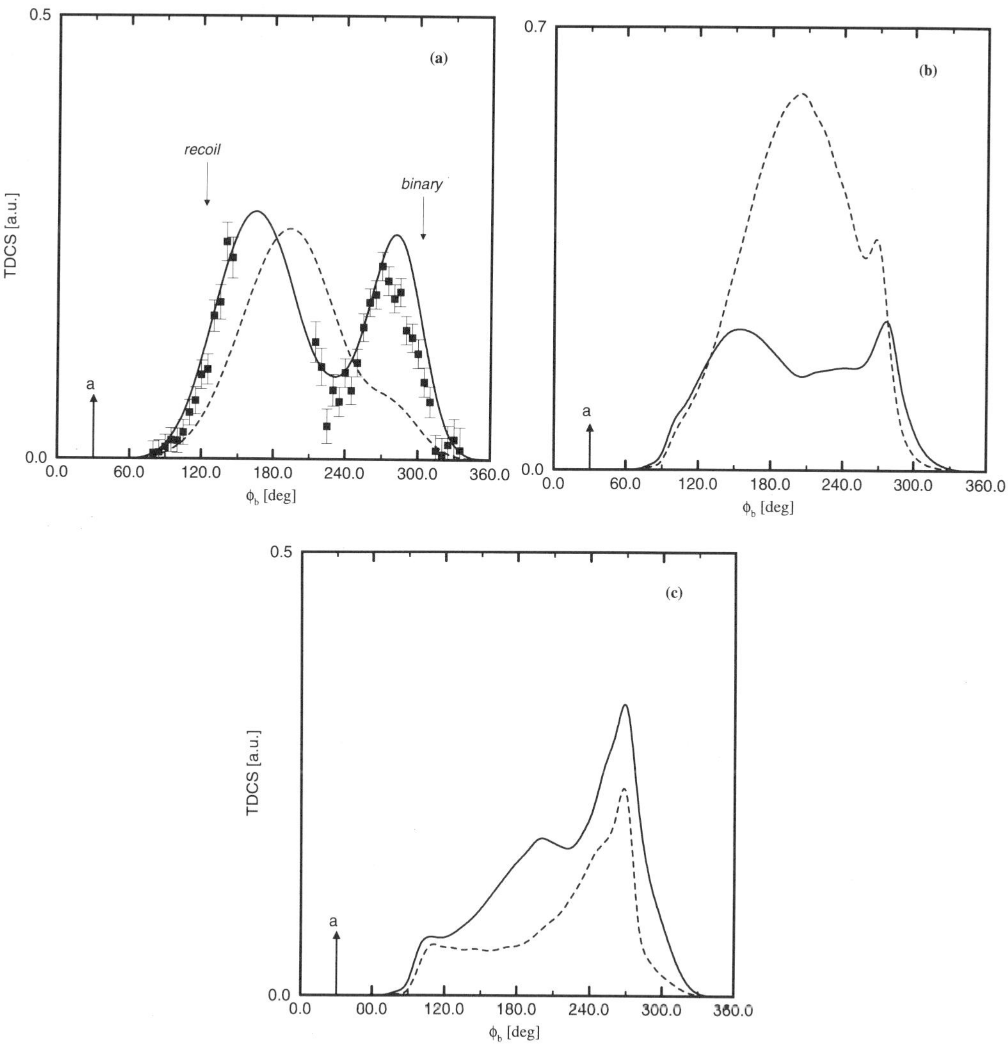

Figure 4a,b,c: (a) shows the TDCS for the electron-impact ionization of atomic hydrogen in the co-planar, asymmetric energy-sharing geometry. The incident energy is $E_i = 27.2\ eV$. the fast scattered electron is detected at an angle of 16° with respect to the incident direction whereas the angular distribution of the secondary electron with fixed energy of $4\ eV$ is measured. The emission angle of the latter electron with respect to the incident direction is denoted by Φ_b. Experiments are courtesy of Ref.[20]. The positions of the binary and recoil peaks are indicated. (b): For the collision geometry chosen in (a) the TDCS has been calculated using the projectile-nucleus scattering amplitude (T_{eN}) only, whereas in (c) the projectile-atomic-electron scattering amplitude (T_{ee})is employed. The solid lines show the predictions of the DS3C theory whereas the dotted curves indicate the 3C results.

projectile encounter, and a less prominent *recoil peak* which is assigned to a second order process in which the atomic electron, after being hit by the projectile, recoils off the nucleus via initial binding. No substantial interference (within our theoretical model) is noticed. This situation is expected to change dramatically at lower energies, as all scattering amplitudes tend to be of the same order. Surprisingly, the angular distributions show the binary-recoil-peak double structure (Fig.4a) observed at higher energies. In fact, the positions of these peaks almost coincide with the classical recoil and binary-peak positions, as indicated on Fig. 4a. However, analyses of the incoherent contributions of the scattering amplitude from the nucleus T_{eN} and that from the bound electron T_{ee} (Fig.4b,c) reveals that the observed cross sections have a much more subtle origin. Due to the distinctive positions at which the dominant peaks in Fig.4b,c occur, we conclude that these peaks in T_{eN} and T_{ee} are largely decided by two-body interactions. That means, the direct electron-electron scattering gives rise to the binary peak indicated in Fig. 4(c) whereas the direct projectile-nucleus scattering largely decides the recoil peak[1], as seen in Fig. 4(b). Although there are differences in the shape of the scattering amplitudes within the 3C and DS3C models the decisive role of the three-body coupling, which is partly taken into account by the DS3C model and neglected by the 3C theory, is the different weighting of these individual amplitudes which leads to completely different interference patterns [compare Fig. 4a]. Hence, from this we come to the conclusion that structures in the individual, incoherent scattering amplitudes can still be traced to two-body scattering processes even at lower energies. The observed cross sections, however, are determined by the interference pattern of these amplitudes which is decided by three-body coupling. Thus, it can be said that, within our theoretical framework, three-body couplings control the interference between individual scattering amplitudes (and thus decisively the observed spectra [16]), however, it does not lead to any new structure of a collisional origin. More extensive data confirming this picture of the underlying physical mechanisms have been published elsewhere [17, 12].

References

[1] H. Ehrhardt, M. Schulz, T. Tekaat and K. Willmann, *Phys. Rev. Lett.* **22**: 89 (1969).

[2] U. Amaldi, A. Egioli, R. Marconero and G. Pizella, *Rev.Sci.Instrum.* **40**: 1001 (1969).

[3] B. Lohmann, I.E. McCarthy, A.T. Stelbovic and E. Weigold, *Phys. Rev.* A **30** 758 (1984).

[4] H. Ehrhardt, G. Knoth, P. Schlemmer and K. Jung, *Phys. Lett.* **110** A: 92 (1985).

[5] A. Pochat, R.J. Tweed, J. Peresse, C.J. Joachain, B. Piraux and F.W.Jr. Byron, *J. Phys. B: At. Mol. Phys.* **16**: L775 (1983).

[6] M. Brauner, J.S. Briggs, H. Klar, J.T. Broad, T. Rösel, K. Jung and H. Ehrhardt, *J. Phys. B: At. Mol. Phys.* **24**: 657 (1991).

[7] L. Frost, P. Freienstein and W. Wagner, *J. Phys. B: At. Mol. Phys.* **23**: L715 (1990).

[8] I.E. McCarthy and E. Weigold, *Rep. Prog. Phys.* **54**: 781 (1991).

[9] A.J. Murray and F.H. Read, *Phys. Rev. Lett.* **69**: 2912 (1992).

[10] A.J. Murray and F.H. Read *Phys. Rev.* A **47**: 3724 (1993).

[11] A.J. Murray and F.H. Read *J. Phys. B: At. Mol. Phys.* **26**: L359 (1993).

[12] J. Berakdar, J. Röder, J.S. Briggs and H. Ehrhardt, *J. Phys. B: At. Mol. Phys.* in press (1996).

[13] G. Garibotti and J.E. Miraglia *Phys.Rev.A* **21**: 572 (1980)

[14] M. Brauner, J.S. Briggs and H. Klar, *J. Phys. B* **22**: 2265 (1989).

[15] J. Berakdar, *Phys. Rev. A.* **53**: 2314 (1996).

[1] The smaller binary and recoil peaks observed in Fig. 3(b) and Fig. 3(a), respectively, are due to successive two-body scatterings contained in the final-state wavefunctions.

[16] J. Berakdar and J.S. Briggs, *Phys. Rev. Lett.* **72**: 3799 (1994); *J. Phys. B* **27**: 4271 (1994); *ibid* **29**: 2289 (1996).

[17] J. Berakdar *Phys. Rev. A* (1997) in press.

[18] J. Berakdar and H. Klar, *J. Phys. B* **26**: 4219 (1993).

[19] C.J. Joachain, *Comments At. Mol. Phys.* **17**: 261 (1986).

[20] J. Röder, *private communication* (1996).

[21] G. Wannier, *Phys. Rev.* **90**: 817 (1953).

[22] T. Rösel, J. Röder, L. Frost, K. Jung, H. Ehrhardt, S. Jones and D.H. Madison *Phys. Rev. A* **46**: 2539 (1992).

[23] F.W. Byron Jr., C.J. Joachain, *Phys. Rep.* **179**: 211 (1989).

ANALYSIS OF INTEGRATED CROSS SECTIONS AND SPIN ASYMMETRIES FOR THE ELECTRON-IMPACT IONIZATION OF ONE- AND TWO-ELECTRON ATOMIC SYSTEMS

J. Berakdar[1], S.J. Buckman [1], P.F. O'Mahony[2] and F. Mota-Furtado[2]

[1]Atomic and Molecular Physics Laboratories,
Research School of Physical Sciences and Engineering,
Institute of Advanced Studies, Australian National University,
Canberra, ACT 0200, Australia

[2]Department of Mathematics, Royal Holloway,
University of London, Egham, Surrey TW20 0EX, UK

INTRODUCTION

A feature inherent to the motion of two electrons in a nuclear field is the threshold behaviour of the total cross section for the ejection of one electron from a hydrogenic target upon electron impact. Based on a classical analysis of the phase space of two continuum electrons Wannier [1] has established that, at threshold, the analytical dependence of the total ionization cross section, hereafter referred to as $\sigma(E)$, on the total excess energy derives from the phase-space volume available for double escape:

$$\sigma(E) \propto E^{\mu/2\ -\ 1/4} \qquad (1)$$

where Z is the nuclear charge of the residual ion and the Wannier index μ is given by

$$\mu = \frac{1}{2}\sqrt{\frac{100\ Z\ - 9}{4\ Z\ -\ 1}}\ . \qquad (2)$$

Meanwhile a considerable body of experimental and theoretical work exists which confirms the Wannier predictions [1, 2, 3, 4, 5, 6, 7, 8, 9, 10, 11, 12, 13]. Wannier considered the reaction as a "half collision". Thus, relation (1) derives from considerations of the final state and does not depend on the initial state from which, or the transition operator via which, this final state is achieved. In particular, the threshold law (1) applies for the ionization of hydrogenic positive ions upon electron impact. With increasing strength of the nuclear field, $1 \gg Z$, the functional form (1) tends to a linear dependence which is the threshold law provided by an independent Coulomb particle model [5]. In this case the electron-nuclear interaction can be considered as the "strong potential" and the electron-electron Coulomb interaction can be

Coincidence Studies of Electron and Photon Impact Ionization
Edited by Whelan and Walters, Plenum Press, New York, 1997

treated in a perturbative way. Because of the half-collision treatment of the scattering process, Wannier-type theories do not provide estimate of the absolute value of $\sigma(E)$ and other integrated reaction properties. Hence, for increasing target charge it is worthwhile to develop a dynamical model of the collision process which is capable of providing information on the behaviour of absolute differential and integrated cross sections. For initially charged targets this will also help assess the influence of initial-state distortions of the projectile motion due to the presence of the long-range projectile-target interaction. Therefore, we consider the electron-impact single ionization of the ground-state electrons of arbitrary one- and two-electron atomic systems. In this work we assume reasonably light targets so that the LS-coupling scheme and a nonrelativistic treatment of the reaction are justified. In the following, where the theory is developed for the ionization of two-electron targets, the same consideration are applicable for the simpler case of electron-impact ionization of one-electron systems.

THEORETICAL FRAME WORK

The most complete information on the collision process is obtained by the simultaneous detection of the vector momenta of the outgoing electrons. Assuming an infinitely heavy nucleus with respect to the electron mass and using atomic units (a.u.), the fully differential cross section at a fixed total energy E of the electron-target system is given by

$$d^6\sigma(E) = \frac{(2\pi)^4}{v_i}|\mathcal{T}(\mathbf{k}_a, \mathbf{k}_b)|^2\delta(E - E_i)d^3k_a d^3k_b \tag{3}$$

where E_i and v_i are respectively the total energy in the initial channel and the velocity of the incoming projectile with respect to the nucleus. The momenta of the escaping electrons (with respect to the nucleus) are denoted by $\mathbf{k}_a, \mathbf{k}_b$. In the prior formulation, the $\mathcal{T}$-matrix element reads

$$\mathcal{T}(\mathbf{k}_a, \mathbf{k}_b) = \langle\ \Psi^-|W_i|\Phi\ \rangle \tag{4}$$

where the asymptotic initial state of the electron-target system is described by the state vector $|\Phi\ \rangle$ which is the eigenstate of the Hamiltonian

$$\left(-\frac{1}{2}\nabla_{\mathbf{r}_a} + \frac{-(Z-2)}{r_a} + h_i - E\right)\langle\ \mathbf{r}_a, \mathbf{r}_b, \mathbf{r}_c|\Phi\ \rangle = 0. \tag{5}$$

In Eq.(5), h_i stands for the Hamiltonian of the undisturbed target in the initial channel and the coordinate $\mathbf{r}_a$ refers to the position of the projectile electron with respect to the nucleus. The solution of Eq.(5) with boundary conditions describing the incoming-electron bound-state scattering system can be given in a closed form

$$\Phi(\mathbf{r}_a, \mathbf{r}_b, \mathbf{r}_c) = (2\pi)^{-3/2}e^{\pi\gamma/2}\Gamma(1+i\gamma)\ \phi_i(\mathbf{r}_b, \mathbf{r}_c)\ e^{i\mathbf{k}_i\cdot\mathbf{r}_a}\ {}_1F_1\left[-i\gamma, 1, i(k_i r_a - \mathbf{k}_i\cdot\mathbf{r}_a)\right] \tag{6}$$

where $\mathbf{k}_i$ is the momentum of the incident electron and the Sommerfeld parameter γ is given by $\gamma = -(Z-2)/v_i$. The confluent hypergeometric function is denoted by ${}_1F_1[a, b, x]$. In Eq.(6) the initial bound state of the target is designated by ϕ_i. Hence, $(h_i - \epsilon_i)\phi_i = 0$ where ϵ_i is the (positive) initial binding energy of the target and therefore $E_i = k_i^2/2 - \epsilon_i$. The state vector $\langle\Psi^-|$ occurring in Eq.(4) constitutes an exact eigenstate of the total Hamiltonian of the scattering system with boundary conditions describing two outgoing electrons moving in the continuum of the final-state ion. In this work we assume the ionized target to be left in its ground state and disregard any distortion of this state due to the continuum electrons. Thus, the wavefunction $\langle\ \mathbf{r}_a, \mathbf{r}_b, \mathbf{r}_c|\Psi^-\ \rangle$ is written in the form

$$\langle\ \mathbf{r}_a, \mathbf{r}_b, \mathbf{r}_c|\Psi^-\ \rangle = \psi^-(\mathbf{r}_a, \mathbf{r}_b)\phi_f(\mathbf{r}_c) \tag{7}$$

where $\phi_f(\mathbf{r}_c)$ is the ground-state wavefunction of the ion in the exit channel with (positive) energy ϵ_f. The three-body continuum wavefunction $\psi^-(\mathbf{r}_a, \mathbf{r}_b)$ describes the motion of two electrons in the field of the charge $Z - 1$. In a independent Coulomb particle model the wavefunction $\psi^-(\mathbf{r}_a, \mathbf{r}_b)$ takes on the form

$$\begin{aligned}\psi^-(\mathbf{r}_a, \mathbf{r}_b) \approx\ & (2\pi)^{-3} \exp(i\, \mathbf{r}_a \cdot \mathbf{k}_a + i\, \mathbf{r}_b \cdot \mathbf{k}_b) \\ & N_a\, {}_1F_1 \left[i\frac{Z_a}{k_a}, 1, -i(k_a r_a + \mathbf{k}_a \cdot \mathbf{r}_a) \right] \\ & N_b\, {}_1F_1 \left[i\frac{Z_b}{k_b}, 1, -i(k_a r_a + \mathbf{k}_b \cdot \mathbf{r}_b) \right] \end{aligned} \tag{8}$$

where $Z_a = -(Z - 1) = Z_b$ are the electron-ion product charges in the final state. The normalization constants N_j; $j = a, b$ are given by

$$N_j = \exp\left(-\frac{\pi Z_j}{2k_j}\right) \Gamma\left(1 - i\frac{Z_j}{k_j}\right); \ j = a, b. \tag{9}$$

Thus the total final-state energy is given by $E = (k_a^2 + k_b^2)/2 - \epsilon_f$. The approximation (9) means that the strength of the final-state electronic correlations are negligibly small compared to the strength of the electron-ion final-state interactions. Therefore, the model of independent Coulomb particles is expected to work best for large nuclear charges. The perturbation operator W_i which appears in Eq.(4) is the part of the Hamiltonian not diagonalized by the initial state (6), i.e.

$$W_i = \frac{-2}{r_a} + \frac{1}{|\mathbf{r}_a - \mathbf{r}_b|} + \frac{1}{|\mathbf{r}_a - \mathbf{r}_c|}. \tag{10}$$

Within the independent Coulomb particle model, the above treatment is applicable to an arbitrary charge-state of the initial target, i.e. for positive, negative and neutral one and two-electron systems. In general, the evaluation of the cross section (3) involves a nine-dimensional integral which can be reduced analytically to a two-dimensional integral on the real axis. The latter integral is then performed numerically. In case the measurement of (3) does not resolve the spin states of the outgoing electrons, the cross sections are statistical averages over singlet (σ^s) and triplet (σ^t) states. Information on the ratio of singlet to triplet scattering can be gleaned from spin-polarized measurements where the spin asymmetry is defined as

$$A(E) = \frac{\sigma^s(E) - \sigma^t(E)}{\sigma^s(E) + 3\sigma^t(E)}. \tag{11}$$

RESULTS AND DISCUSSION

According to a Wannier-type treatment [14, 6] all $L-$states show the same excess-energy dependence close to the double-escape threshold and nearly all $LS\pi$-states (total angular momentum, total spin and parity of the two-electron states), in particular singlet and triplet states, possess the same threshold law. The consequence of this is a constant spin asymmetry $A(E)$ near to threshold, although the absolute value of A can not be given within this model for the reasons mentioned above. In Fig.1(a) the spin asymmetry for the ionization of atomic hydrogen is depicted. In this case the incoming Coulomb-wave in the wavefunction (6) reduces to unity. The measurements are well described by the independent Coulomb particle model (8) as well as by the convergent-close coupling method, which provides a full numerical solution of the scattering problem [15]. In addition, in accordance with the Wannier theory predictions, $A(E)$ is a flat function of E over a considerable energy range above threshold. With increasing strength of the nuclear field [Fig.1(b)] we observe the emergence of a peak structure in the spin asymmetry. This is of interest as, near to threshold, an error analysis of the measured spin asymmetry for atomic hydrogen [16] has revealed a slightly positive slope of $A(E)$.

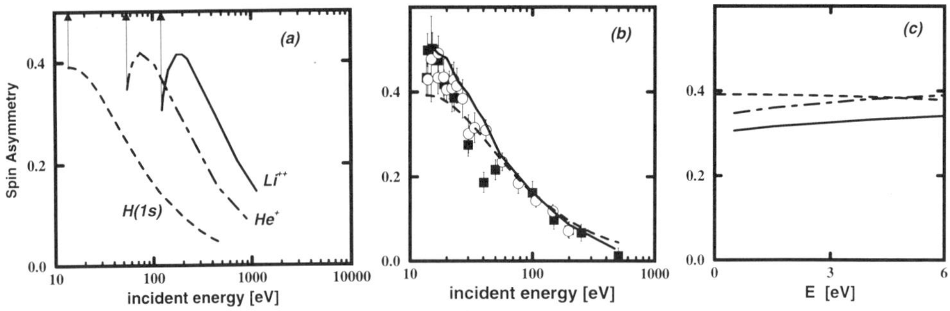

Figure 1. (a) shows the spin asymmetry, as defined by Eq.(11), for the the electron-impact ionization of atomic hydrogen as a function of the incident energy. The solid curve is the result of the convergent-close coupling method [15] whereas the dashed curve is calculated from the present model. The experimental data are due to Ref.[20] (full squares) and Ref.[21] (open circles). In (b) the results for the electron-impact ionization of He^+ (dot-dashed curve) and Li^{++} (solid curve) are also included. The ionization thresholds of $H(1s)$, He^+ and Li^{++} are indicated by arrows. Figure 1(c) shows $A(E)$ as a function of the excess energy E at small energies of the escaping electrons.

Figures 1(a,b) suggest that this positive slope develops to a peak for increasing nuclear charge and thus it is reasonable to suggest that the positive slope in the measurements of $A(E)$ for the ionization of atomic hydrogen has its origin in the electron-nucleus final-state interactions since the electron-electron final-state interaction is disregarded in our treatment [22]. Although there is no direct experimental evidence to support the prediction of Fig.1(b), it is remarkable that the shape of spin-asymmetry measurements for electron-lithium [17] and electron-sodium [17] ionization clearly resemble the structure shown in Fig.1(b) with a peak in the low excess-energy region. Although comparison of the latter experimental data with the calculations shown in Fig.1(b) is not possible, the similarities in their shape lead us to the suggestion that at low excess energy, the main contribution to the measured spin asymmetry in electron-lithium and sodium ionizing collisions [17] originates from small impact parameter scattering where the ionized electrons experience the net nuclear charge. It should be mentioned here that the peaks occurring in Fig.1(b) are not in contradiction of the Wannier-theory predictions of a constant $A(E)$ since, at lower excess energies, the spin asymmetries shown in Fig.1(b) are quite flat, even though a positive slope is observable for increasing nuclear charge [Fig.1(c)].

A further important feature of the spin asymmetries depicted in Fig.1(b) is the high-energy behaviour. With increasing electron-nuclear coupling the spin-asymmetry increases for a given high excess energies. A possible explanation of this behaviour is found when expressing the spin-asymmetry in the form

$$A = \frac{|f|^2 + |g|^2 - |f-g|^2}{|f|^2 + |g|^2 + |f-g|^2} \tag{12}$$

where f and g are the direct and exchange scattering amplitudes, respectively. In the case of extremely high incident energies, the direct ionization process dominates the exchange process, i.e. $\lim_{E\to\infty} |g|(E)/|f|(E) \to 0$. Thus, it is anticipated that $\lim_{E\to\infty} A(E) \to 0$. For increasing nuclear charge the size of the target in the initial state decreases. Therefore, ionization occurs predominantely at small impact parameters in which case exchange becomes more probable. Hence, it is conceivable that $\lim_{E\to\infty} A(E) \to 0$ is reached faster for smaller nuclear charge of the target in the initial channel.

In conclusion we present in this work a model for describing the electron-impact ionization of one and two-electron systems with arbitrary nuclear charge. Spin-asymmetries in the total cross sections have been analysed and their dependence on the strength of the nuclear field has been elucidated. The present model neglects the final-state interaction of the escaping electrons. The influence of this interaction has been studied in details [23, 19]. Furthermore, we have recently conducted a study [22] investigating the energy-sharing and angular correlations of the outgoing electrons.

We would like to thank Erich Weigold for helpful suggestions and comments on the manuscript. Financial support by the the Alexander von Humboldt Foundation and the Australian National University is gratefully acknowledged.

References

[1] G. Wannier Phys.Rev. **90**, 817 (1953).

[2] R.K. Peterkop J.Phys.B. **4**, 513 (1971).

[3] S. Cvejanović and F.H. Read, J.Phys.B **7**, 1841 (1974).

[4] F.H. Read in T.D. Märk and G.H. Dunn ed. (1984) *Electron Impact Ionisation* (New York:Springer)

[5] A.R.P. Rau Phys.Rev.A **4**, 207 (1971).

[6] H. Klar and W.Schlecht J.Phys.B **9**, 1699 (1976).

[7] H. Klar J.Phys.B. **14**, 3255 (1981).

[8] A.R.P. Rau Phys.Rep. **110**, 369 (1984).

[9] J.M. Feagin J.Phys.B. **17**, 2433 (1984).

[10] K. Kossmann, V. Schmidt and T. Andersen Phys.Rev.Lett **60**. 1266 (1988).

[11] P. Lablanquie, K. Ito, P. Morin, I. Nenner and J.H.D. Eland Z.Phys.D. **16**, 77 (1990).

[12] J-M. Rost J.Phys.B. **28** 3003 (1995).

[13] J.H. Macek and S.Yu. Ovchinnikov Phys.Rev.Lett. **74**, 4631 (1995).

[14] C.H. Greene and A.R.P. Rau Phys.Rev.Lett. **48**, 533 (1982).

[15] I. Bray and A.T. Stelbovics, Phys.Rev.Lett. **70**, 746 (1993).

[16] X.Q. Guo, D.M. Crowe, M.S. Lubell, F.C. Tang, A. Vasilakis, J. Slevin and M. Eminyan Phys.Rev.Lett. **15**, 1857 (1990).

[17] G. Baum, M. Moede, W. Raith and W. Schröder J.Phys.B **18**, 531 (1985).

[18] S. Cvejanović, R.C. Shiell and T.J. Reddish J.Phys.B. **28**, L707 (1995).

[19] J. Berakdar Aust.J.Phys. **49**, 1095 (1996)

[20] G.D. Fletcher, M.J. Alguard, T.J. Gay, V.W. Hughes, P.F. Wainwright, M.S. Lubell and W. Raith Phys.Rev.A **31**, 2854 (1985).

[21] D.M. Crowe, X.Q. Guo, M.S. Lubell, J. Slevin and M. Eminyan J.Phys.B. **23**, L325 (1990).

[22] J. Berakdar and S.J. Buckman *Phys. Rev. A* **54**, 5431 (1996).

[23] J. Berakdar, P.F. O'Mahony and F. Mota-Fortado Z. Phys. D in press (1996).

DOUBLE IONIZATION MECHANISMS FROM TRIPLE COINCIDENCE EXPERIMENTS

A. Lahmam-Bennani,[1] A. Duguet,[1] B. El Marji,[1] C. Schröter,[2] and L. Spielberger[3]

[1]Laboratoire des Collisions Atomiques et Moléculaires (Unité associée au CNRS N° 281) - Université de Paris-Sud, 91405 Orsay Cedex, France

[2]Istituto di Metodologie Avanzate Inorganiche del CNR - Area della Ricerca di Roma, Via Salaria km 29, CP 10, 00016 Monterotondo, Italy

[3]Institut Für Kernphysik, August Euler-Str.6, 60486 Frankfurt, Germany

INTRODUCTION

Differential studies of the double ionization (DI) process, both under photon and charged particle impact, have known a remarkable development in the last few years, due to major technical advances in multianalysis and multidetection systems, and the use of multiparameters coincidence spectrometers. An important issue of these studies is to investigate the electronic correlations, i.e. the correlated motion of the electrons, both in the initial state and in the final state of the collision. For instance, it is the initial state electron correlation which is responsible[1] for the photon impact DI, and this largely prevails also under electron impact. Moreover, the asymptotic angular distribution of the products is largely dominated[2] by the final state correlations, that is the Coulombic interactions between the outgoing particles. Therefore, an (e,3e) experiment where three electrons and one doubly charged ion are present in the continuum is expected to yield unique information towards understanding the solution of the 3-and the 4- body problems for which exact solutions are not yet known. Main objective of this work is placed along these lines, with a special emphasis being put on understanding which particular mechanism is responsible for the electron impact double ejection of two electrons from the target into the continuum. In other words, we would like to answer the question : is the electron impact DI due to a direct process, or does it proceed via the electron correlations ? We will see that the (e,3e) experiments may enable disentangling between various competing DI processes.

A schematic diagram of a complete DI experiment is shown in figure 1. The incident electron labeled 0 ionizes an atomic target and is mostly forward scattered (labeled a). The two ejected or "atomic" electrons, though indistinguishable are labeled b and c for convenience. To fully determine the kinematics, one needs to measure all three final electrons energies and angles and detect them in a triple coincidence, an (e,3e) experiment. The measured quantity is then a five-fold differential cross section (5DCS), $d^5\sigma/dE_a dE_b d\Omega_a d\Omega_b d\Omega_c$, analogous to the corresponding single ionization (e,2e) triple differential cross section. Note that the 5DCS is sometimes termed 8DCS by some authors, to account for the fact that each Ω has a polar and an azimuthal dependence, θ and φ.

Coincidence Studies of Electron and Photon Impact Ionization
Edited by Whelan and Walters, Plenum Press, New York, 1997

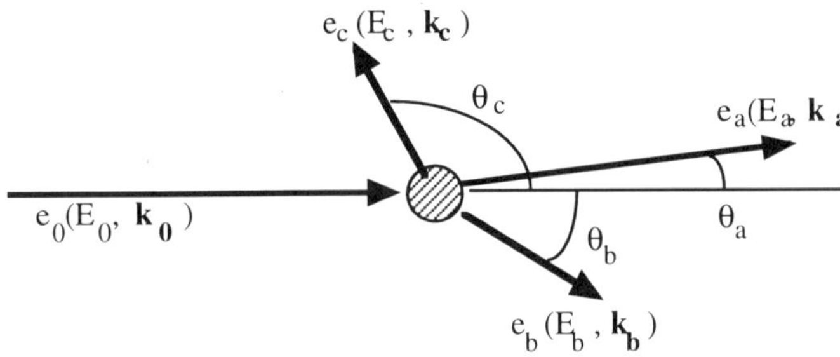

Figure 1. Schematic diagram of a coplanar (e,3e) experiment

In the case of outer shell DI, conservation of energy and momentum requires

$$E_0 = E_a + E_b + E_c + E^{++}$$

and

$$\mathbf{k_0} = \mathbf{k_a} + \mathbf{k_b} + \mathbf{k_c} + \mathbf{q_r}$$

where E_i and $\mathbf{k_i}$ (i=0, a, b, c) are the incoming and outgoing electron energies and momenta, $\mathbf{K} = \mathbf{k_0}\text{-}\mathbf{k_a}$ is the total momentum transfer to the target, E^{++} is the energy needed to doubly ionize the target, and $\mathbf{q_r}$ is the ion recoil momentum. From the measurement of all four energies E_i, one knows from which orbital(s) the two atomic electrons have been removed, leaving the ion in a given well defined state.

EXPERIMENTAL

Apparatus

The first reported (e,3e) spectrometer[3,4] was a "monodetection" system where the three electrons were detected at three preselected angles. This system suffered from its extremely low collection efficiency which led very long accumulation times. For this reason, a second generation spectrometer was built[5], including dual multiangle detection for the two slow electrons. The system is based on a double toroidal analyzer (Fig. 2), designed according to the criteria given by Leckey and Riley[6]. Main difference, however, is that these authors were looking at electrons which all had the same energy, and hence they used a full toroid. Our experiments require observing two electrons with energies E_b and E_c which are in general different from each other. Therefore, the toroid was cut in two halves : electrons emitted between 0 and 180° with respect to the incident beam direction are deflected downwards, whereas those emitted between 180 and 360° are deflected upwards. Each group is imaged as a half-ring on a position sensitive detector, PSD. The angular information contained in the collision plane ($\mathbf{k_0}$, $\mathbf{k_a}$), i.e. the ejection angles θ_b and θ_c, is preserved upon arrival on the PSD's due to focusing properties of toroidal analyzers[7]. Each PSD assembly consists of three multichannel plates followed by a resistive anode. One thus have a measure of the (X,Y) coordinates of each impinging electron, which are then transformed into in-plane angles, θ_b or θ_c.

The fast a-electron is detected as previously[3,4] in a 127° cylindrical analyzer. The technique for triple coincidences recording and for the data handling is basically identical to that given by Dupré et al[8] and need not be repeated here.

Results and Discussion.

(e,3e) cross sections were measured for argon by fixing the scattered electron energy and angle, E_a=5500 eV and θ_a=+0.45° (hereafter, positive angles are counted counter-clockwise), whereas the excess energy transferred to the target is evenly shared among the two ejected electrons, $E_b=E_c$=10 eV. The incident energy is consequently adjusted (E_0=5563 eV) to meet the $3p^{-2}\,{}^3P$ DI of argon.

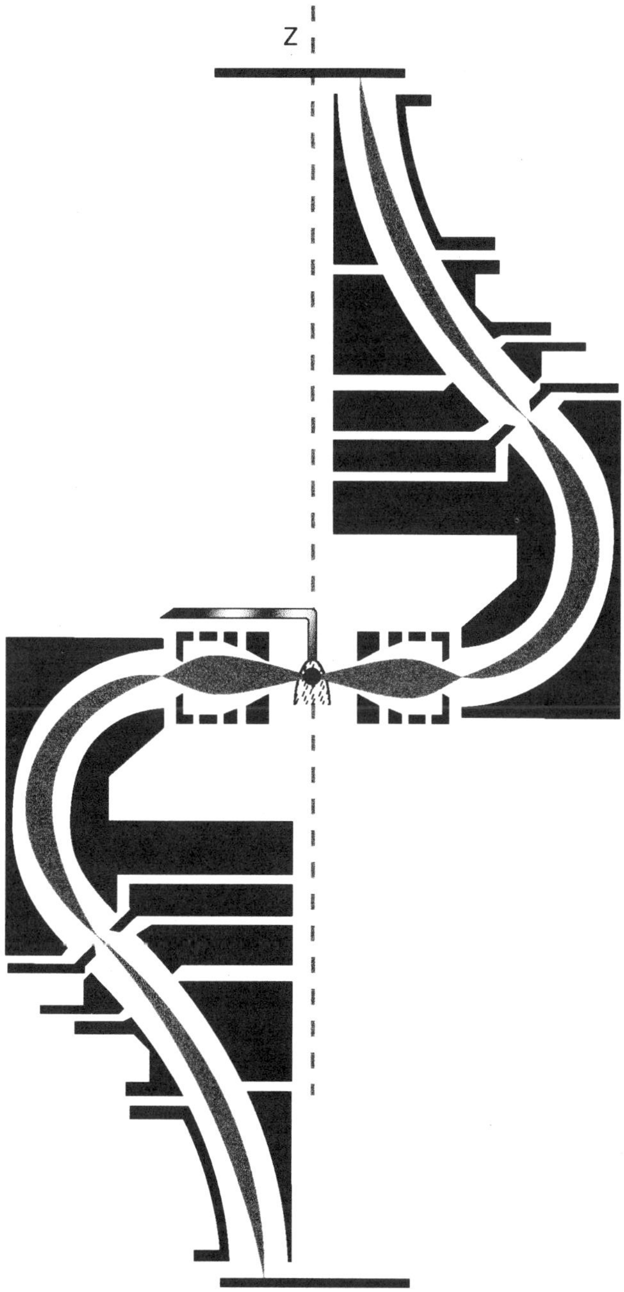

Figure 2. Schematic representation of the double toroidal analyzer used for the multi-angle detection of the two ejected electrons. The primary electron beam, issued from an electron gun (not shown) is coming perpendicular to the figure, at the centre, where it crosses the target gas beam. The system is of cylindrical symmetry around the vertical z-axis. Electrons ejected in the x-y horizontal plane of collision are imaged as two half-rings on the position sensitive detectors located at the top and the bottom, respectively.

During the off-line analysis of the data, the "ring" of triply coincident events on each PSD is divided into 10 sectors, each ± 7° width. Hence, 2x10 (e,3e) angular distributions with fixed θ_a and θ_b, and variable θ_c (or fixed θ_a and θ_c, but variable θ_b) are simultaneously obtained. Two examples of such distributions are displayed in figure 3. One ejected electron is detected at a fixed direction, shown by the arrow, while the other one is mapped in the opposite half plane. Two important observations can be made here. First, in contrast to the (e,2e) single ionization where the ejected electron angular distribution is known to be peaked in the **K** direction (so-called binary lobe), this direction is not anymore a preferential direction of emission in DI, the maximum of the distribution being shifted some 50 degrees backwards. Second, the **K** direction is clearly not a symmetry axis for the measured distributions. Such a symmetry around **K** is predicted[9] by any first order theoretical model where only single interaction processes of the projectile with the target are included. Therefore, the symmetry breaking in Fig. 3 is a clear indication that second or higher order effects must play an important role in the collision process. In fact, it is possible to show, through polynomial fits to these data and many similar ones, that the measured angular distributions always show a shoulder -if not a secondary maximum- in the **K** direction This implies that these distributions are made up by the sum of two contributions, as tentatively drawn in Fig. 3, the first order contribution centered around **K** and the second order contribution which is the dominant one. The backward shift of this contribution means that a large amount of momentum is transferred to the ion which plays an important role in the collision process. The most likely origin of this contribution is a two step mechanism (so-called TS2) where the incident electron collides successively with two target electrons, resulting in their quasi-simultaneous double ejection.

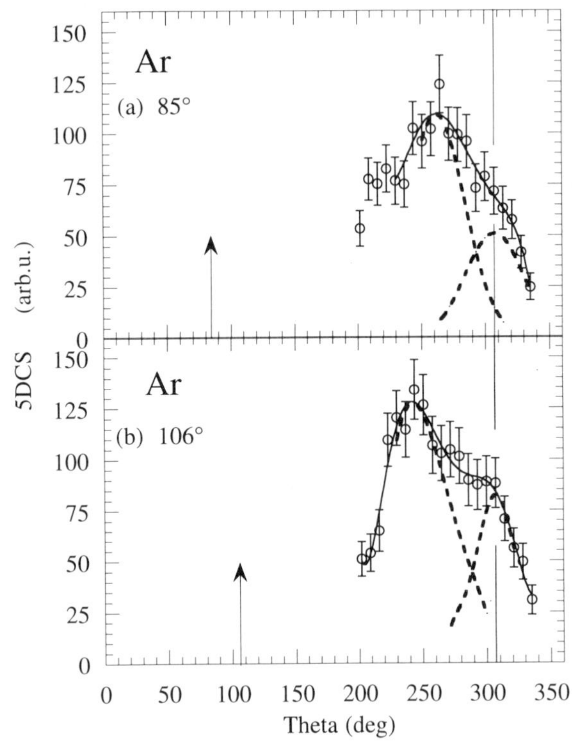

Figure 3. The relative (e,3e) cross section for double ionization of argon. Kinematical parameters are E_0=5563 eV, E_a=5500 eV, E_b=E_c=10 eV, θ_a=+0.45 deg. One electron is detected along a fixed direction shown by the arrow and the labelling, i.e. (a) 85° and (b) 106°. The second electron is mapped in the opposite half-plane. The +**K** direction is shown by the vertical thin line. Error bars are one standard deviation statistical error. The full curve is a polynomial fit to the data, while the dashed curves are tentative deconvolution of the experimental results into first and second order contributions, see text.

It might look surprising that a second order process is found here to dominate over the first order one. This seems to violate the commonly accepted idea (see for instance McGuire[10]) that at ~5 keV electron impact energy, the DI process is essentially due to a shake-off (SO) process, the two-step (TS) contribution being negligible. However, the contradiction is only apparent. Indeed, the energy distribution function among the two emitted electrons is of a U-type[11], where the most significant part of the cross section corresponds to highly asymmetric sharings. When integrated, it is this part which largely dominates the total cross section, and which is found both in the e,(3-1)e differential experiments of El Marji et al[12] and in McGuire's[10] theoretical predictions for the total cross sections to be due to a SO process. The part of the U-differential cross section close to the equal energy-sharing corresponds to a small fraction of the total cross section, and it is this part which is found in the present (e,3e) experiments to be mostly due to a TS process. This fraction is small enough so that the total cross section is not sensitive to its effect.

CONCLUSION

We have presented a selection of the results obtained in (e,3e) experiments performed on argon at 5.5 keV impact energy. It is found that, under the present kinematics, the DI process is dominated by a two-step process. This is shown to be due to the fully differential nature of the (e,3e) cross section, where only equal energy sharing events are observed here. However, such conclusion does not mean that other (first-order) processes are not present. It is only that the choice we have made of detecting both electrons in opposite half planes strongly favours, from the kinematics, one process among the other.

ACKNOWLEDGMENTS

This work was partially supported by the European Community Contract No. CHRX-CT93-0350.

REFERENCES

1. F.W. Byron Jr., and C.J. Joachain, 1967, *Phys.Rev. A* 64:1
2. C. Dal Cappello and H. Le Rouzo, 1991, *Phy.Rev. A* 43:1395
3. A. Lahmam-Bennani, C. Dupré, and A. Duguet, 1989, *Phys.Rev.Lett.* 63:1582
4. A. Lahmam-Bennani, 1991, *J.Phys.B:At.Mol.Opt.Phys.* 24:2401
5. B. El Marji, A. Duguet, A. Lahmam-Bennani, M. Lecas, 1994, An (e,2e) - (e,3e) spectrometer based on a double toroidal analyzer with multidetection system, in: *Proceedings of the EuroConference on Coincident Studies of Electron and Photon Impact Ionization,* Colmar, France
6. R.C.G. Leckey and J.D. Riley, 1985, *Appl.Surf.Sci.* 22/23:196
7. F. Toffoletto, R.C.G. Leckey, and J.D. Riley, 1985, *Nucl.Instrum.Methods B* 12:282
8. C. Dupré, A. Lahmam-Bennani, and A. Duguet, 1991, *Meas.Sci.Technol.* 2:327
9. C. Dal Cappello and B. Joulakian, 1997, private communication, *to be published*
10. J.H. McGuire, 1982, *Phys.Rev.Lett.* 49:1153
11. Z.-J. Teng, and R. Shakeshaft, 1994, *Phys.Rev. A* 49:3597
12. B. El Marji, A. Lahmam-Bennani, A. Duguet, and T.J. Reddish, 1996, *J.PhysB:At.Mol.Opt.Phys.* 29:L157

DOUBLE IONIZATION OF HELIUM BY ELECTRON IMPACT: A STUDY OF THE TWO-STEP MECHANISM

Claude Dal Cappello, Rachid El M'Khanter
and Pascal Lamy

Laboratoire de Physique Moléculaire et des Collisions
Institut de Physique et d'Electronique de Metz
1, rue Arago, 57078 Metz Cedex 03, FRANCE

INTRODUCTION

The study of the double ionization of atoms or molecules by electron impact is a powerful tool in investigating finer details of the ionization process. This double ionization by electron impact is due to different processes :

- direct process consisting to a simultaneous double ionization of two electrons via different mechanisms
- indirect process resulting of an ionization of a single inner-shell electron followed by Auger decay
- resonant processes in collisions with positive ions referred to as-called resonant excitation-autoionization process which lead to a final emission of three electrons (the incident one is captured by the target ion and ejected with two target electrons).

We restrict here to the double ionization of helium by electron impact which only involves direct process.

This last process is generally described[1] by three mechanisms which need the first and second Born approximation. The so-called shake-off (SO) process is a single interaction between the incident electron and one target electron which leads to a first ejected electron, followed by a relaxation process due the sudden change in potential which gives a second ejection. The first mechanism is well described by a first Born approximation where the interaction potential between the incident electron and the target is only present here.

The so-called two-step1 (TS1) process is a first interaction between the incident electron and one target electron, which leads to a first ejected electron, followed by a second interaction between this first ejected electron and another target electron. The result of this second interaction is a second ejection of a target electron. This second mechanism must be described by a second Born approximation where the intermediate electron is the first ejected electron.

The so-called two-step2 (TS2) process consists to a first interaction between the incoming electron and one target electron which leads to a first ejected electron, followed by a second interaction between the scattered electron and another target electron.

This third mechanisms requires the second Born approximation too. In this case the intermediate electron is the scattered electron.

The study of the double ionization is also a good way to investigate the correlation in the atoms or molecules. The SO process leads directly to the double Fourier transform of the initial wave function of the target[2-4]. If the two-step mechanisms are involved the information about the correlation in the initial state of the target is destroyed[3]. It is an urgent task to know the kinematical conditions to get spectroscopic informations about the target.

In a recent paper Popov et al[3] showed that the TS1 mechanism is important when the asymmetric kinematics are used (the scattered electron is fast and both ejected electrons are slow). They conclude that symmetric kinematics (the scattered and one ejected electron have close energies and are fast, the second ejected electron is slow) are better to study the correlation in an atom or a molecule. They also write that the TS2 mechanism is negligible for high incident energy. But this last sentence was written after a lot of approximations[3]. We propose here to study the TS2 mechanism without any approximations in the case of the double ionization of helium by fast incident electrons. We want to show that only this mechanism is able to explain the brocken symmetry about the direction of the momentum transfer (it is the difference between the momenta of the incoming electron and of the scattered electron) recently observed in the double ionization of krypton[5]. We also show that this mechanism is not negligible when we compare it to the SO process.

THEORY

We restrict here to the double ionization of helium by high incident electrons (from 1 keV to 5 keV). In this case we describe the incoming electron and the scattered electron by plane waves. So we arrive at the standard Born formulation of this collision problem. The fivefold differential cross sections (FDCS) is given by (atomic units are used throughout) :

$$\frac{d^5\sigma}{d\Omega_s d\Omega_1 d\Omega_2 dE_1 dE_2} = \frac{k_s k_1 k_2}{k_i}\left[\left|M(SO)+M(TS1)+M(TS2)\right|^2\right] \tag{1}$$

where $d\Omega_s, d\Omega_1, d\Omega_2$ denote, respectively, the elements of solid angles for the scattered and the two ejected electrons, and dE_1 and dE_2 represent the energy intervals of the two ejected electrons. Here M (SO), M (TS1) and M (TS2) represent the matrix element when we consider, respectively, the SO, TS1 and TS2 mechanisms. So interference effects can arise. The shake-off process leads to :

$$M(SO) = \left(\frac{-1}{2\pi}\right)\left\langle \psi_f^-(\vec{k}_1,\vec{k}_2;\vec{r}_1,\vec{r}_2)\, e^{i\vec{k}_s\cdot\vec{r}_o} \left| \frac{-2}{r_o} + \frac{1}{r_{o1}} + \frac{1}{r_{o2}} \right| \psi_i(\vec{r}_1,\vec{r}_2)\, e^{i\vec{k}_i\cdot\vec{r}_o} \right\rangle \tag{2}$$

where $\vec{k}_i, \vec{k}_s, \vec{k}_1$ and $\vec{k}_2$ denote, respectively, the momenta of incident, scattered, first ejected and second ejected electrons. The initial-state wave function of helium $\psi_i(\vec{r}_1,\vec{r}_2)$ is those of Silvermann and al[6] which includes radial and angular correlations. The final-state wave function is the approximate BBK wave function[7] :

$$\psi_f(\vec{k}_1,\vec{k}_2;\vec{r}_1,\vec{r}_2) = \frac{1}{\sqrt{2}}\left[\varphi_c^-(\vec{k}_1,\vec{r}_1)\,\varphi_c^-(\vec{k}_2,\vec{r}_2) + \varphi_c^-(\vec{k}_2,\vec{r}_1)\,\varphi_c^-(\vec{k}_1,\vec{r}_2)\right]\varphi(\left|\vec{k}_1-\vec{k}_2\right|) \tag{3}$$

where φ_c is a Coulomb wave function :

$$\varphi_c^- = \frac{e^{i\vec{k}.\vec{r}}}{(2\pi)^{3/2}}\ e^{\pi Z/2k}\ \Gamma(1+i\,Z/k)\ {}_1F_1(-i\,Z/k,1,-i(kr+\vec{k}.\vec{r})) \tag{4}$$

and $\varphi(|\vec{k}_1-\vec{k}_2|)$ the repulsive Gamow factor :

$$\varphi(|\vec{k}_1-\vec{k}_2|) = e^{-\pi\chi_{12}}\ \Gamma(1-i\chi_{12});\ \chi_{12} = \frac{1}{|\vec{k}_1-\vec{k}_2|} \tag{5}$$

If the initial-state wave function is a single product as $\psi_i(\vec{r}_1,\vec{r}_2)=\varphi_i(\vec{r}_1)\,\varphi_i(\vec{r}_2)$ if we neglect the interaction between the incoming electron and the nucleus and if the gamow factor is also neglected we arrive at :

$$M(SO) \simeq \sqrt{2}\,\frac{2}{k^2}\Big[\big\langle\varphi_c(k_1)\big|e^{i\vec{k}.\vec{r}}\big|\varphi_i\big\rangle\big\langle\varphi_c(k_2)\big|\varphi_i\big\rangle + \big\langle\varphi_c(k_2)\big|e^{i\vec{k}.\vec{r}}\big|\varphi_i\big\rangle\big\langle\varphi_c(k_1)\big|\varphi_i\big\rangle\Big] \tag{6}$$

which exactly represents an interaction between an incident electron and one target electron $\big\langle\varphi_c\big|e^{i\vec{k}.\vec{r}}\big|\varphi_i\big\rangle$ multiplied by a relaxation term $\langle\varphi_c|\varphi_i\rangle$. This "pure" shake-off amplitude[8] is sufficient to describe a double ionization with symmetric kinematics (the scattered electron and one ejected electron are fast while the other ejected electron is slow).
But if we consider asymmetric kinematics the formula (6) is not sufficient and the repulsive Ganow factor must be taken into account in order to describe the two slow ejected electrons.
The TS1 process requires the second Born approximation :

$$M(TS1) = \left(\frac{-1}{\pi}\right)\sum_n \int \frac{d\vec{k}_a}{(2\pi)^3\big[k_i^2-k_a^2-k_s^2-2I_n\big]}\left\langle\psi_f^-(\vec{k}_1,\vec{k}_2;\vec{r}_1,\vec{r}_2)\right|\frac{1}{r_{12}}\left|\psi_n^+(\vec{k}_a;\vec{r}_1,\vec{r}_2)\right\rangle$$

$$\times\left\langle\psi_n^-(\vec{k}_a;\vec{r}_1,\vec{r}_2)\,e^{i\vec{k}_s.\vec{r}_o}\right|\frac{-2}{r_o}+\frac{1}{r_{o1}}+\frac{1}{r_{o2}}\left|\psi_i(\vec{r}_1,\vec{r}_2)\,e^{i\vec{k}_i.\vec{r}_o}\right\rangle \tag{7}$$

where $$\psi_n^\pm(\vec{k}_a;\vec{r}_1,\vec{r}_2) = \frac{1}{\sqrt{2}}\big[\varphi_c^\pm(\vec{k}_a,\vec{r}_1)\,\varphi_n(\vec{r}_2)+\varphi_c^\pm(\vec{k}_a,\vec{r}_2)\,\varphi_n(\vec{r}_1)\big] \tag{8}$$

$$\varphi_c^+(\vec{k},\vec{r}) = \frac{e^{i\vec{k}.\vec{r}}}{(2\pi)^{3/2}}\ e^{\pi Z/2k}\ \Gamma(1-i\,Z/k)\ {}_1F_1(i\,Z/k,1,i(kr-\vec{k}.\vec{r})) \tag{9}$$

and $\varphi_n(\vec{r})$ is a wave function describing the He^+ atom.
We restrict here to n = 1 and n = 2. The term I_n is the necessary energy to eject one target electron and to get the fundamental (or an excited) state of He^+ (n = 1 or n = 2).
The first term :

$$\left\langle\psi_n^-(\vec{k}_a;\vec{r}_1,\vec{r}_2)\,e^{i\vec{k}_s.\vec{r}_o}\right|\frac{-2}{r_o}+\frac{1}{r_{o1}}+\frac{1}{r_{o2}}\left|\psi_i(\vec{r}_1,\vec{r}_2)\,e^{i\vec{k}_i.\vec{r}_o}\right\rangle \tag{10}$$

corresponds to a single ionization and (if n = 2) to a simultaneous excitation of the He^+

atom. This first ejected electron (described by a Coulomb wave $\varphi_c(\vec{k}_a, \vec{r})$) collides the other target electron and ejects it. This second collision is represented by the term :

$$\left\langle \psi_f^-(\vec{k}_1, \vec{k}_2; \vec{r}_1, \vec{r}_2) \left| \frac{1}{r_{12}} \right| \psi_n^+(\vec{k}_a; \vec{r}_1, \vec{r}_2) \right\rangle \tag{11}$$

This last term[11] is difficult to calculate because it involves the product of three Coulomb waves.

These calculations are in progress and we only remind here the principal result obtained when we reduce the Coulomb waves to the plane waves[3]. We observe a big peak in the direction of the momenta $\vec{k}_i - \vec{k}_s - \vec{k}_1$ when the second ejected electron is detected at variable angles while the first ejected electron is detected with a fixed angle.

Despite the use of the second Born approximation the symmetry around the momentum transfer is kept.

The TS2 process requires the use of the second Born approximation too :

$$M(TS2) = \left(\frac{-1}{\pi}\right) \sum_n \int \frac{d\vec{k}_b}{(2\pi)^3 (k_i^2 - k_b^2 - k_1^2 - 2I_n)} \left\langle \psi_f^-(\vec{k}_2, \vec{r}_2)\, e^{i\vec{k}_s \cdot \vec{r}_o} \left| \frac{1}{r_{o2}} \right| e^{i\vec{k}_b \cdot \vec{r}_o}\, \varphi_N(\vec{r}_2) \right\rangle$$

$$\times\, \varphi(|\vec{k}_1 - \vec{k}_2|) \left\langle \psi_n^-(\vec{k}_1; \vec{r}_1, \vec{r}_2)\, e^{i\vec{k}_b \cdot \vec{r}_o} \left| \frac{-2}{r_o} + \frac{1}{r_{o1}} + \frac{1}{r_{o2}} \right| \psi_i(\vec{r}_1, \vec{r}_2)\, e^{i\vec{k}_i \cdot \vec{r}_o} \right\rangle$$

$$+ \left(\frac{-1}{\pi}\right) \sum_n \int \frac{d\vec{k}_b}{(2\pi)^3 \left[k_i^2 - k_b^2 - k_2^2 - 2I_n\right]} \left\langle \psi_f^-(\vec{k}_1, \vec{r}_1)\, e^{i\vec{k}_s \cdot \vec{r}_o} \left| \frac{1}{r_{o1}} \right| e^{i\vec{k}_b \cdot \vec{r}_o}\, \varphi_N(\vec{r}_1) \right\rangle$$

$$\times\, \varphi(|\vec{k}_1 - \vec{k}_2|) \left\langle \psi_n^-(\vec{k}_2; \vec{r}_1, \vec{r}_2)\, e^{i\vec{k}_b \cdot \vec{r}_o} \left| \frac{-2}{r_o} + \frac{1}{r_{o1}} + \frac{1}{r_{o2}} \right| \psi_i(\vec{r}_1, \vec{r}_2)\, e^{i\vec{k}_i \cdot \vec{r}_o} \right\rangle \tag{12}$$

This amplitude represents the successive interaction between the incoming electron and the target. The term :

$$\left\langle \psi_n^-(\vec{k}_i; \vec{r}_1, \vec{r}_2)\, e^{i\vec{k}_b \cdot \vec{r}_o} \left| \frac{-2}{r_o} + \frac{1}{r_{o1}} + \frac{1}{r_{o2}} \right| \psi_i(\vec{r}_1, \vec{r}_2)\, e^{i\vec{k}_i \cdot \vec{r}_o} \right\rangle \tag{13}$$

with i = 1 or i = 2 corresponds (as in the TS1 process) to a single ionization (n = 1) or a single ionization and a simultaneous excitation of the He^+ atom (if n = 2).

Then the intermediate scattered electron (described here by the plane wave $e^{i\vec{k}_b \cdot \vec{r}_o}$) collides the other target electron. This second collision is represented by :

$$\left\langle \psi_n^-(\vec{k}_i, \vec{r}_i)\, e^{i\vec{k}_s \cdot \vec{r}_o} \left| \frac{1}{r_{oi}} \right| e^{i\vec{k}_b \cdot \vec{r}_o}\, \varphi_N(\vec{r}_i) \right\rangle \tag{14}$$

i = 1 or i = 2

In this case the symmetry around the momentum transfer is broken. It is due to the successive interactions of the incoming electron where two intermediate momentum transfer appear : $\vec{k}_i - \vec{k}_b$ and $\vec{k}_b - \vec{k}_s$.

We remember that the same result is obtained by the use of the second Born approximation to describe (e,2e) collisions at low incident energy[9].

RESULTS

The recent measurements of Lahmam-Bennani et al[5] have been made for high incident energies and very low scattering angles while the energies of the two ejected electrons are low in comparison with the incident energy. These measurements are coplanar and the three outgoing electrons are detected in coincidence.

The first ejected electron is detected along the momentum-transfer direction while the second is detected at variable angles. These experiments[5] show a broken symmetry around the momentum transfer when both ejected electrons have the same energy. These measurements have been made for the krypton atom and we propose here to study the case of the helium atom as a first step.

We investigate two cases an incident energy of 3473 eV and 5473 eV. Both ejected electrons have the same energy (10 eV) in each case. The scattering angle is fixed to $\theta_s = 1°$. The first ejected electron is always detected along the momentum-transfer direction.

Figures 1 and 4 exhibit the FDCS as a function of the direction $\vec{k}_2$ of one ejected electron. The polar axis is the direction of the incident electron. In figure 1 and 3 we have drawn each contribution to the FDCS : the matrix element M (SO) and the matrix element M (TS2) without any interference. For each figure we notice that M (SO) and M (TS2) have the same order of magnitude. We also see the broken symmetry along the direction of the momentum transfer. The inclusion of the contributions n = 2 give no big change in the results.

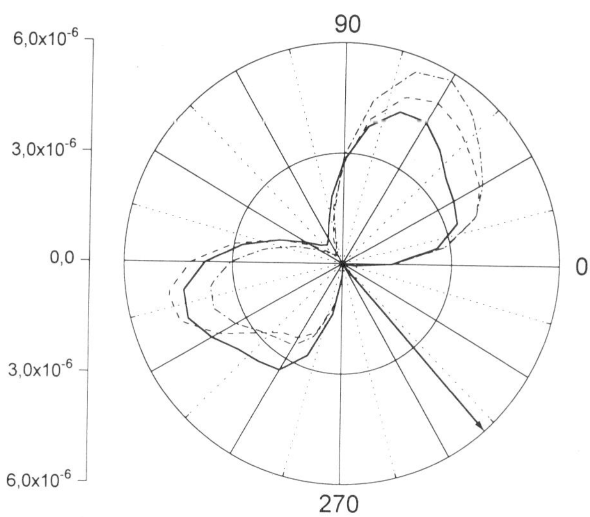

Figure 1. The FDCS in atomic units for the double ionization of helium. Electron energies are 3374, 10 and 10 eV, respectively for scattered and ejected electrons. The scattering angle $\theta_s = 1°$. The first ejected electron is detected along the momentum-transfer direction while the second is detected at variable angles. The polar axis is the direction of the incident electron.

_______ the SO contribution

- - - - - - - the TS2 contribution (n = 1)

—·—·— the TS2 contribution (n = 1 and n = 2)

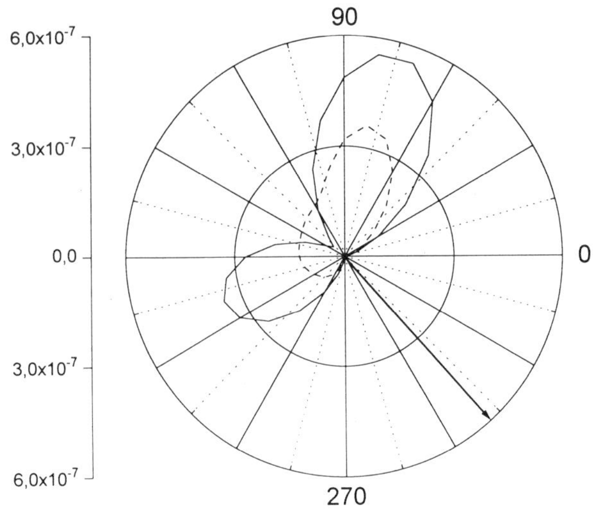

Figure 2. The FDCS in atomic units
____ the SO and TS2 contributions (n = 1)
- - - - the SO and TS2 contributions (n = 1 and n = 2)

Figures 2 and 4 show the results of the FDCS as the sum of M (SO) and M (TS2). Here the interference term is included and we notice the importance of this term. The asymmetry along the momentum transfer is larger here when we take into account for the interference term. The inclusion of the excited intermediate states of He^+ (n = 2) generally reduces the final amplitude.

These preliminary results show, for the first time, that the TS2 mechanism is not negligible even at high incident energies. The interference term has a very big contribution here. It is now an urgent task to include the exact contribution of TS1 mechanism This contribution must be added to those given by SO and TS2 mechanisms. But we also demonstrate that only the TS2 mechanism is able to explain the asymmetry along the momentum-transfer direction.

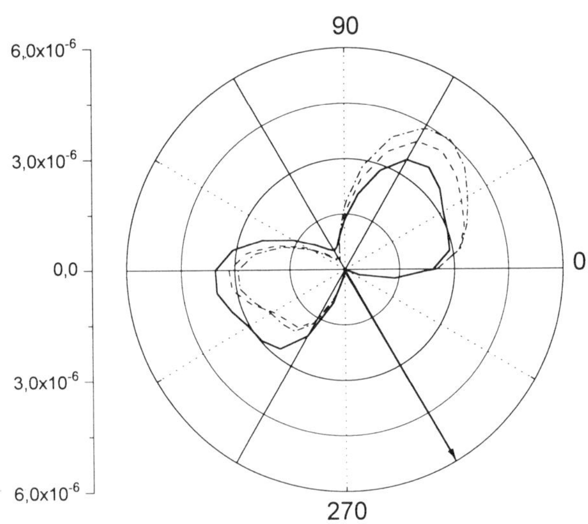

Figure 3. Same as figure 1 but with 5373 eV for the scattered electron

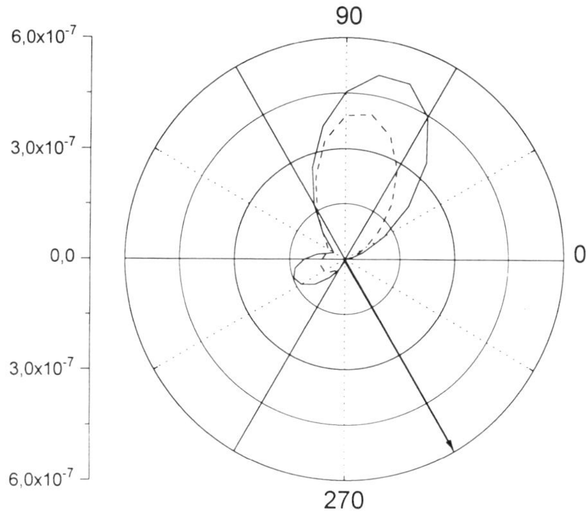

Figure 4. Same as figure 2 but with 5373 eV for the scattered electron

ACKNOWLEDGMENTS

We would like to thank P. Sierputowicz and P. Senot for technical assistance. Discussions with Pr. R. Tweed, A. Lahmam-Bennani and J. Langlois were very useful. This work was partially supported by the EEC Contrat Science N° SCI* 0175-C (EDB).

REFERENCES

1 - R. Tweed, Double processes in electron-Helium collisions, Z. Phys. D 23 : 309 (1992)

2 - C. Dal Cappello and H. Le Rouzo, Angular distributions in the double ionization of helium by electron impact , Phys. Rev. A 43 : 1395 (1991)

3 - Yu. V. Popov, C. Dal Cappello, B. Joulakian and N. M. Kuzmina, On the mechanisms of (e,3e) collisions with fast projectiles, J. Phys. B : At. Mol. Opt. Phys. 27 : 1599 (1996) (see also this volume)

4 - Yu. V. Popov, C. Dal Cappello and K. Kuzakov, (e,3e) electronic momentum spectroscopy : perspectives and advantages, J. Phys. B : At. Mol. Opt. Phys. (accepted)

5 - A. Lahmam-Bennani, A. Duguet, B. El Marji, C. Schröter and L. Spielberger, Double ionization mechanism from triple coincidence experiments (see this volume)

6 - J.N. Silverman, O. Platas and F.A. Matsen, Simple configuration interaction wave functions, J. Chem. Phys. 32 : 1402 (1960)

7 - C. Dal Cappello, H. Hda and A.C. Roy, Angular distributions in the double ionization of the noble gases by electron impact, Phys. Rev. A 51 : 3735 (1995)

8 - R. El M'Khanter, C. Dal Cappello, A. Lahmam-Bennani and Yu. V. Popov, On the mechanisms ,for double ionization : a study of (e,3-1 e) experiments on argon, J. Phys. B 29 : 1101 (1996)

9 - F.W. Byron Jr., C.J. Joachain and B. Piraux, Triple differential cross sections for the ionization of atomic hydrogen by fast electrons, Phys. Lett. 9 : 427 (1983)

PARTITIONING OF MOMENTUM IN ELECTRON-IMPACT DOUBLE IONIZATION

John H. Moore, Michael J. Ford

Department of Chemistry
University of Maryland
College Park, MD 20742

Michael A. Coplan, John W. Cooper

Institute for Physical Science and Technology
University of Maryland
College Park, MD 20742

John P. Doering

Department of Chemistry
Johns Hopkins University
Baltimore, MD 21218

INTRODUCTION

A complete description of electron-impact double ionization requires the determination of the momenta of the scattered projectile and two ejected electrons. The experiment is referred to as (e,3e) implying a collision in which an incident electron gives rise to three electrons that are detected in coincidence. The cross section for a given incident-electron energy and a specified final state of the doubly-charge ion is eightfold differential and is typically reported as a function of the solid angle of acceptance of the detectors of the three outgoing electrons and the energy bandpass of two of these detectors (the energy of the third electron being fixed by conservation of energy).

The (e,3e) experiment has the potential to elucidate in a straight-forward manner several aspects of the so-called "many-body problem" such as the mechanism of electron-impact double ionization as well as the nature of the post-collision interaction between the

Coincidence Studies of Electron and Photon Impact Ionization
Edited by Whelan and Walters, Plenum Press, New York, 1997

outgoing electrons and between these electrons and the doubly-charged ion. In addition, (e,3e) measurements may provide information on the correlated motion of electrons within the atom. The nature of the impulse delivered by the incident electron to the two subsequently-ejected electrons as well as to the ionic core is fundamental to understanding both double ionization and atomic electron correlation. These interactions can be assessed by observing the way in which the available projectile momentum is partitioned between the products of the collision.

Owing to the large number of variables in an (e,3e) measurement, it is difficult to survey broad aspects of electron-impact double ionization. For this reason we have investigated momentum partitioning in experiments in which one of the outgoing electrons is ignored, thus effectively performing a partial integration of the eight-fold differential cross section. Data obtained by this means over a large range of experimental variables demonstrate that double ionization cannot be described by a simple extension of that which is known about electron-impact single ionization.

THE EXPERIMENT

Momentum Vectors

Electron-impact double ionization of an atom may be described by the vector diagram shown in Fig. 1 where $\mathbf{p_o}$ represents the momentum of the incident electron that is scattered with momentum $\mathbf{p_s}$, and $\mathbf{p_1}$ and $\mathbf{p_2}$ represent the momenta of the two ejected

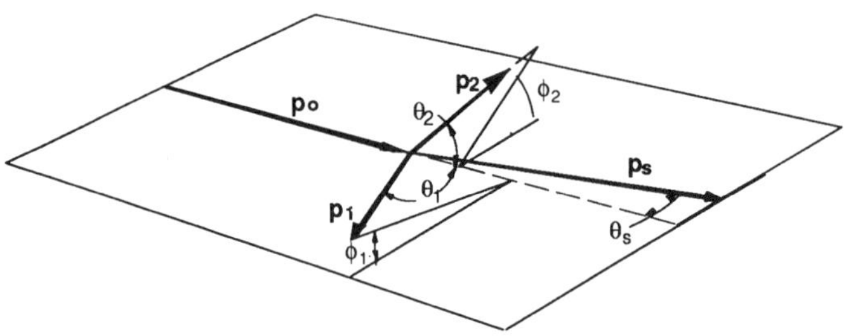

Figure 1. Electron-impact double ionization in terms of the momentum vectors of the incident electron ($\mathbf{p_o}$), the scattered electron ($\mathbf{p_s}$), and two ejected electrons ($\mathbf{p_1}$ and $\mathbf{p_2}$).

electrons. Momentum conservation requires these vectors to satisfy the relation

$$\mathbf{p_o} + \mathbf{p_{ion}^{i}} + \mathbf{q_1} + \mathbf{q_2} = \mathbf{p_s} + \mathbf{p_1} + \mathbf{p_2} + \mathbf{p_{ion}^{f}}$$

where $\mathbf{q_1}$ and $\mathbf{q_2}$ are the momenta of the two target atomic electrons, $\mathbf{p_{ion}^{i}}$ is the momentum of the ionic core an instant before the collision, and $\mathbf{p_{ion}^{f}}$ is the momentum of the ionic core following the collision. The *momentum transfer* in the collision is the difference between the final and initial momenta of the projectile electron:

$$\mathbf{K} = \mathbf{p_o} - \mathbf{p_s}$$

On the assumption that the target atom is stationary,

$$\mathbf{p}^{i}_{ion} + \mathbf{q}_1 + \mathbf{q}_2 = 0,$$

and $\mathbf{p}^{i}_{ion}$ is the *recoil momentum* suffered by the ionic core upon the loss of the two target electrons:

$$\mathbf{p}^{i}_{ion} = -(\mathbf{q}_1 + \mathbf{q}_2)$$

The net target-electron momentum, $(\mathbf{q}_1 + \mathbf{q}_2)$, provides a means for assessing any correlation between these two electrons in the atom. If the (e,3e) reaction is carried out in such a way that the residual ion retains only its recoil momentum (i.e., $\mathbf{p}^{i}_{ion} = \mathbf{p}^{f}_{ion}$), then the net target-electron momentum, and hence its distribution, can be determined from experimentally measured quantities:

$$\mathbf{q}_1 + \mathbf{q}_2 = \mathbf{p}_s - \mathbf{p}_o + \mathbf{p}_1 + \mathbf{p}_2$$

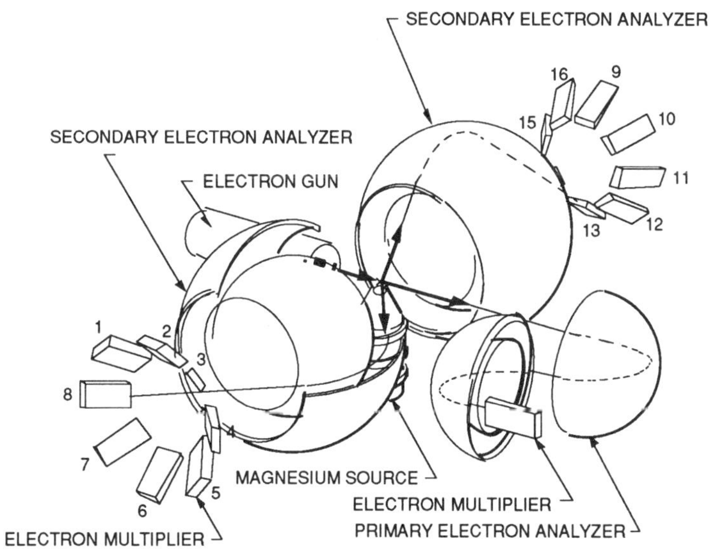

Figure 2. Schematic illustration of multichannel (e,3e) apparatus.

Apparatus

The measurement of the (e,3e) reaction cross section is especially demanding. The eightfold differential cross section, $\frac{d^8\sigma}{d\Omega_s d\Omega_1 d\Omega_2 dE_1 dE_2}$, is small, typically of the order of $10^{-8}\ a_o^2$. To distinguish double ionization from all other electron-scattering and ionization events that occur when a beam of electrons is directed into an atomic target, the scattered and the two ejected electrons must be detected in triple coincidence with nanosecond resolution. Many measurements must be made to survey any significant part of the multidimensional parameter space. Because of the nature of the background, acceptable statistics in a triple coincidence measurement can only be obtained with very low incident electron currents. In practice, the observable triple coincidence rate is of the order of 10^{-2} s^{-1}. To confront these realities, we have built an apparatus[1] that uses a multidetector array so that as many as 64 triple-coincidence measurements can be carried out simultaneously (Fig. 2).

Partially Integrated Cross Sections

Magnesium, a quasi-two-electron atom ($1s^2 2s^2 2p^6 3s^2$), is the target in our experiment. The valence electrons in this atom are expected to display a high degree of correlation. Our first (e,3e) measurements were of the angular correlation of ejected electrons from double ionization *via* the $L_{2,3}M_1M_1$ Auger process. These measurements were carried out under small-momentum-transfer conditions (i.e., θ_s small) where the cross section is largest. It was immediately clear that a simpler experiment would be necessary to survey a nonresonant ionization process over a large range of variables. We therefore undertook measurements in which the scattered electron goes undetected, thereby obtaining the differential cross section integrated with respect to **K**, the momentum transfer.

With integration over two independent variables, the cross section is sixfold differential: $\dfrac{d^6\sigma}{d\Omega_1 d\Omega_2 dE_1 dE_2}$. To focus on momentum partitioning, the data are treated as a function of **P**, the net momentum of the ejected pair of electrons:

$$\mathbf{P} \equiv \mathbf{p}_1 + \mathbf{p}_2$$

Measured cross sections are plotted as a function of the magnitude, $|\mathbf{P}|$, and direction, $\theta_\mathbf{P}$, of the **P** vector.

RESULTS AND DISCUSSION

Measurements have been carried out with incident energies, E_o, of 1000 eV and 422 eV and ejected electron energies, E_1 and E_2, between 80 eV and 120 eV, this energy range chosen to avoid detection of Auger electrons that have energies between 35 eV to 70 eV. From the data presented in Fig. 3, it can be seen that for E_o = 1000 eV the cross section is largest for the largest values of $|\mathbf{P}|$ and the smallest values of $\theta_\mathbf{P}$.

It has been suggested[2] that double ionization cross sections should display a feature analogous to the Bethe Ridge[3] observed for single ionization cross sections. The Bethe Ridge appears on a plot of the triply-differential cross section as a function of $|\mathbf{K}|^2$ and E_n and corresponds to ionization *via* a binary collision between a projectile electron and a free target electron. No momentum is transferred to the residual core. The ridge follows the line given by $|\mathbf{K}|^2 = E_n$ or, alternatively, $\mathbf{p}^i_{ion} = \mathbf{p}^f_{ion} = 0$. For ionization from an atom, the ridge is broadened, reflecting the momentum of the electron in the atom. The profile of the Bethe Ridge gives the distribution of ion recoil momentum.

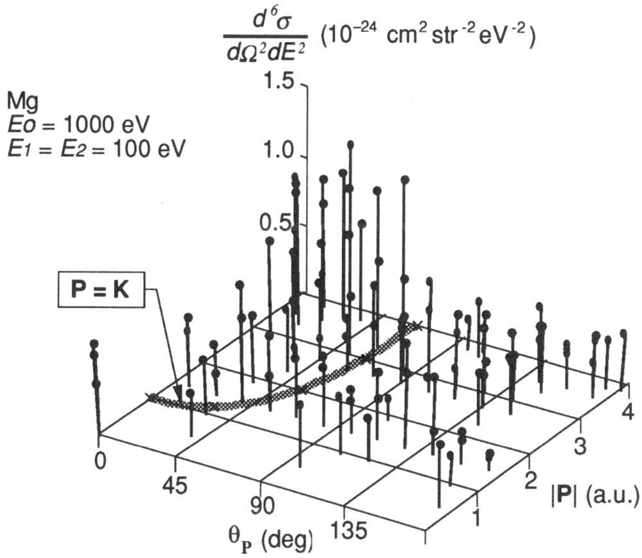

Figure 3. The cross section for double ionization of magnesium integrated over the momentum transfer for an incident-electron energy E_o = 1000 eV and ejected-electron energies of 100 eV. The locus of (|P|, θ_P) values for which **P** = **K** is indicated by the curve in the base plane.

For double ionization, the feature analogous to the Bethe Ridge has been dubbed the "Bethe Sphere".[2] The conditions for knockout of two electrons with no momentum transfer to the core can be related to the observed net momentum, **P**. Combining the equations for conservation of momentum gives

$$\mathbf{P} = \mathbf{p_1} + \mathbf{p_2} = \mathbf{K} - \mathbf{p}_{\mathbf{ion}}^{\mathbf{f}}$$

The requirement $\mathbf{p}_{\mathbf{ion}}^{\mathbf{i}} = \mathbf{p}_{\mathbf{ion}}^{\mathbf{f}} = 0$ implies that the location of the Bethe Sphere is given by **P** = **K**. That is, in the impulsive knockout of two electrons, the momentum transfer goes

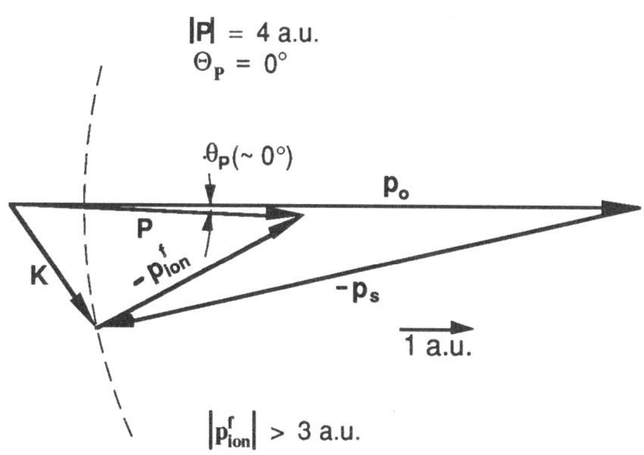

Figure 4. Diagram showing the vector sums $\mathbf{K} = \mathbf{p_o} - \mathbf{p_s}$ and $\mathbf{P} = \mathbf{K} - \mathbf{p}_{\mathbf{ion}}^{\mathbf{f}}$ for an incident electron with energy E_o = 1000 eV scattered with energy loss 222 eV (corresponding to the data in Fig. 3). The scattering angle θ_s is not known but the negative of the scattered electron momentum vector must terminate on the dashed arc. For |P| = 4 a.u. and θ_P = 0°, the magnitude of $\mathbf{p}_{\mathbf{ion}}^{\mathbf{f}}$ must be greater than 3 a.u.

entirely into the momentum of the ejected electrons. The line **P** = **K** is shown in the base plane of Fig. 3.

There is no feature corresponding to the Bethe Sphere apparent in our data. In fact, to the contrary, there appears to be a large probability of the ion's carrying away a great deal of momentum. Consider the case where |**P**| = 4 au and $\theta_P = 0°$, in the region where the cross section is largest. From the vector diagram in Fig. 4, it can be seen that the magnitude of the final ion momentum $|\mathbf{p}^f_{ion}|$ must be greater than 3 a.u. at this point.

It is also remarkable that the cross section is relatively large for values of θ_P approaching 180°, i.e., when the net momentum vector is oriented backwards relative to the incident-beam direction. This is particularly evident in Fig. 5 where we show a cut through the data in Fig. 3 along the line **P** = 3.83 a.u. The location of the Bethe Sphere condition is indicated along with the result of a Born calculation for the double ionization of the two Mg 3s electrons. The calculation clearly does not anticipate the cross section amplitude in the backwards direction. The occurrence of backwards emission is even more evident at lower incident-electron energies as shown in Fig. 6 for an incident-electron energy $E_o = 422$ eV. Here the cross section for ejection with a net momentum |**P**| = 3.83 is nearly as large in the backward direction ($\theta_P = 180°$) as in the forward direction ($\theta_P = 0°$). It is also worth noting that the magnitude of the final ion momentum exceeds 5 a.u. at this point. These observations are reminiscent of single ionization at intermediate-to-low incident energies wherein backwards emission of the ionized electron is attributed to reflection from the core.

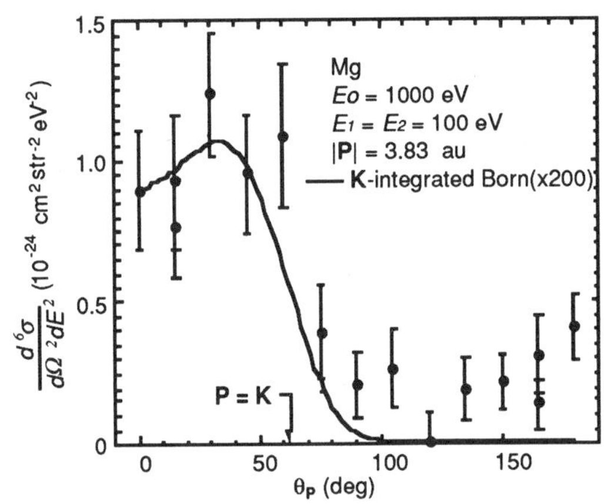

Figure 5. The cross section for the double ionization of magnesium for the case in which the magnitude of the net ejected-electron momentum |**P**| = 3.83 a.u. (This corresponds to an angle of 90° between the two ejected-electron momentum vectors.) The solid line shows the calculated cross section integrated over momentum transfer

There are at least two possible explanations for our observation of large residual ion momentum following electron-impact double ionization. Either there is a large momentum transfer to the core or the ionization process selects atoms in which the core initially possesses a large momentum. From measurements of the momentum distribution of a 3s electron in the outer shell of an atom,[4] one expects the initial ion momentum $\mathbf{p}^i_{ion}$ to be relatively small--on average less than 1 a.u. Our observations then imply a very large momentum transfer from the projectile electron to the ionic core in agreement with the ion-momentum studies of Jagutzki, et al.[5] This is certainly quite different from the case of single ionization at comparable collision energies. Alternatively, the data could be interpreted to imply that double ionization favors atoms in which the initial core momentum, $\mathbf{p}^{i}_{\mathbf{ion}} = \mathbf{q}_1 + \mathbf{q}_2$, is extraordinarily large so that $\mathbf{p}^{i}_{\mathbf{ion}} \approx \mathbf{p}^{f}_{\mathbf{ion}}$.

An additional challenge for interpretation is presented by ionization events in which the net electron momentum moves in the backward direction, contrary to the momentum-transfer vector that kinematically must always be oriented into the forward half-sphere. Preliminary Born Approximation calculations do not account for the cross section amplitude in the backwards direction, however, it must be recognized that these calculations account only for ionization of two 3s electrons to leave the ion in the ground state. It is possible that the angular distribution of ejected electrons may be quite different for ionizations leaving an excited ion (e.g., $2p^43s^2$, $2p^53s$). Calculations taking these possibilities into consideration are underway. In addition, to investigate the mechanism of backward ejection, we are doing experiments at lower incident-electron energies to systematically exclude ionization events involving inner shell electrons.

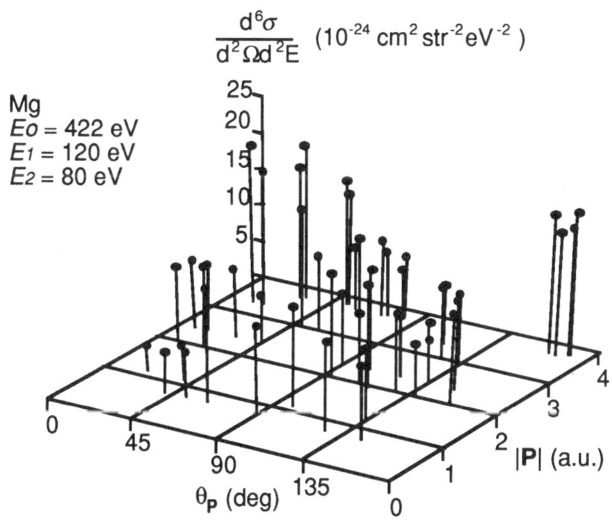

Figure 6. The cross section for double ionization of magnesium integrated over the momentum transfer for an incident-electron energy $E_o = 422$ eV and ejected-electron energies $E_1 = 120$ eV and $E_2 = 80$ eV. The cross section for |P| near 4 a.u. and θ_P near 180° corresponds to both ejected electrons moving in the backwards direction.

ACKNOWLEDGMENTS

This work is supported by National Science Foundation grant PHY-95-15516. We thank Profs. R. S. Berry and R. M. Stehman for the program used in the calculations.

REFERENCES

1. M. J. Ford, J. P. Doering, J. H. Moore, and M. A. Coplan, Rev. Sci. Instrum. **66**, 3137 (1995).
2. J. Berakdar and H. Klar, J. Phys. B **26**, 4219 (1993).
3. M. Inokuti, Rev. Mod. Phys. **43**, 297 (1971).
4. Y. Zheng, I. E. McCarthy, E. Weigold and D. Zhang, Phys. Rev. Lett. **64**, 1358 (1990).
5. O. Jagutzki, L. Spielberger, R. Dorner, S. Nuttgens, V. Mergel, H. Schmidt-Bocking, J. Ullrich, R. E. Olson, and U. Buck, Z. Phys. D. - Atoms, Molecules and Clusters **36**, 5 (1996).

5DCS AND 4DCS CALCULATIONS FOR SYMMETRIC (e,3e) IMPACT IONIZATION

Yuri V. Popov[1], Claude Dal Cappello[2],
and Konstantin Kuzakov[1]

[1]Nuclear Physics Institute, Moscow State University,
Moscow 119899, Russia
[2]Institute de Physique et d'Electronique de Metz,
57078 Metz Cedex 3, France

INTRODUCTION

The $(e,2e)$ electron momentum spectroscopy (EMS) of atoms and atomic targets has become now the powerful method of investigation in the quantum chemistry and the physics of surfaces. This $(e,2e)$ EMS is the single ionisation of a target by fast electrons with both outgoing electrons having close energies and angles. Such geometry of the experiment allows to study one electron wave function in the target[1,2].

The successes of last several years in the carrying out of $(e,3e)$ coincident experiments[3–5] allow to speak about the reality of the $(e,3e)$ EMS-method for electron-electron correlations' studies. The validity of the double ionisation for spectroscopy was discussed for the first time in the paper of Smirnov et al.[6] The characteristic features of any EMS-experiments are fast projectile electrons and big momentum transfers. The last requirement differ essentially the use of $(e,3e)$ and $(\gamma,2e)$ experiments for the spectroscopic purposes.

The conservation of energy and momentum for the $(e,3e)$ scattering gives:

$$E_0 + \varepsilon_0^A = E_a + E_b + E_c + \varepsilon_0^{A^{++}}, \tag{1}$$

$$\vec{p}_0 = \vec{p}_a + \vec{p}_b + \vec{p}_c + \vec{q}. \tag{2}$$

Here $(E_0,\vec{p}_0)$ are the energy and the momentum of the incident electron, $(E_i,\vec{p}_i)$ $(i = a,b,c)$ are the energies and the momenta of the outgoing electrons, ε_0^A and $\varepsilon_0^{A^{++}}$ are the atomic and ionic ground energies, $\vec{q}$ is the recoil momentum, or the residual ion momentum. We can formally consider the momentum transfer here as $\vec{Q} = \vec{p}_0 - \vec{p}_a$, but for big E_0, E_a and E_b with $E_a \sim E_b \sim \frac{E_0}{2}$ the vector $\vec{\Delta} = \vec{p}_0 - \vec{p}_a - \vec{p}_b$ plays the important role being small enough within the definite angular domain.

The advantage of $(e,3e)$ EMS-method concludes in the possibility to take into account the final state electron-electron correlations in terms of the simple semiclassical approach[7], in contrast to the asymmetric kinematics with one fast scattered electron in the final state. Although the five-fold differential cross section (5DCS) is bigger at such kinematics and the ratio signal/background is better we are not able to make the definite theoretical conclusions on the target structure due to the strong influence of ionization mechanisms and the final state ee-correlations[8,9]. Unfortunately the first $(e,3e)$ experimental settings up realize the asymmetric kinematical conditions with small momentum transfer although the unit of Ford et al.[5] can be in principle adapted to the EMS geometry.

The purpose of this paper is to present some arguments and calculations to show the advantages of the $(e,3e)$ experiments to study the details of targets' ground states. We have begun to realize this purpose in our previous paper[10], and here the new calculations are performed with the helium target.

THEORY

We briefly repeat here the main equations of $(e,3e)$ EMS-method. The 5DCS corresponding to the EMS-amplitude takes the form:

$$\frac{d^5\sigma}{dE_b dE_c d\Omega_a d\Omega_b d\Omega_c} = \frac{2p_a p_b}{(2\pi)^3 p_0}\left|\frac{t_{ee}}{2\pi}\right|^2 \rho(\vec{p}_c,\vec{\Delta}) \tag{3}$$

with

$$\left|\frac{t_{ee}}{2\pi}\right|^2 = \frac{4}{Q^4} f(x)C(y);$$

$$\rho(\vec{p}_c,\vec{\Delta}) = \frac{p_c}{(2\pi)^3}\left|\iint d\vec{r}_1\, d\vec{r}_2 \varphi^{-*}(\vec{p}_c,\vec{r}_1)e^{i\vec{\Delta}\vec{r}_2}\Phi_0(\vec{r}_1,\vec{r}_2)\right|^2 \tag{4}$$

$$\left\{\begin{array}{lll} f(x) = \dfrac{2\pi x}{e^{2\pi x}-1}; & C(y) = [1+y^4-y^2\cos(2x\ln y)] & \\ x = |\vec{p}_a-\vec{p}_b|^{-1}; & y = \dfrac{|\vec{p}_0-\vec{p}_a|}{|\vec{p}_0-\vec{p}_b|}; & \vec{\Delta} = \vec{p}_0-\vec{p}_a-\vec{p}_b \end{array}\right\}$$

The form (3) is equivalent to $(e,2e)$ binary triple differential cross section (TDCS), but $\rho(\vec{p}_c,\vec{\Delta})$ now corresponds to the Fourier transform of the overlap function $\chi(\vec{p},\vec{r})$

$$\chi(\vec{p},\vec{r}) = \int d\vec{r}\,'\varphi^{-*}(\vec{p},\vec{r}\,')\Phi_0(\vec{r}\,',\vec{r})$$

which is the projection of the ground state wave function Φ_0 on the continuum Coulomb states.

At high energy we describe the incident, scattered and fast ejected electrons by plane waves keeping in mind to implement later the semiclassical corrections mentioned. The second ejected electron which has low energy is described by a coulomb wave. Let us denote

$$I(\vec{p}_c,\vec{\Delta}) = \iint d\vec{r}_1\, d\vec{r}_2 \varphi^{-*}(\vec{p}_c,\vec{r}_1)e^{i\vec{\Delta}\vec{r}_2}\Phi_0(\vec{r}_1,\vec{r}_2) \tag{5}$$

and expand $\varphi^-(\vec{p}_c,\vec{r}_1)$, $\exp(i\vec{\Delta}\vec{r}_2)$ and $\Phi_0(\vec{r}_1,\vec{r}_2)$ in terms of partial waves:

$$\varphi^-(\vec{p}_c,\vec{r}) = 4\pi\left(\frac{2\pi\eta}{e^{2\pi\eta}-1}\right)^{1/2}\sum_{l=0}^{\infty} i^l e^{-i\sigma_l}R_l(p_c,r)\langle Y_l(\vec{p}_c)Y_l(\vec{r})\rangle, \tag{6}$$

$$\exp(i\vec{\Delta}\vec{r}) = 4\pi\sum_{l=0}^{\infty} i^l j_l(\Delta r)\langle Y_l(\vec{\Delta})Y_l(\vec{r}\,')\rangle, \tag{7}$$

$$\Phi_0(\vec{r},\vec{r}\,') = 4\pi \sum_{l=0}^{\infty} \Phi_{0l}(r,r')\langle Y_l(\vec{r})Y_l(\vec{r}\,')\rangle. \tag{8}$$

The integral in eq. (5) simplifies now:

$$I(\vec{p}_c,\vec{\Delta}) = (4\pi)^2 \left(\frac{2\pi\eta}{e^{2\pi\eta}-1}\right)^{1/2} \sum_{l=0}^{\infty} e^{i\sigma_l}(2l+1)P_l(\cos\theta_{\vec{p}_c\vec{\Delta}}) \int_0^{\infty} r_1^2 j_l(\Delta r_1)\,dr_1 \int_0^{\infty} r_2^2 R_l(p_c,r_2)\Phi_{0l}(r_1,r_2)\,dr_2 \tag{9}$$

The (5DCS) can be analytically integrated over the solid angle of the slow ejected electron. The (4DCS) of $(e,3-1\,e)$ experiments reads now

$$\frac{d^4\sigma}{dE_b\,dE_c\,d\Omega_a\,d\Omega_b} = 2^9\frac{p_a p_b}{p_0 Q^4} f(x)C(y)(1-e^{-\frac{4\pi}{p_c}})^{-1} \sum_{l=0}^{\infty}(2l+1)\left|\int_0^{\infty} r_1^2 j_l(\Delta r_1)\,dr_1 \int_0^{\infty} r_2^2 R_l(p_c,r_2)\Phi_{0l}(r_1,r_2)\,dr_2\right|^2. \tag{10}$$

For comparison (TDCS) for $(e,2e)$ binary collisions looks like

$$\frac{d^3\sigma}{dE_b\,d\Omega_a\,d\Omega_b} = 2^8\pi\frac{p_a p_b}{p_0 Q^4} f(x)C(y)\left|\int_0^{\infty} r_1^2 j_0(\Delta r_1)\,dr_1 \int_0^{\infty} r_2^2 \varphi_0(r_2)\Phi_{00}(r_1,r_2)\,dr_2\right|^2 \tag{11}$$

with $\varphi_0(r) = \sqrt{8/\pi}e^{-2r}$. Both types of the cross-sections (3) and (10) are the functions of the angle $\theta_a = \theta_b = \theta$ between the projectile electron and one of the fast final electrons if one considers the coplanar fully symmetric geometry.

Besides that in the $(e,3e)$ case we have additional dimensions connected with the vector $\vec{p}_c$, so we can expect to obtain more information about the object of investigation compare to the $(e,2e)$ case. Even the less informative $(e,3-1e)$ experiments allow us to conclude about the relative waits of the zero and higher partial contributions.

RESULTS AND DISCUSSIONS

In the case of helium a big number of ground state wave functions can be found in the special literature. Most of them have a big zero partial component with inclusion of relatively small angular terms. In order to examine the influence of radial and/or angular correlations the following types of different wave functions are considered here:

a) the RHF wave function containing neither radial nor angular correlations[11]:

$$\begin{cases} \Phi_0^{\mathrm{RHF}}(\vec{r}_1,\vec{r}_2) = \phi(r_1)\phi(r_2) \\ \phi(r) = \sum_{i=1}^{5} \alpha_i \varphi_i^{1s}(r) \end{cases} \tag{12}$$

b) the wave function of Silvermann, Platas and Matsen[12] that includes both radial and angular correlations:

$$\Phi_0^{\mathrm{SPM}}(\vec{r}_1,\vec{r}_2) = (1+\lambda^2)^{-1/2}\Big\{N[\varphi^{1s}(r_1)\varphi^{1s'}(r_2) + \varphi^{1s'}(r_1)\varphi^{1s}(r_2)] + \frac{4\pi\lambda}{\sqrt{3}}\varphi^{2p}(r_1)\varphi^{2p}(r_2)\langle Y_1(\vec{r}_1)Y_1(\vec{r}_2)\rangle\Big\} \tag{13}$$

c) one of the Hylleraas-type (H) wave functions obtained by Bonham and Kohl[13]:

$$\Phi_0^{\mathrm{H}}(\vec{r}_1, \vec{r}_2) = N[\varphi^{1s}(r_1)\varphi^{1s'}(r_2) + \varphi^{1s'}(r_1)\varphi^{1s}(r_2)](1 + C_0 e^{-\lambda r_{12}}) \tag{14}$$

In eqs. (12)–(14)

$$\varphi^{1s}(r) = \sqrt{\frac{\gamma^3}{\pi}} e^{-\gamma r}; \quad \varphi^{2p}(r) = \sqrt{\frac{\delta^5}{3\pi}} r e^{-\delta r}.$$

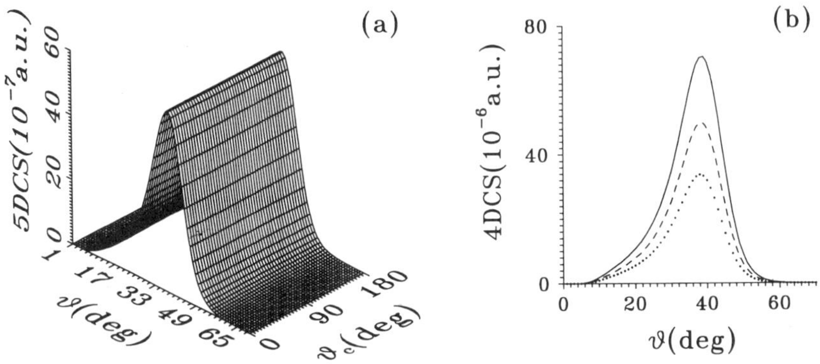

Figure 1. The 5DCS (a) and the 4DCS (b) for the RHF wave function (12). Kinematical conditions: $E_0 = 580\,eV$, $E_a = E_b$; (a) $E_c = 1\,eV$; (b) ——— $E_c = 1\,eV$, - - - - - - $E_c = 5\,eV$, · · · · · $E_c = 10\,eV$.

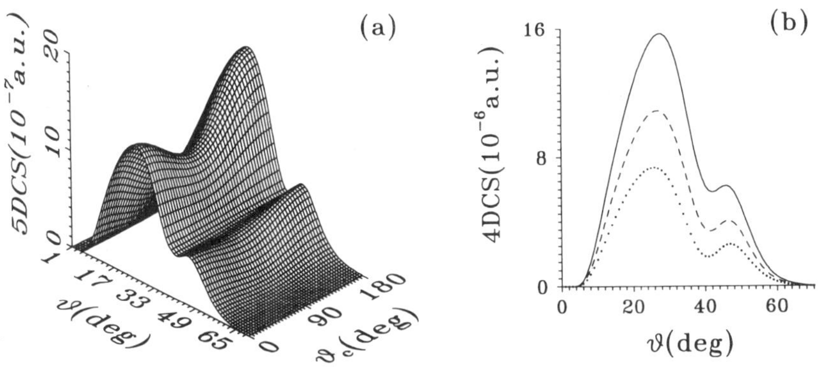

Figure 2. The same like Fig. 1, but for the SPM wave function (13).

Figures 1–3(a) display the dependence of 5DCS on both the scattering angle θ varying around 45° in respect to the initial electron direction and the angle θ_c of the slow ejected electron. The behavior of 5DCS along the θ_c-axis provides us with the information about angular correlations.

Figures 1–3(b) exhibit the 4DCS as the functions of θ. One can see that they keep the main features of 5DCS' curves for H and SPM wave functions including the unexpectable two-peak structure which occurs because of the radial correlations. Such situation is due to the bulk of the zero partial component in the helium wave function (compare Figs. 2b and 3b with Figs. 4(a,b)).

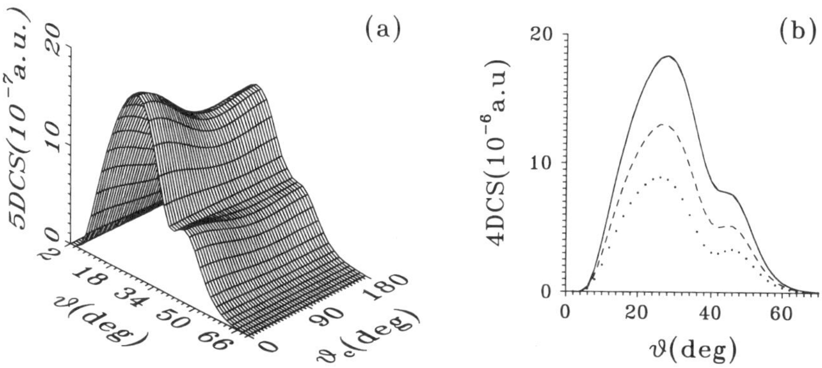

Figure 3. The same like Fig. 1, but for the H wave function (14).

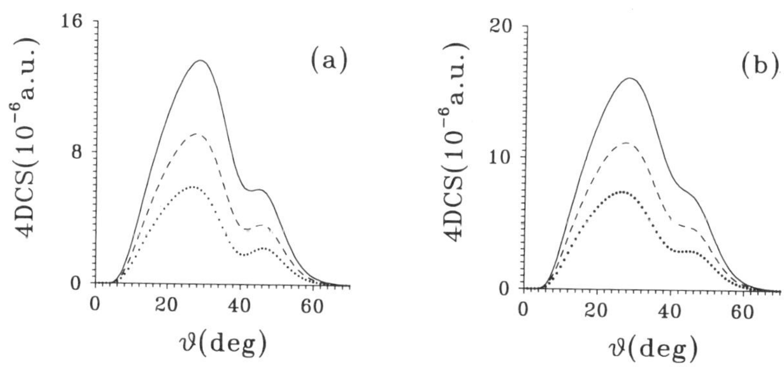

Figure 4. The contribution of the zero partial component $l = 0$ to the 4DCS in (10). (a) SPM; (b) H. The kinematical conditions are the same like Fig. 1.

One sees the characteristic manifestation of radial correlations in the two peaks' curve in Figures 2b and 3b. The position of the local minimum between two peaks corresponds to the position of the maximum in Fig. 1b. i.e. the single Hartree-Fock peak seems to split into the double peaks' structure.

At last Figure 5 demonstrates the Bethe-ridge for $(e, 3e)$. In this kinematical case $\vec{q} = 0$, i.e. $\vec{\Delta} = \vec{p}_c$, and φ_c is the angle between two triangles formed by vectors $(\vec{p}_0, \vec{p}_a, \vec{Q})$ and $(\vec{Q}, \vec{p}_b, \vec{p}_c)$. The profile along the φ_c-axis reflects the behavior of the two-electron Mott cross section in (3). Firstly, one can note very big value of the 5DCS in the Bethe-ridge vicinity (compare Figs. 1 and 5 at the same values of the angle θ_a). This is ordinary result which can be used by experimentators. Secondly, the profiles along the E_c-axis indicate again the considerable difference both in shape and absolute scale.

Note that all Φ_0-models discussing here give practically the same TDCS curves.

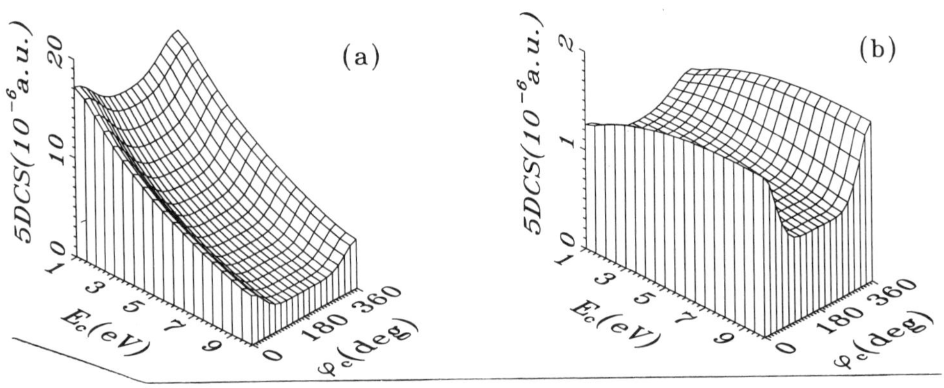

Figure 5. The Bethe-ridge kinematics ($\vec{q} = 0$). $E_0 = 580\, eV$, $E_a = 250\, eV$, $\theta_a = 40°$. (a) RHF; (b) H.

Let us take a look at the set of reactions: $He(e,3e)He^+$, $He(e,2e)He^{+*}$ and $He(e,3e)He^{++}$, carried out in the EMS geometry (He^{+*} means the excitation of the residual ion). Their electron momentum densities $\rho(\vec{\Delta})$ have the same principle structure like in (4), but the overlap functions χ are different: they are the projections of the function Φ_0 on the different parts of the Coulomb spectrum. Each reaction gives us some information on the atomic structure, but obviously the information in case of $(e,3e)$ EMS geometry is grater than that obtained from $(e,2e)$-experiments. For the first time we can see the effect of the Coulomb continuum spectrum which is usually omitted in most of practical calculations.

CONCLUSION

We have calculated 5DCS and 4DCS for electron-impact double ionization of helium in a new proposed geomtry. We consider in this case the scattered and one of the ejected electrons with close energies while the second ejected electron has lower energy. These $(e,3e)$ and $(e,3-1e)$ experiments strongly depend on the correlation in the initial state whereas the $(e,2e)$ experiments are not sensitive to this correlation.

Acknowledgments

We are very grateful to Dr. C. Whelan for the fruitful discussions and help.

REFERENCES

1. V.G. Neudatchin, G.A. Novoseltzeva, and Yu.F. Smirnov, Quasielastic knock out of an electron by a fast electron from atoms, molecules and very thin crystalline films, *JETP.* 28:540 (1969).
2. I.E. Mc Carthy and E. Weigold, Electron momentum spectroscopy of atoms and molecules, *Rep. Prog. Phys.* 54:789 (1991).

3. A. Lahmam-Bennani, C. Dupré, and A. Duguet, Electron-impact double ionization of argon studied by double and triple coincidence techniques: the first (e,3e) experiment, *Phys. Rev. Lett.* 63:1582 (1989).
4. A. Lahmam-Bennani, A. Duguet, A.M. Grisogono, and M. Lecas, (e,3e) absolute five-fold differential cross sections for double ionization of krypton, *J. Phys. B: At. Mol. Opt.* 25:2873 (1992).
5. M.J. Ford, J.P. Doering, M.A. Coplan, J.W. Cooper, and J.H. Moore, (e,3e) observation of the angular correlation between ejected and Auger electrons in the double ionization of magnezium, *Phys. Rev. A.* 51:418 (1995); (see also this volume).
6. Yu.F. Smirnov, A.V. Pavlitchenkov, V.G. Levin, and V.G. Neudatchin, A study of the two-electron Fourier amplitudes of atomic and molecular wave functions using the $(\gamma, 2e)$ and (e,3e) processes of high energies, *J. Phys. B: At. Mol. Phys.* 11:3587 (1978).
7. L. Avaldi, R. Camilloni, Yu.V. Popov, and G. Stefani, Coulomb interaction in the final state of electron impact ionization: effects on the triple differential cross section, *Phys. Rev. A.* 33:851 (1986).
8. Yu.V. Popov, C.Dal Cappello, B. Joulakian, and N.M. Kuzmina, On the mechanisms of (e,3e) collisions with fast projectiles, *J. Phys. B: At. Mol. Opt. Phys.* 27:1599 (1994).
9. Yu.V. Popov, C. Dal Cappello, B. Joulakian, and I.V. Farnakeev, Double ionization of atoms by electron impact as a source of information on electron-electron correlations in targets, *JETP.* 80:179 (1995).
10. Yu.V. Popov, C. Dal Cappello, and K. Kuzakov, (e,3e) electronic momentum spectroscopy: pespectives and advantages, *J. Phys. B: At. Mol. Opt. Phys.* (accepted for publication).
11. E. Clementi and C. Roetti, Roothan-Hartree-Fock atomic wavefunctions: basis functions and their coefficients for ground and certain exited states of neutral and ionized atoms, Z<54, *At. Data and Nucl. Data Tables.* 14:177 (1974).
12. J.N. Silverman, O. Platas, and F.A. Matsen, Simple configuration-interaction wave-functions. I.Two-electron ions: a numerical study, *J. Chem. Phys.* 32:1402 (1960).
13. R.A. Bonham and D.A. Kohl, Simple correlated wavefunctions for the ground state of heliumlike atoms, *J. Chem. Phys.* 45:2471 (1966).

EXACT AND APPROXIMATE METHODS OF THE RIGOROUS COULOMB SCATTERING THEORY

Igor V. Farnakeev[1], Vladimir L. Shablov[1], and Yuri V. Popov[2]

[1]Obninsk Institute of Nuclear Power Engineering, Obninsk, 249020 Russia
[2]Institute of Nuclear Physics
Moscow State University, Moscow 119899, Russia

INTRODUCTION

Intensive experimental studies of (e,2e) and (e,3e) processes have caused a new wave of interest to the approximate methods of the rigorous many-body Coulomb scattering theory. The vast "market" of such methods was replenished recently by the convergent close-coupling method of Bray[1], the hyperradial-adiabatic approach of Matveenko and Fukuda[2], the parabolic - hyperspherical approach of Berakdar[3], and many others.

However there have been no theoretical schemes up to now leading to the convergent integral equations for Coulomb scattering amplitudes. The definite rules must be formulated which determine the correct transition of the amplitudes on the energy shell in an initial or final channels[4]. It is the crucial point of the Coulomb scattering theory based mainly on the short-range potential approach. We would like to mention only one phenomenon in this respect when both the homogeneous and the inhomogeneous Lippman-Schwinger integral equations are valid simultaneously[5]. This in turn means that most of first Born approximations has no sense without a preliminary investigation of the residual terms.

Why does the Coulomb scattering theory run into the problems? It happens because we try to apply the mathematical apparatus developed for short-range potentials to long-range ones. In the nuclear potential case we have two characteristic scales: the nuclear force radius and a free space. This circumstance is in the foundation of the mathematical methods of the scattering theory. However there is no free space at all in the Coulomb case, so we do not have plane waves as a basis of the theory.

The goal of this paper is to give a summary of the methods of the multichannel Coulomb scattering theory which allow to overcome these methodical difficulties. Two fundamental problems are considered. Firstly, different correct approaches to the definition of the Coulomb scattering wave functions as well as the multiparticle scattering amplitudes are formulated. Secondly, we set up requirements within the rigorous theory that the approximate wave functions and reaction amplitudes should satisfy.

MUHLERIN-ZINNES METHOD

If there were no Coulomb interaction the scattering process would be described by wave operators $\Omega_{\pm\alpha}$

$$\Omega_{\pm\alpha} = s - \lim_{t\to\mp\infty} e^{iHt} e^{-iH_\alpha t} |\phi_\alpha\rangle, \tag{1}$$

where H is the total Hamiltonian of the N - particle system, H_α is the channel Hamiltonian and $|\phi_\alpha\rangle$ describes the internal state of n_α fragments colliding in the channel α. In case of systems with the Coulomb interaction the limit (1) does not exist and the definition of the wave operators must be altered.

The correct definition of the Coulomb wave operators was done by Dollard[6] and later by Muhlerin and Zinnes[7], and Chandler et al.[8,9]. The modified Dollard wave operators look like

$$\Omega_{\pm\alpha} = s - \lim_{t\to\mp\infty} e^{iHt} U_\alpha^\pm(t) e^{-iH_\alpha t} |\phi_\alpha\rangle, \tag{2}$$

In the momentum space

$$U_\alpha^\pm(t) = \exp(\pm it(\eta_\alpha \ln|t| + A_\alpha)),$$

The value η_α is the Coulomb parameter of the channel α

$$\eta_\alpha = \sum_{\beta,\gamma\in\alpha} \frac{q_\beta q_\gamma \mu_{\beta\gamma}}{k_{\beta\gamma}} = \sum_{\beta,\gamma\in\alpha} \eta_{\beta\gamma},$$

where β and γ are two fragments with the charges q_β and q_γ colliding in the channel α, $\mu_{\beta\gamma}$ is its reduced mass, $\vec{k}_{\beta\gamma}$ is its relative momentum, and

$$A_\alpha = \sum_{\beta,\gamma\in\alpha} \eta_{\beta\gamma} \ln \frac{2k_{\beta\gamma}^2}{\mu_{\beta\gamma}}.$$

Muhlerin and Zinnes proposed to insert the asymptotic operators $\mathbf{V}_\alpha^\pm$ instead of $U_\alpha^\pm(t)$ in (2). Their matrix elements look like plane waves with the well-known Coulomb distortions

$$\langle\vec{r}_\alpha|\mathbf{V}_\alpha^\pm|\vec{p}_\alpha\rangle = \exp\left(i < \vec{r}_\alpha, \vec{p}_\alpha > \pm i \sum_{\beta,\gamma\in\alpha} \eta_{\beta\gamma} \ln(k_{\beta\gamma} r_{\beta\gamma} \mp \vec{k}_{\beta\gamma}\vec{r}_{\beta\gamma})\right), \tag{3}$$

where $\vec{r}_\alpha$ and $\vec{p}_\alpha$ are the sets of the spatial and momentum Jakobi coordinates, $\vec{r}_{\beta\gamma}$ is the relative coordinate of the fragments β and γ, and $\exp(i < \vec{r}_\alpha, \vec{p}_\alpha >)$ denotes the plane wave describing the free motion of the colliding fragments. The operator $\mathbf{V}_\alpha^\pm$ is asymptotically equivalent to the Dollard operator (see Refs.7-9), i.e.

$$\lim_{t\to\mp\infty} ||(U_\alpha^\pm - \mathbf{V}_\alpha^\pm) e^{-iH_\alpha t}|| = 0 \tag{4}$$

The norm in (4) is defined in the Hilbert space $\mathcal{J}_\alpha$. This result implies that the Dollard asymptotic operator can be replaced in (2) by the Muhlerin-Zinnes asymptotic operator without altering the value of the limit. Examination of the proof of (4) shows that the logarithmically distorted plane waves describe properly the asymptotic boundary conditions in all directions of the space except the direction in which fragments have zero relative velocities. Therefore, the result (4) leads to the following representation of the Coulomb wave operators

$$\Omega_{\pm\alpha} = s - \lim_{t\to\mp\infty} e^{iHt}\mathbf{V}_{\alpha}^{\pm}e^{-iH_\alpha t}|\phi_\alpha\rangle. \qquad (5)$$

The expression (5) is more convenient as the starting point for the development of the time-independent Coulomb scattering theory rather than the initial Dollard's approach (2).

One should point out, that the logarithmic distortion of plane waves in (5) can be replaced by the corresponding two-body Coulomb wave functions, and we obtain instead of (3)

$$\langle\vec{r}_\alpha|\mathbf{V}_{\alpha}^{\pm}|\vec{p}_\alpha\rangle = \exp(i<\vec{r}_\alpha,\vec{p}_\alpha>)\times$$
$$\times\prod_{\beta,\gamma\in\alpha} e^{-\frac{\pi}{2}\eta_{\beta\gamma}}\Gamma(1\pm i\eta_{\beta\gamma})\Phi(\mp i\eta_{\beta\gamma},1;\mp ik_{\beta\gamma}r_{\beta\gamma} - i\vec{k}_{\beta\gamma}\vec{r}_{\beta\gamma}) \qquad (6)$$

Using the identity

$$e^{-iHt} = \int_{-\infty}^{\infty}\frac{dx}{-2i\pi}e^{-ixt}G(x+i0), \qquad (t\geq 0)$$

where $G(Z)=(Z-H)^{-1}$ is the resolvent of the Hamiltonian, we obtain from (5)

$$\Omega_{+\alpha}|\vec{p}_\alpha\rangle = \int_{-\infty}^{\infty}\int_{-\infty}^{\infty}\frac{dx}{-2i\pi}e^{-i(x-E_\alpha)t}[I+G(x+i0)V^\alpha]G_\alpha(x+i0)\mathbf{V}_{\alpha}^{+}|\phi_\alpha,\vec{p}_\alpha\rangle. \qquad (7)$$

In (7) $G_\alpha(Z)$ is the channel Green function, $V^\alpha = H - H_\alpha$, E_α is the total energy of the system in the channel α. Substituting the identity

$$G_\alpha(Z)\mathbf{V}_{\alpha}^{+}|\phi_\alpha,\vec{p}_\alpha\rangle = \frac{1}{Z-E_\alpha}(\mathbf{V}_{\alpha}^{+} + G_\alpha[H_\alpha,\mathbf{V}_{\alpha}^{+}])|\phi_\alpha,\vec{p}_\alpha\rangle \qquad (8)$$

into (7) and taking the limit

$$\lim_{t\to+\infty}\frac{e^{-ixt}}{x+i0} = -2i\pi\delta(x)$$

we arrive at the result of integration

$$\Omega_{+\alpha}|\vec{p}_\alpha\rangle = (I+G(E_\alpha+i0)V^\alpha)\mathbf{V}_{\alpha}^{+}|\phi_\alpha,\vec{p}_\alpha\rangle + G(E_\alpha+i0)[H_\alpha,\mathbf{V}_{\alpha}^{+}]|\phi_\alpha,\vec{p}_\alpha\rangle. \qquad (9)$$

The case $(t\leq 0)$ can be considered in analogy.

Note, that the wave-operator can be also written in the equivalent form of the Povzner limit[10]

$$\Omega_{\pm\alpha}|\vec{p}_\alpha\rangle = \lim_{\varepsilon\to 0}\pm i\varepsilon G(E_\alpha\pm i\varepsilon)\mathbf{V}_{\alpha}^{\pm}|\phi_\alpha,\vec{p}_\alpha\rangle \qquad (10)$$

that can be derived from (5) applying the Abel limit method[11].

Now we are ready to write down the expression for the wave function

$$|\Psi_{\pm\alpha}^{+}(\vec{p}_\alpha)\rangle = \Omega_{\pm\alpha}|\vec{p}_\alpha\rangle = \mathbf{V}_{\alpha}^{\pm}|\phi_\alpha,\vec{p}_\alpha\rangle + G(E_\alpha\pm i0)(V^\alpha\mathbf{V}_{\alpha}^{\pm} + [H_\alpha,\mathbf{V}_{\alpha}^{\pm}])|\phi_\alpha,\vec{p}_\alpha\rangle \qquad (11)$$

The result (11) was obtained earlier[12] and is equivalent to that of Progovecki and Zorbas[13].

The formulae (10) and (11) allow to evaluate after simple transformations the expression for the S-matrix element

$$\langle \vec{p}_\beta | S_{\beta\alpha} | \vec{p}_\alpha^{\,0} \rangle = \langle \vec{p}_\beta | \Omega^\dagger_{-\beta} \Omega_{+\alpha} | \vec{p}_\alpha^{\,0} \rangle$$

$$= \lim_{\varepsilon \to 0} \frac{i\varepsilon}{E_\beta - E_\alpha + i\varepsilon} \langle \phi_\beta, \vec{p}_\beta | (\mathbf{V}_\beta^-)^\dagger \mathbf{V}_\alpha^+ | \phi_\alpha, \vec{p}_\alpha^{\,0} \rangle +$$

$$\lim_{\varepsilon \to 0} \frac{i\varepsilon}{(E_\beta - E_\alpha + i\varepsilon)} \frac{1}{(E_\alpha - E_\beta + i0)} \langle \vec{p}_\beta | \Omega^\dagger_{-\beta} (V^\alpha \mathbf{V}_\alpha^+ + [H_\alpha, \mathbf{V}_\alpha^+]) | \phi_\alpha, \vec{p}_\alpha^{\,0} \rangle = A_1 + A_2. \quad (12)$$

It has been shown[14] that the first term in (12) equals zero if the initial or final reaction channels contain charged particles. The second term includes the energy conservation law δ - function multiplied by the Coulomb scattering amplitude $t_{\alpha\beta}(\vec{p}_\beta, \vec{p}_\alpha^{\,0}; E + i0)$ and can be presented in the prior form:

$$A_2 = -2i\pi\delta(E_\alpha - E_\beta) \langle \vec{p}_\beta | \Omega^\dagger_{-\beta} (V^\alpha \mathbf{V}_\alpha^+ + [H_\alpha, \mathbf{V}_\alpha^+]) | \phi_\alpha, \vec{p}_\alpha^{\,0} \rangle, \quad (13)$$

or in the post form

$$A_2 = -2i\pi\delta(E_\beta - E_\alpha) \langle \phi_\beta, \vec{p}_\beta | (V^\beta \mathbf{V}_\beta^- + [H_\beta, \mathbf{V}_\beta^-])^\dagger \Omega_{+\alpha} | \vec{p}_\alpha^{\,0} \rangle \quad (14)$$

BBK approximation

We consider now one consequence resulting from the equations obtained and concerning the (e,2e) reactions on atoms. As a rule the initial channel contains only two fragments: target + projectile electron. Assume that at least one of these fragments is neutral. In this particular case $\mathbf{V}_\alpha^\pm = I$. Hence, the prior amplitude in (13) takes the form

$$t_{\alpha\beta}(\vec{p}_\beta, \vec{p}_\alpha^{\,0}, E_\alpha + i0) = \langle \Psi_\beta^-(\vec{p}_\beta) | V^\alpha | \phi_\alpha, \vec{p}_\alpha^{\,0} \rangle. \quad (15)$$

Remind that $|\Psi_\beta^\pm(\vec{p}_\beta)\rangle = \Omega_{\pm\beta} |\vec{p}_\beta\rangle$ denotes the exact scattering wave function in the channel β.

Substituting into (15) the first term from (11) we obtain the approximate expression for the prior amplitude which in the case (3) is equivalent to the well known model[15] obeying the nickname BBK in the special literature

$$t_{\alpha\beta}^{BBK}(\vec{p}_\beta, \vec{p}_\alpha^{\,0}, E_\alpha + i0) = \langle \phi_\beta, \vec{p}_\beta | (\mathbf{V}_\beta^-)^\dagger V^\alpha | \phi_\alpha, \vec{p}_\alpha^{\,0} \rangle. \quad (16)$$

When the correlation phenomena are investigated the post-form of the reaction amplitude looks to be more preferable. In this case the scattering amplitude takes the form

$$t_{\alpha\beta}^*(\vec{p}_\beta, \vec{p}_\alpha^0, E_\alpha + i0) = \langle \phi_\alpha, \vec{p}_\alpha^{\,0} | W_\beta | \phi_\beta, \vec{p}_\beta \rangle \quad (17)$$

with the transition operator W_β:

$$W_\beta = V^\beta \mathbf{V}_\beta^- + [H_\beta, \mathbf{V}_\beta^-].$$

In the case of (e,2e) reactions it looks like the transition operator given in the velocity form of special kind.

For the BBK-model the transition operator W_β reads

$$\langle \vec{r}_1, \vec{r}_2 | W | \vec{p}_1, \vec{p}_2 \rangle = \prod_{\alpha=1,2,12} e^{-\frac{\pi}{2}\eta_\alpha} \Gamma(1 - i\eta_\alpha) \exp(i\vec{p}_1\vec{r}_1 + i\vec{p}_2\vec{r}_2) \times$$

$$\times(\Phi_2\vec{\nabla}_1\Phi_1\vec{\nabla}_1\Phi_{12}+\Phi_1\vec{\nabla}_2\Phi_2\vec{\nabla}_2\Phi_{12}) \tag{18}$$

where Φ_1 denotes $\Phi(+i\eta_1,1;-ip_1r_1-i\vec{p}_1\vec{r}_1)$ for brevity and so on. Owing to the relation:

$$\vec{\nabla}_{\vec{r}}\Phi(i\eta,1;-ikr-i\vec{k}\vec{r})=\frac{k}{r}\vec{\nabla}_{\vec{k}}\Phi(i\eta,1;-ikr-i\vec{k}\vec{r})$$

the post amplitude takes the form

$$t^*(\vec{p}_0,\vec{p}_1,\vec{p}_2,E-i0)=\prod_{\alpha=1,2,12}e^{-\frac{\pi}{2}\eta_\alpha}\Gamma(1-i\eta_\alpha)p_1p_2\times$$

$$\times(\vec{\nabla}_{\vec{p}_1}\vec{\nabla}_{\vec{p}_{12}}R_1-\vec{\nabla}_{\vec{p}_2}\vec{\nabla}_{\vec{p}_{12}}R_2)$$

where

$$R_1=\int\frac{d\vec{r}_1}{r_1}\int\frac{d\vec{r}_2}{r_{12}}e^{i(\vec{p}_1-\vec{p}_0)\vec{r}_1+i\vec{p}_2\vec{r}_2}\Phi(i\eta_1,1;-i\vec{p}_1\vec{r}_1-ip_1r_1)\times$$
$$\times\Phi(i\eta_2,1;-i\vec{p}_2\vec{r}_2-ip_2r_2)\Phi(i\eta_{12},1;-i\vec{p}_{12}\vec{r}_{12}-ip_{12}r_{12})\chi_{fi}(\vec{r}_2).$$

and

$$R_2=\int\frac{d\vec{r}_1}{r_{12}}\int\frac{d\vec{r}_2}{r_2}e^{i(\vec{p}_1-\vec{p}_0)\vec{r}_1+i\vec{p}_2\vec{r}_2}\Phi(i\eta_1,1;-i\vec{p}_1\vec{r}_1-ip_1r_1)\times$$
$$\times\Phi(i\eta_2,1;-i\vec{p}_2\vec{r}_2-ip_2r_2)\Phi(i\eta_{12},1;-i\vec{p}_{12}\vec{r}_{12}-ip_{12}r_{12})\chi_{fi}(\vec{r}_2).$$

In the previous integrals $\vec{p}_{12}=\frac{1}{2}(\vec{p}_1-\vec{p}_2)$ and $\chi_{fi}(\vec{r})$ is the overlap of the atomic and the ionic bound states.

The investigation carried out shows the existence of alternative models which follow the post form of reaction amplitudes. Moreover, one can construct on the basis of the equation (11) any other models with the correct Coulomb asymptotics.

RENORMALIZATION OPERATORS

The Coulomb scattering amplitude $t_{\alpha\beta}$ given by (13) and (14) is described by means of convergent integrals. It turns into a nuclear scattering amplitude if $\mathbf{V}^{\pm}_{\alpha,\beta}=I_{\alpha,\beta}$, i.e. in the absence of the Coulomb forces. However in the letter case we can write down the integral equations for the scattering amplitudes. It is convenient in many applications. Formally the same equations are valid for the Coulomb potentials but only out of energy shell, when the total energy is not equal to the energies in the initial and/or in the final channels. If we want to describe the real scattering process we must apply the renormalization procedure (see, for example, Ref.4).

The formula (11) is very convenient to derive the integral equations. We rewrite it in the following equivalent form

$$|\Psi^{\pm}_\alpha(\vec{p}_\alpha)\rangle=F^{\pm}_\alpha(E_\alpha\pm i0)|\phi_\alpha,\vec{p}_\alpha\rangle+G(E_\alpha\pm i0)V^\alpha F^{\pm}_\alpha(E_\alpha\pm i0)|\phi_\alpha,\vec{p}_\alpha\rangle \tag{19}$$

with

$$F^{\pm}_\alpha(E_\alpha\pm i\varepsilon)=\mathbf{V}^{\pm}_\alpha+G_\alpha(E_\alpha\pm i\varepsilon)[H_\alpha,\mathbf{V}^{\pm}_\alpha]=\pm i\varepsilon G_\alpha(E_\alpha\pm i\varepsilon)\mathbf{V}^{\pm}_\alpha \tag{20}$$

The properties of the operators $F^{\pm}_\alpha$ were investigated in Ref.12. It was shown in particular that these operators play the role of the renormalization Coulomb operators in the Dollard's approach. Actually rewriting (4) as

$$\lim_{t\to\mp\infty}\langle\phi_\alpha|e^{iH_\alpha t}(U^{\pm}_\alpha(t))^\dagger\mathbf{V}^{\pm}_\alpha e^{-iH_\alpha t}|\phi_\alpha\rangle=I_\alpha,$$

where I_α is the unit operator in $\mathcal{J}_\alpha$, and applying the Abel limit method we are convinced that the renormalization Coulomb operators satisfy the integral equation

$$\Gamma(1 \mp i\hat{\eta}_\alpha) e^{\frac{\pi}{2}\hat{\eta}_\alpha \mp i\hat{A}_\alpha} (E_\alpha \pm i\varepsilon - H_\alpha)^{\pm i\hat{\eta}_\alpha} F_\alpha^\pm (E_\alpha \pm i\varepsilon)|\phi_\alpha\rangle = |\phi_\alpha\rangle \quad (\varepsilon \to 0), \tag{21}$$

where $\hat{\eta}_\alpha$ and $\hat{A}_\alpha$ are the multiplication operators by η_α and A_α in the momentum space.

The solution of (21) was found in Ref.12. It has been shown in particular that the operators $F_\alpha^\pm$ are equivalent to that obtained earlier by Prugovecki and Zorbas (see Ref. 13). The main feature of these operators is following: if the operator $\hat{\eta}_\alpha$ differs from zero, the matrix element $\langle f|F_\alpha^\pm(E_\alpha + i0)|\phi_\alpha, \vec{p}_\alpha\rangle$ is equal zero for any smooth function $|f\rangle$.

The expression (19) leads to the modified Lippman-Schwinger equation

$$|\Psi_\alpha^\pm(\vec{p}_\alpha)\rangle = F_\alpha^\pm(E_\alpha \pm i0)|\phi_\alpha(\vec{p}_\alpha)\rangle + G_\alpha(E_\alpha \pm i0)V^\alpha|\Psi_\alpha^\pm(\vec{p}_\alpha)\rangle, \tag{22}$$

which according to the previous result can be reduced into the "homogeneous" Lippman-Schwinger equation, i.e.

$$\langle f|\Psi_\alpha^\pm(\vec{p}_\alpha)\rangle = \langle f|G_\alpha(E_\alpha + i0)V^\alpha|\Psi_\alpha^\pm(\vec{p}_\alpha)\rangle.$$

This unexpectable result was obtained earlier in Refs.5,10,13 etc.

The expression (19) is the base for the time-independent Coulomb scattering theory in the momentum space while the expression (11) is more suitable for a coordinate description of the scattering. For example, the representation (11) contains explicitly all types of the Coulomb asymptotics: the operator $\mathbf{V}_\alpha^\pm$ generates the proper logarithmic phase (3), while the total Green function gives the asymptotics in the form of ingoing or outgoing spherical waves. The formulae (11) and (19) allow to erect the bridge between the momentum and coordinate representations of the Coulomb scattering theory, that was an insoluble problem for a long time.

The following expression for the reaction amplitude can be obtained from (15)[16,17] (see also Refs.6,12)

$$t_{\beta\alpha}(\vec{p}_\beta, \vec{p}_\alpha^{\,0}, E + i0) = \lim_{\varepsilon \to 0} \langle \vec{p}_\beta|(\mathbf{Z}_\beta^-(E - i\varepsilon))^\dagger T_{\beta\alpha}(E + i\varepsilon)\mathbf{Z}_\alpha^+(E + i\varepsilon)|\vec{p}_\alpha^{\,0}\rangle \tag{23}$$

$$(E = E_\alpha = E_\beta)$$

In (23) the operator $T_{\beta\alpha}(Z)$ takes the form

$$T_{\beta\alpha}(Z) = \langle \phi_\beta|V^\beta + V^\beta G(Z)V^\alpha|\phi_\alpha\rangle, \tag{24}$$

and is equivalent to the reaction amplitude in the short-range potentials case. If the Coulomb interaction presents the momentum space matrix elements of this operator contain the specific Coulomb divergences discussed in Refs.8,12, which must be removed to obtain the physical amplitude. The renormalization operators $\mathbf{Z}_\alpha^\pm$ which are given by the expression

$$\mathbf{Z}_\alpha^\pm(E \pm i\varepsilon) = \langle \phi_\alpha|F_\alpha^\pm(E \pm i\varepsilon)|\phi_\alpha\rangle \tag{25}$$

just do this.

Using (19) we can obtain the following results

$$t_{\beta\alpha}(\vec{p}_\beta, \vec{p}_\alpha^{\,0}, E + i0) = \int \frac{d\vec{p}_\alpha^{\,\prime}}{(2\pi)^{3(n_\alpha - 1)}} \langle \vec{p}_\beta|\Omega_{-\beta}^\dagger V^\alpha|\phi_\alpha, \vec{p}_\alpha^{\,\prime}\rangle\langle \vec{p}_\alpha^{\,\prime}|\mathbf{Z}_\alpha^+(E + i0)|\vec{p}_\alpha^{\,0}\rangle \tag{26}$$

$$t_{\beta\alpha}(\vec{p}_\beta, \vec{p}_\alpha^{\,0}, E + i0) = \int \frac{d\vec{p}_\beta^{\,\prime}}{(2\pi)^{3(n_\beta - 1)}} \langle \vec{p}_\beta|(\mathbf{Z}_\beta^-(E - i0))^\dagger|\vec{p}_\beta^{\,\prime}\rangle\langle \phi_\beta, \vec{p}_\beta^{\,\prime}|V^\beta\Omega_{+\alpha}|\vec{p}_\alpha^{\,0}\rangle. \tag{27}$$

The formula (26) is the prior-form for the reaction amplitude, while (27) is its post-form.

It is important to point out that the versions (13)-(14) and (26)-(27) are equivalent. To prove this statement we start from the expression (11) which is written here in the post-form

$$t_{\beta\alpha}(\vec{p}_\beta, \vec{p}_\alpha^{\,0}, E + i0) = \langle \vec{p}_\beta | \Omega_{-\beta}^\dagger V^\alpha F_\alpha^+(E + i0) | \phi_\alpha, \vec{p}_\alpha^{\,0} \rangle. \tag{28}$$

Using (20) we can reduce (28) to the sum

$$t_{\beta\alpha}(\vec{p}_\beta, \vec{p}_\alpha^{\,0}, E + i0) = C_1 + C_2$$

with

$$C_1 = \langle \phi_\beta, \vec{p}_\beta | (F_\beta^-(E - i0))^\dagger (V^\alpha - V^\beta) F_\alpha^+(E + i0) | \phi_\alpha, \vec{p}_\alpha^{\,0} \rangle +$$
$$\langle \phi_\beta, \vec{p}_\beta | (F_\beta^-(E - i0))^\dagger ((V^\beta - V^\alpha) - [H_\alpha, \mathbf{V}_\alpha^+]) | \phi_\alpha, \vec{p}_\alpha^{\,0} \rangle \tag{29}$$

and

$$C_2 = \langle \vec{p}_\beta | (\Omega_{-\beta})^\dagger (V^\alpha \mathbf{V}_\alpha^+ + [H_\alpha, \mathbf{V}_\alpha^+]) | \phi_\alpha, \vec{p}_\alpha^{\,0} \rangle. \tag{30}$$

Replacing in (29) the operator V^α by $(H - H_\alpha)$, $(V^\alpha - V^\beta)$ by $((E - H_\alpha) - (E - H_\beta))$ etc., we obtain $C_1 = 0$. Therefore,

$$t_{\beta\alpha}(\vec{p}_\beta, \vec{p}_\alpha^{\,0}, E + i0) = C_2,$$

that was needed to prove.

CONCLUSION

A summary of methods of the rigorous many-body Coulomb scattering theory is given. The Muhlerin - Zinnes' approach which is proven to be equivalent to the basic Dollard's approach seems to be the most preferable as the starting point of the time - independent scattering theory. It gives not only the correct convergent form of the wave functions but logically connects the Schrödinger and the Lippman - Schwinger equations. Some popular and new approximations can be simply derived from the equations obtained.

REFERENCES

1. I. Bray, Convergent close-coupling method for the calculation of electron scattering on hydrogenlike targets, *Phys.Rev.A* 49:1066 (1994).
2. A. V. Matveenko, and H. Fukuda, Hyperradial-adiabatic approach to Coulomb three-body systems, *J.Phys B: At.Mol.Opt.Phys* 29:1575 (1996).
3. J. Berakdar, Parabolic-hyperspherical approach to the fragmentation of three-particle Coulomb systems, *Phys.Rev.A* 54:1480 (1996).
4. Yu. Popov, Investigation of a three-charged-particle break-up scattering amplitude, *J.Phys.B: At.Mol.Phys.* 14:2449 (1981).
5. A. M. Mukhamedzhanov, Integral equations for Coulomb scattering wave functions and Coulomb asymptotic states, *Theor.Math.Phys. (Rus.)* 62:105 (1985).
6. J.D. Dollard, Asymptotic convergency and Coulomb interaction, *J.Math.Phys* 5:729 (1964).
7. D. Muhlerin, and I.I. Zinnes, Coulomb scattering. I. Single channel. *J.Math.Phys.* 11:1402 (1970).

8. C. Chandler, and A. G. Gibson, Time-dependent Coulomb scattering theory, *J.Math.Phys.* 15:291 (1974).
9. C. Chandler, The Coulomb problem. A selective review, *Nucl.Phys.A* 353:1290 (1981).
10. S.P. Merkuriev, and L.D. Faddeev, *Quantum Scattering Theory for the Systems of Few Particles*, Nauka, Moscow (1985).
11. M. Reed, and B. Simon, *Methods of Modern Mathematical Physics. V. Scattering Theory*, Academic Press, New York (1979).
12. Yu.V. Popov, V.L. Shablov, and Yu.Yu. Shitkov, The relation between time-dependent and time-independent scattering theories for the system of particles with Coulom interaction, *Fundamental and Applied Math. (Rus.)* 2:925 (1996).
13. E. Prugovecki, and J. Zorbas, Many-body modified Lippmann-Schwinger equation for Coulomb-like potentials, *Nucl.Phys.A* 213:541 (1973).
14. A.A. Kvitsinsky, V.V. Kostrykin, and S.P. Mercuriev, Scattering theory for quantum three-body systems at fixed total-angular momentum, *J.Elem.Part. and Atom.Nucl. (Rus.)* 21:1301 (1986).
15. M. Brauner, J.S. Briggs, and H. Klar, Triply differential cross section for ionisation of hydrogen atoms by electrons and positrons, *J.Phys.B: At.Mol.Opt.Phys.* 22:2265 (1989).
16. A.M. Veselova, The definition of the scattering amplitudes in the problems of two and three charged particles, *Theor.Math.Phys. (Rus.)* 35:180 (1978).
17. V.V. Komarov, A.M. Popova, and V.L. Shablov, *Scattering of Few Quantum Particles*, Moscow University Press, Moscow (1993).

STRONG AND WEAK STATEMENTS IN THE THEORY OF DIPOLAR (e,2e) IMPACT IONIZATION

Yuri V.Popov[1], Victor A. Knyr[2], Leonid V. Stotland[2], and Viacheslav V. Nasyrov[2]

[1]Nuclear Physics Institute, Moscow State University, Moscow 119899, Russia
[2]Department of Physics, Khabarovsk State Technical University, Khabarovsk, 680035, Russia

INTRODUCTION

The correct theory must be predictable. This is an axiom. We beleive that a single ionization of targets by fast projectile electrons, or (e,2e)-experiments can supply us with a new information about delicate and subtle atomic effects, and an aim of the theory is to identify them. The dipolar kinematics of (e,2e) experiments in this respect are the most debatable. The matter of fact that even in the case of fast incident electrons the matrix element includes the overlap of initial and final targets' wave functions which in turn are very complex objects. Using more or less crude models of these wave functions for calculations we are not only able to predict something but even differ between models. Thus we must look for special kinematical conditions to make definite theoretical conclusions.

In attempts to realize this program we paid attention on the dipolar high energy $He(e,2e)He^+$ experiments carried out by the Lahmam-Bennani's scientific team[1] . In the experiments considered the projectile beam had the energy $E_0 \sim 4 \div 8$ keV, the scattered electron was practically of the same energy. Such high incident energies allow to use the first Born approximation (FBA). Helium is the simplest two-electron target, and this circumstance attracts the special interest of theoreticians. In the experiments the residual ion He^+ left both in the ground (n=1) and in the excited (n=2) states. Our particular interest is focused on small momentum transfers $Q \sim 0.1 \div 0.2$ a.u. and small ejected electron energies $E_b \sim 4 \div 10$ eV.

This choice is not occasional. Two characteristic features pay one's attention in the case n=1, when the residual ion remains in the ground state. Firstly, most of calculations of triple differential cross sections (TDCS) performed in this kinematical domain demonstrate the results coincident with each other. Secondly, the calculations in the vicinity of the recoil peak are situated about $15 \div 20\%$ lower of the experimental points, which in turn reproduce almost near values of the binary and the recoil peaks.

Coincidence Studies of Electron and Photon Impact Ionization
Edited by Whelan and Walters, Plenum Press, New York, 1997

The calculations have been presented both in Refs.1,2 and later by Kheifets et al.[3] (we shall call this model KBMS) and by Knyr et al.[4] (we shall call it KNSP).

The theories can be conditionally separated into two groups. The first one usually operates with some simple or multiconfiguration Hartree-Fock wave function as a ground state and a distorted or the coulomb wave for the slow ejected electron. The models named OCW and V1/U/SEP in Refs.1,2 and the KBMS model belongs to this set. The second group is based on the preliminary reduction of the six-dimensional Schrodinger equation to an infinite system of one-dimensional coupled equations. The system can consist of differential equations like in the R-matrix approach of Mota Furtado and O'Mahoni[5] , or algebraic equations like in the J-matrix approach of Knyr et al. This group of theories as a rule gives better calculation of electron-electron correlations in the final state.

Thus, the aim of the analysis presented is to answer the question: Would it possible to obtain any new substantial information on a target or ionization mechanisms studying high incident energy dipolar (e,2e) experiments in the limit of vanishing momentum transfer and small ejected electron energy?

THEORY

The first Born TDCS for a helium target is given by:

$$\frac{d^3\sigma_n}{d\Omega_a d\Omega_b dE_b} = \frac{16 p_a p_b}{(2\pi)^3 p_0 Q^4} \sum_{lm} | < \phi^-_{(\mu)} | \hat{T}(\vec{Q}) | \phi_0 > |^2, \tag{1}$$

where $(E_0, \vec{p}_0)$, $(E_a, \vec{p}_a)$, $(E_b, \vec{p}_b)$ are the energies and the momenta of the incident, scattered and ejected electrons respectively, $\vec{Q} = \vec{p}_0 - \vec{p}_a$ is the momentum transfer, $|\phi^-_{(\mu)} >$ and $|\phi_0 >$ are the singlet wave functions of the two-electrons final and initial states, $\hat{T}(\vec{Q})$ the transition operator, index $(\mu) = (nlm)$ denotes the reaction channel.

Different forms of the transition operator are acceptable: length form (L), velocity form (V), acceleration form etc. If the initial and the final state wave functions are exact then the transition operator form plays no role, but it is not so for models, especially if functions in the couple are obtained by different methods. The V-form as a rule gives better results for more crude final state wave function than the L-form[6] The couple of wave functions yielding the near results of TDCS calculations for different forms of the transition operator can be called optimal.

At present the couple of wave functions computed by Knyr et al. seems to be the optimal one. Various KNSP-calculations are presented in Figures 1 and 2 both in the L- and in the V- form. One easely sees that even for $n = 2$ the L- and the V-calculations almost coincide with each other. For instance, the L- and V-KBMS calculations demonstrate even the different TDCS shapes (see Figures in Ref.3).

As we mentioned in the Introduction, all couples of the wave functions presented now "on the market of calculations" in the kinematic region chosen display practically coincident with each other results for all angles Θ_b if Q and E_b are rather small. This circumstance shows that this kinematical domain is weakly sensitive to details of the wave functions. To understand the reason of this effect we write down:

$$< \phi^-_{(\mu)} | \hat{T}(\vec{Q}) | \phi_0 > = F(\vec{Q}, \vec{p}_b) = \int d\vec{r}_1 \exp(i\vec{Q}\vec{r}_1) I(\vec{r}_1, \vec{p}_b), \mu = (100) \tag{2}$$

where

$$I(\vec{r}_1, \vec{p}_b) = \int d\vec{r}_2 \phi^{-*}_{(\mu)}(\vec{p}_b; \vec{r}_1, \vec{r}_2) \phi_0(\vec{r}_1, \vec{r}_2) \tag{3}$$

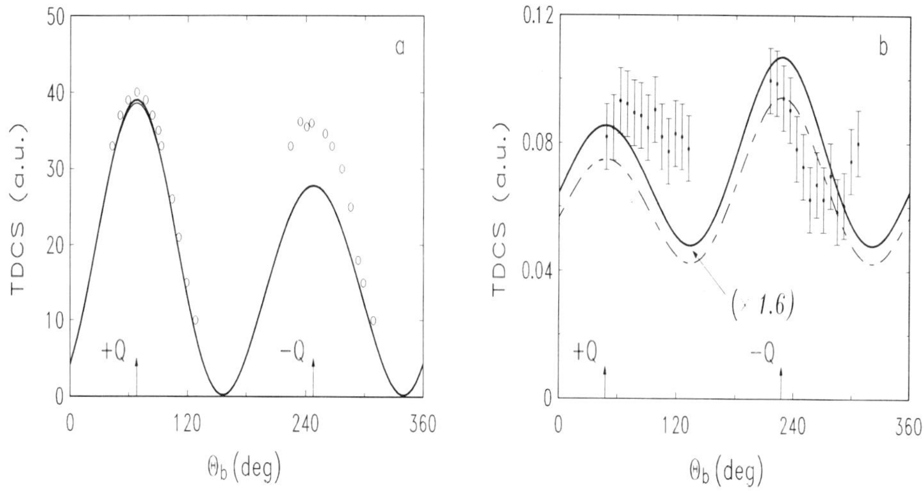

Figure 1. Triple differential cross section $d^3\sigma_n/d\Omega_a d\Omega_b dE_b$ (in atomic units) for the $He(e,2e)He^+$ reaction as a function of the angle Θ_b. The kinematic conditions are as follows: $E_a = 5500$ eV; E_b=5 eV; a) n=1, Q=0.134 a.u.; b)n=2, Q=0.177 a.u. The experimental points are taken from Ref 2. The solid curve represents L-form of the transition operator, the solid-dotted curve represents its V-form. They practically coincide on Fig. 1a.

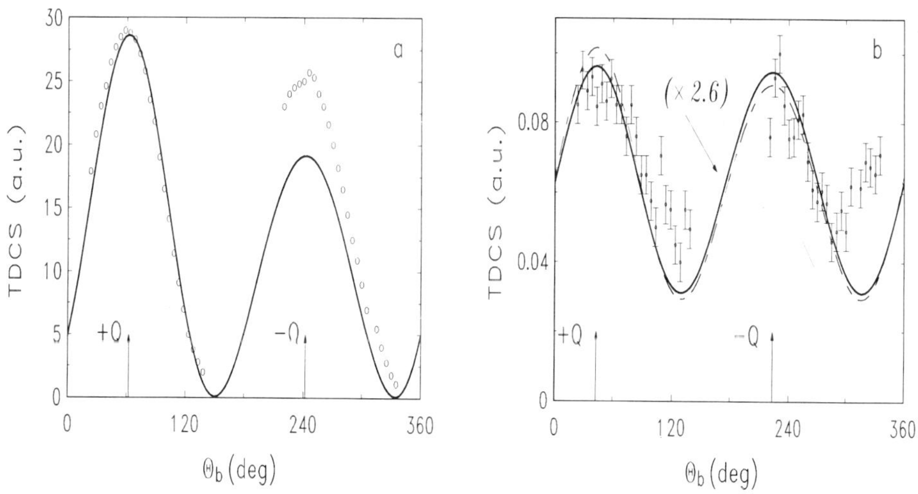

Figure 2. The same like Fig.1 but E_b=10 eV; a) n=1, Q=0.129 a.u.; b) n=2, Q=0.177 a.u.

The orthogonality condition gives $\int d\vec{r} I(\vec{r},\vec{p}) = 0$. The partial decomposition of the function $I(\vec{r},\vec{p})$ takes the form:

$$I(\vec{r},\vec{p}) = 4\pi \sum_{l=0}^{\infty} I_l(r,p) < Y_l(\vec{r}) Y_l(\vec{p}) >$$

Inserting this decomposition into (2) and taking into account that Q is small we obtain from (1)

$$\frac{d^3\sigma}{d\Omega_a d\Omega_b dE_b} = \frac{2^5}{\pi}\frac{p_a p_b}{p_0 Q^2} |Q a_0 + a_1 \cos\Theta + Q a_2 \cos^2\Theta + O(Q^2)|^2 \qquad (4)$$

In(4) Θ is the angle between the vectors $\vec{Q}$ and $\vec{p}_b$. The connection between a_i and I_l is transparent from (2).

It is seen from Figures 1a and 2a that a_0 is practically zero for small p_b. If a_1 and a_2 are of the same order of magnitude then at small Q we have from (4)

$$\frac{d^3\sigma}{d\Omega_a d\Omega_b dE_b} \simeq \frac{2^5}{\pi}\frac{p_a p_b}{p_0 Q^2}|a_1 + Q a_2 \cos\Theta|^2 \cos^2\Theta \tag{5}$$

This result we just observe in the experiment. It is also in agreement with the arguments put foreward earlier[7] and reproduce the photoionization limit as well.

The following expressions for the coefficients a_0, a_1 and a_2 are valid:

$$a_0 = \frac{i}{16\pi} < \phi^-_{(\mu)}|\rho^2|\phi_0 > \tag{6}$$

$$a_1 = \frac{1}{4\pi} < \phi^-_{(\mu)}|z|\phi_0 > \tag{7}$$

$$a_2 = \frac{i}{8\pi} < \phi^-_{(\mu)}|z^2|\phi_0 > \tag{8}$$

In (6)–(8) $\vec{\rho}$ is the component of the vector $\vec{r}_1$ or $\vec{r}_2$ (it does not metter due to symmetries) perpendicular to the vector $\vec{p}_b$, and z is the parallel one.

The formulae (6)–(8) have very instructive physical sense. The experimental fact a_0=0 and the relations (7) and (8) mean that the outgoing electron behaves like a billiard ball moving strictly along the z-direction. This ball either goes out just after the collision with the fast projectile electron (the binary peak), or interacts first with an almost firm wall and then leaves the atom (the recoil peak). The smallness of a_2 means that an atomic structure looks like "core + electron" rather than "nucleus + two electrons".

DISCUSSIONS AND CONCLUSIONS

The specific feature of the most helium ground state wave functions is the dominant contribution of the $1s^2$ - state, on the level of $93 \div 98$ %. Physically the dominance of the $1s^2$-state means that both electrons occupy the same orbit ($r_1 = r_2$), and the role of correlations reduces to some widening of its radius compare to the radius of an electron in a hydrogen-like ion. For example, the only absolute maximum of the KNSP zero partial ground state component $|r_1 r_2 \phi_{00}(r_1, r_2)|^2$ is reached at $r_1 = r_2 = 0.6$ a.u., that corresponds to $Z_{eff} = 1.67$. On the other hand one can expect that only the lowest partial waves contribute to the final state at relatively small ejected energies. In particular case of the simple OCW model the final state is the product of the continuum coulomb function with Z=1 and the bound coulomb function with Z=2. This is a typical manifestation of the radial correlations. The lowest partial components of more developed final state models have in principle the same structure. The varieties display at higher harmonics.

The presented simple arguments make clear why different couples of the wave functions including the best one give the coincident results of the calculations at small Q. The matrix element (2) is nothing more than the overlap between the OCW final state and $1s^2$ initial wave function. This is a strong statement.

To reach now the coincidence between the theoretical and experimental results we must change the coefficients a_1 and a_2 in (5), i.e. to reduce or exclude the influence of the $1s^2$ - state. Another words, we must include at least radial correlations to $\phi_{00}(r_1, r)$ which are evidently absent in the Hartree-Fock model of light atoms. One possible way to do this is to use for estimations the Hylleraas-type wave functions calculated by

Bonham and Kohl[8] which are of $1s1s'$–form. The preliminary calculations demonstrate however no effect in this case. Another way can be found in the recent paper of Draeger et al.[9].

The $1s1s'$–structure of the helium ground state has a clear physical sense. If the electron number 1 occupies the lowest energy level in the coulomb field of an ion He^{++} with Z=2, then the electron number 2 "sees" the charge Z=1 and disposes at bigger distance from the nucleus. Such structure is not of the Hartree-Fock type, and, perhaps, the Hartree-Fock model is not suitable for light atoms at all, but this is a weak statement.

REFERENCES

1. A.Duguet, M.Cherid, A.Lahmam-Bennani, A.Franz, and H.Klar, High–accuracy (e,2e) cross sections for helium: reference data in the first Born approximation, *J.Phys.B: At.Mol.Phys.*, 20:6145 (1987).
2. C.Dupre, A.Lahmam-Bennani, A.Duguet, F.Mota-Furtado, P.F.O'Mahony, and C.Dal Cappello, (e,2e) triple differential cross sections for the simultaneous ionization and excitation of helium, *J.Phys.B: At.Mol.Opt.Phys.*, 25:259 (1992).
3. A.S.Kheifets, I.Bray, I.E. McCarthy, and Bo Shang, Theoretical triple differential cross section of the helium atom ionization with excitation to the n=2 ion state, *Phys.Rev.A*, 50:4700 (1994).
4. V.A.Knyr, V.V. Nasyrov, L.Ya.Stotland, and Yu.V. Popov, Application of the J-matrix method to problems of single ionization of atoms by fast electrons, *JETP*, 22:190(1996).
5. F.Mota-Furtado and P.F.O'Mahony, The R-matrix method in second Born calculations of electron impact ionization of atoms: case study for helium, *J.Phys.B: At.Mol.Opt.Phys.*, 21:137 (1988).
6. Yu.V.Popov and E.K. Shabalina, Notes on dipolar (e,2e) reactions, *J.Phys.B:At.Mol.Opt.Phys.*, 21:L855 (1986).
7. G.Stefani, M.V. Marabello, L.Avaldi, and R.Camilloni Does asymmetric (e,2e) approach the dipolar limit at intermediate energies?, in: *(e,2e) and Related Processes*, C.T.Whelan, H.R.J.Walters, A.Lahmam-Bennani, and H.Ehrhardt, ed., Kluwer Academic Publishers, Dordrecht/Boston/London (1993).
8. R.A.Bonham and D.A.Kohl, Simple correlated wavefunctions for the ground state of heliumlike atoms, *J.Chem.Phys.*, 45:2471 (1966).
9. M.Draeger, G.Handke, W.Ihra, and H.Friedrich, One- and two-electron excitations of helium in the s-wave model, *Phys.Rev.A*, 50:3793 (1994).

THE TRIPLE DIFFERENTIAL CROSS SECTIONS FOR (γ,2e) IN HELIUM AT INTERMEDIATE ENERGIES

J Wightman*, T J Reddish*, S Cvejanovic†
* Physics Department, University of Newcastle, Newcastle upon Tyne, NE1 7RU, U.K.
† On leave from : Institute of Physics, University of Belgrade, P O Box 57,11001, Belgrade, Yugoslavia.

INTRODUCTION

Photodouble Ionization (PDI) is of fundamental interest because it results in an ion and two escaping electrons which are subject to long-range Coulomb forces. The most sensitive and direct way of studying this 3-body Coulomb problem is to use synchrotron radiation and measure the angular distributions (or Triple Differential Cross Sections, TDCS) of the two outgoing electrons using coincidence techniques. PDI in helium is the archetypal system for these studies because a bare nucleus is left behind and no process other than 'direct' double ionisation can occur.

A photoelectron spectrometer has been developed to investigate PDI using novel toroidal analysers which energy select the photoelectrons while preserving the initial angle of emission[1,2]. The angular distributions of the electrons are mapped directly onto the detectors and the resulting efficiency gain helps to compensate for the low cross section of the (γ,2e) process. In this way, the TDCS's in helium have been measured for (5,5), (10,10) and (20,20)eV energy conditions at the Daresbury Synchrotron Radiation Source (SRS).

THEORY

In recent years, there have been several theoretical approaches in describing the TDCS for PDI in helium [3-12], producing results which broadly agree with each other, and with the available experimental data [13-24]. The angular properties of the TDCS's were first deduced by Huetz *et al* [3], and successfully applied to the special case of equal energy sharing in helium[15,16,18]. In this description, the TDCS's of helium are described by two terms (gerade a^g and ungerade a^u) which are symmetric or antisymmetric with respect to electron energy exchange $E_1 \leftrightarrow E_2$:

$$TDCS_{\hat{\varepsilon}} = \left| a^g(E_1,E_2,\theta_{12})(\cos(\hat{\varepsilon}.\hat{k}_1)+(\cos(\hat{\varepsilon}.\hat{k}_2))+a^u(E_1,E_2,\theta_{12})(\cos(\hat{\varepsilon}.\hat{k}_1)-(\cos(\hat{\varepsilon}.\hat{k}_2))\right|^2 \quad (1)$$

where $\theta_{12} = \hat{k}_1.\hat{k}_2$, and $\hat{\varepsilon},\hat{k}_1,\hat{k}_2$ are the unit vectors which refer to the direction of polarisation and the wavevectors of the two outgoing electrons, respectively. The gerade term dominates the TDCS in the low energy "threshold" region whereas at "intermediate" energies both terms contribute [18]. When $E_1 = E_2$ the ungerade term is identically zero for all excess energies and, for our studies in the perpendicular plane geometry, equation (1) reduces to:

$$TDCS_{\hat{\varepsilon}} = C(E,\theta_{12})(\cos\theta_1 + \cos\theta_2)^2 \quad (2)$$

where the angular correlation term

$$C(E,\theta_{12}) = \left| a^g(E/2,E/2,\theta_{12}) \right|^2 \quad (3)$$

and θ_1, θ_2 are the emission angles with respect to the polarisation axis of the two electrons in the detection plane. Thus, choosing equal energy conditions separates the second kinematical - or independent particle - term from the correlation term which contains the effects of the 3-body Coulomb interaction. The form of the kinematical term results in a node in the TDCS at a relative angle of $\theta_{12} = 180°$, and can be understood as the tendency for the uncorrelated electrons to emerge along the direction of polarisation.

The correlation term (C) is predominantly a function of θ_{12}, due to the repulsive Coulomb force between the two electrons. The Wannier threshold model, and its extensions [3,25-28], often use a Gaussian function for the correlation term with a half width 'angle' which is only dependent on the excess energy, via a simple power law:

$$C(E,\theta_{12}) \propto G(180°-\theta_{12}) = \exp\left[-4\ln 2(180°-\theta_{12})^2 / \theta_{1/2}^2\right] \quad (4)$$

where $\theta_{1/2}(E) = \theta_0 E^W$. Traditionally the energy exponent (w) has a value of 0.25 with differing θ_0's [25]. Although the 'threshold' law cannot strictly be applied at high excess energies, it is convenient to continue to adopt a Gaussian form as a first approximation for C, and use the above general form for $\theta_{1/2}$. All the available experimental data, which is summarised in Figure 3, suggests a much smaller value for w indicating that the correlation term continues to dominate the TDCS's for high excess energies, due to the long-range Coulomb force.

EXPERIMENTAL METHOD AND RESULTS

These measurements employed a photoelectron-photoelectron multi-coincidence spectrometer [2] in which the two electrons, of specified energies, are detected over a wide range of emission angles, as shown in Figure 1. Electrons emerge from a central interaction region defined by the intersection of the photon and gas beams. Those emitted in the plane orthogonal to the photon beam that enter the independent analysers are energy-resolved and then focussed onto 2-D position-sensitive detectors, while preserving their initial angles of emission,. The multi-coincidence capability can be realised as electrons arriving anywhere on one detector can be correlated with electrons detected simultaneously anywhere on the other detector. This enables independent angular distributions (or TDCS) to be measured concurrently. The entire TDCS can be mapped as a function of the emission angle of one electron by rotating the whole spectrometer around the direction of the photon beam.

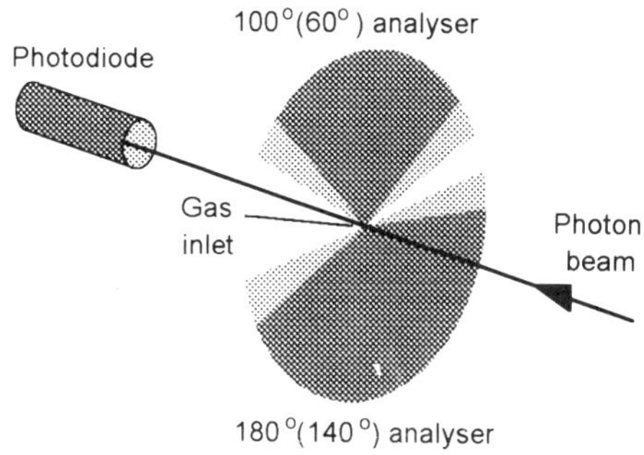

Figure 1 : A schematic diagram showing the fields of view, in the perpendicular plane, of the two partial toroidal analysers. The mechanical acceptance angles are 100° and 180°, but theses are reduced to 60° and 140°, respectively, due to electric field termination effects within the electron optics. The central interaction region is defined by the intersection of the gas and photon beams. The photon flux is monitored by a photodiode.

We have recently measured TDCS's in helium for equal, intermediate energy conditions at the SRS. Synchrotron radiation is predominately linearly polarised in the horizontal plane, and in our case the Stokes S_1 parameter was determined consistently to be 0.83 for the photon energy range covered in these measurements. The data collection time took several days per spectrometer orientation, including angular calibrations and polarisation measurements using the He^+ N = 1,2 states [2].

The measured TDCS's are compared with the calculated distributions using the parametrization of Huetz *et al* [3]. A sample of the results for 10 and 40eV excess energies, with equal energy conditions, is shown in Figure 2. In the polar format the number of (true) coincidences, integrated over 10° windows, is plotted radially. The angular distribution of electron e_2 is plotted for a particular emission angle ($\theta_1 = \alpha$) of e_1 and the electric field vector lies along the horizontal axis. Although the data points cover a restricted angular range of ~120°, the agreement with the theoretical curves obtained from equations (1,2) is very good, and is consistent for all the θ_1 angles. The characteristic two lobes in the theoretical TDCS's are symmetric at 180° and 90° (as is the available data at 90°) and are asymmetric between these limits. A deep minimum is observed at θ_{12} = 180°, as required by the selection rules [3,8].

The (20,20)eV results are at the highest excess energy for equal energy sharing to date. They provide a useful indication as to the strength of the electron correlation in this intermediate energy region, as indicated by our preliminary $\theta_{1/2}$ value of 88° ± 7°. Our measurements support and extend earlier experimental studies and will serve to further the understanding of the energy dependence and functional form of the correlation function; a summary of all the available data is shown in Figure 3.

Further work will extend these studies to unequal energy sharing conditions and to molecules.

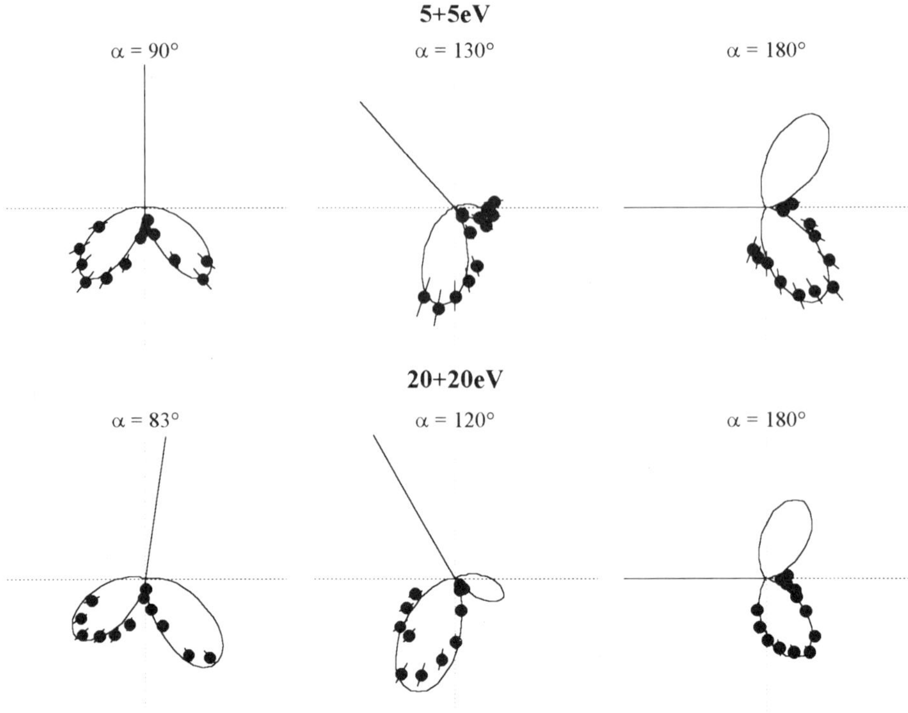

Figure 2 The polar plot of three triple differential cross sections (TDCS) for both (5+5eV) and (20+20eV) electrons in helium with the angle between one ejected electron (α) and the light polarisation direction (horizontal) as indicated. The TDCS's are arbitrarily normalised to the theory of Huetz *et al* [3] (solid curve) with $\theta_{1/2}$'s of 80^0 (±5°) and 87^0 (±7°) for excess energies of 10 and 40eV, respectively, and having taken into account the degree of linear polarisation (S_1 = 0.83) [13].

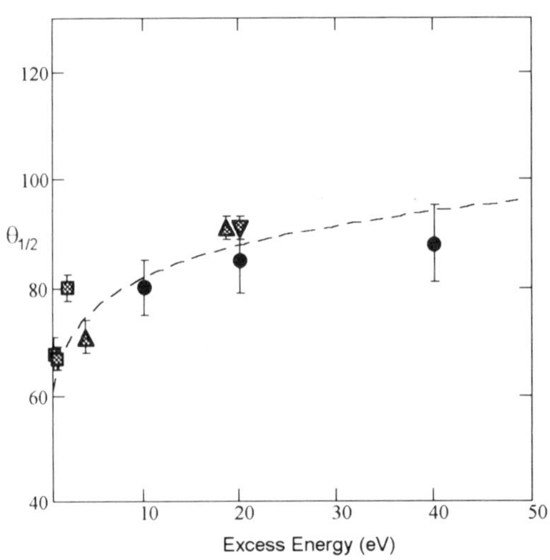

Figure 3: A graph which summarises the available helium data for the excess energy dependence of the (Gaussian) correlation function's half angle. The slow variation with energy above photodouble ionisation threshold is indicative of the strong electron correlation associated with long-range Coulomb force.

● Reddish et al [24]
⊠ Dawber et al [19]
▼ Schwarzkopf and Schmidt [22]
▲ Malegat et al [4]

Dashed line: $\theta_{1/2}$ = 65° $E^{0.1}$ (with Gaussian form for Correlation Function).

ACKNOWLEDGEMENTS

We wish to acknowledge EPSRC for their continuing financial support and synchrotron beamtime. We would also like to thank Dr Mike MacDonald for his valuable assistance while at Daresbury and former graduate students: Graham Richmond, Gareth Bagley, and Ralph Shiell, for their individual contributions to the project.

REFERENCES

1. F. Toffoletto, R.C.G. Leckey, and J.D. Riley, *Nuc. Inst. & Meth.*, B12:282 (1985).
2. T.J. Reddish, G. Richmond, G.W. Bagley, J.P. Wightman and S. Cvejanovic, submitted to *Review of Scientific Instruments* (1996).
3. A. Huetz, P. Selles, D. Waymel, and J. Mazeau, *J. Phys. B.: At. Mol. Opt. Phys.*, 24:1917 (1991).
4. L. Malegat, P. Selles, and A. Huetz, submitted to *Phys. Rev. A* (1996).
5. F. Maulbetsch, and J.S. Briggs, *J. Phys. B.: At. Mol. Opt. Phys.*, 26:1679 (1993).
6. F. Maulbetsch, and J.S. Briggs, *J. Phys. B.: At. Mol. Opt. Phys.*, 26:L647 (1993).
7. F. Maulbetsch, and J.S. Briggs, *J. Phys. B.: At. Mol. Opt. Phys.*, 27:4095 (1994).
8. F. Maulbetsch, and J.S. Briggs, *J. Phys. B.: At. Mol. Opt. Phys.*, 28:551 (1995).
9. F. Maulbetsch, M. Pont, J.S. Briggs, and R.J. Shakeshaft, *J. Phys. B.: At. Mol. Opt. Phys.*, 28:L341 (1995).
10. M. Pont, and R. Shakeshaft, *Phys. Rev. A*, 51:R2676 (1995).
11. M. Pont, and R. Shakeshaft, *J. Phys. B.: At. Mol. Opt. Phys.*, 28:L571 (1995).
12. M. Pont, R. Shakeshaft, F. Maulbetsch, and J.S. Briggs, *Phys. Rev. A*, 53:3671 (1996).
13. O. Schwarzkopf, B. Krässig, J. Elmiger, and V. Schmidt, *Phys. Rev. Letts.*, 70:3008 (1993).
14. O. Schwarzkopf, B. Krässig, and V. Schmidt, *J. de Physique II*, 3:C6169 (1993).
15. D. Waymel, L. Andric, J. Mazeau, P. Selles, and A. Huetz, *J. Phys. B.: At. Mol. Opt. Phys.*, 26:L123 (1993).
16. A. Huetz, P. Lablanquie, L. Andric, P. Selles, and J. Mazeau, *J. Phys. B.: At. Mol. Opt. Phys.*, 27:L13 (1994).
17. O. Schwarzkopf, B. Krässig, V. Schmidt, F. Maulbetsch, and J.S. Briggs, *J. Phys. B.: At. Mol. Opt. Phys.*, 27:L347 (1994).
18. P. Lablanquie, J. Mazeau, L. Andric, P. Selles, and A. Huetz, *Phys. Rev. Letts.*, 74:2192 (1995).
19. G. Dawber, L. Avaldi, A.G. McConkey, H. Rojas, M.A. MacDonald, and G.C. King, *J. Phys. B.: At. Mol. Opt. Phys.*, 28:L271 (1995).
20. A. Huetz, A. Andric, A. Jean, P. Lablanquie, P. Selles, and J. Mazeau, *Conference Proceedings 19th ICPEAC (AIP)* p139 (1995).
21. S.J. Schaphorst, B. Krässig, O. Schwarzkopf, N. Scherer, and V. Schmidt, J. Phys. B.: *At. Mol. Opt. Phys.*, 28:L233 (1995).
22. O. Schwarzkopf, and V. Schmidt, *J. Phys. B.: At. Mol. Opt. Phys.*, 28:2847 (1995) and *J. Phys. B : At. Mol. Opt. Phys.*, Corrigenda p1877 (1996).
23. S.J. Schaphorst, A. Jean, O. Schwarzkopf, P. Lablanquie, L. Andric, A. Huetz and V. Schmidt, *J. Phys. B.: At. Mol. Opt. Phys.*, 29:1901 (1996).
24. T.J. Reddish, J.P. Wightman and S. Cvejanovic, *Proceedings 18th Symposium of the Physics of Ionised Gases* (SPIG) Kotor, Yugoslavia (to be published).
25. A.K. Kazansky, and V.N. Ostrovsky, *J. Phys. B.: At. Mol. Opt. Phys.*, 26:2231 (1993).
26. A.K. Kazansky, and V.N. Ostrovsky, *J. Phys. B.: At. Mol. Opt. Phys.*, 27:447 (1994).
27. A.K. Kazansky, and V.N. Ostrovsky, *J. Phys. B.: At. Mol. Opt. Phys.*, 28:1453 (1995).
28. A.K. Kazansky, and V.N. Ostrovsky, *Phys. Rev. A*, 51:3712 (1995).

TOTAL AND SINGLE DIFFERENTIAL CROSS SECTIONS FOR He DOUBLE PHOTOIONIZATION CALCULATED BY THE CCC METHOD

A. S. Kheifets and Igor Bray

Electronic Structure of Materials Centre, Flinders
University of South Australia, Adelaide 5001, Australia

INTRODUCTION

The helium double photoionization is a fundamental three-body Coulomb problem and, as such, has been at the forefront of atomic physics for the past thirty years. It has attracted considerable theoretical and experimental attention and substantial progress has been made in understanding the complex process of the breakup into three charged particles.

The first application of the the convergent close-coupling (CCC) method to the helium double photoionization problem[1] described accurately the fraction of the double photoionization events in a wide photon energy range. It was also suggested that the method would be well suited for calculating differential cross-sections as well. In this paper we present the CCC formalism and develop it to produce the single-differential, i.e. resolved with respect to the photoelectron energy, double photionization cross-sections.

FORMALISM

In our formalism we consider the double photoionization as a two-stage process. Stage one is single ionization which is followed by electron-impact ionization of the resultant He^+ ion. The one-electron states of the He^+ ion are described by a Laguerre square-integrable basis with excitation of the positive energy pseudostates corresponding to ionization of the He^+ ion, and therefore the double ionized channels for photon-impact ionization of helium. The CCC method has been extensively tested for the e-He^+ scattering system, and has yielded quantitative agreement with the measurements of electron-impact total ionization[2], suggesting that we should be able to obtain accurate double photoionization cross sections for helium. The primary limitation on the accuracy of our approach is a fair description of the helium atom single photoionization leading to the He^+ ion ground and various excited states. We believe this can be achieved by employing a highly correlated helium ground state wave function together with an accurate solution of the e-He^+ excitation and ionization problem using the CCC method.

Coincidence Studies of Electron and Photon Impact Ionization
Edited by Whelan and Walters, Plenum Press, New York, 1997

We use the multi-channel expansion for the final state wave function of the system He^+ ion plus ejected electron:

$$|\Psi_j^{(-)}(\mathbf{k}_b)\rangle = |j\mathbf{k}_b^{(-)}\rangle + \sum_i \int d^3k \frac{\langle \mathbf{k}^{(+)} i|T|j\mathbf{k}_b^{(-)}\rangle}{E - k^2/2 - \epsilon_i + i0} |i\mathbf{k}^{(+)}\rangle, \tag{1}$$

with boundary conditions corresponding to an outgoing wave in a given channel $|j\mathbf{k}_b^{(-)}\rangle$ and incoming waves in all other channels $|i\mathbf{k}^{(+)}\rangle$. Here $E = k_b^2/2 + \epsilon_j$ is the final state energy. The channel wave function $|j\mathbf{k}_b^{(-)}\rangle$ is the product of a one-electron orbital $\bar{\phi}_j$, obtained by diagonalizing the He^+ Hamiltonian in a Laguerre basis, and a Coulomb Z=1 outgoing wave $\chi^{(-)}(\mathbf{k}_b)$. The half off-shell T-matrix in Eq. (1) is the solution of the corresponding Lippmann-Schwinger integral equation [3]. We separate the contribution from the final channels $|j\mathbf{k}_b^{(-)}\rangle$ into single and double ionization according to the energy of the ϵ_j which is positive for the double ionized channels and negative for the singly ionized channels.

The photoionization cross section, as a function of the photon energy ω, corresponding to a particular bound electron state j is given by

$$\sigma_j(\omega) = \frac{4\pi^2}{\omega c} \sum_{m_j} \int d^3k_b \, |\langle \Psi_j^{(-)}(\mathbf{k}_b)\, |\mathcal{D}|\, \Psi_0\rangle|^2 \, \delta(\omega - E_f + E_0) \, , \tag{2}$$

where $c \simeq 137$ is the speed of light in atomic units.

For the ground state wave function Ψ_0 in Eq. (2) we used a 7-term multiconfiguration Hartree-Fock expansion [4]. This gave us a reasonably good account of the ground state correlations judging by the ground state energy $E_0 = -2.90181$ a.u. recovering 95.4% of the correlational energy. Another indication of the quality of our ground state wave function is the fraction of the double photoionization events in the large photon energy limit. According to Åberg [5], this ratio depends only on the ground state wave function. Our ratio of 1.67% is in a good agreement with the measurement of Spielberger *et al.* [6] who carefully separated the Compton scattering from photoionization and reported the value of 1.72%.

The dipole electromagnetic operator $\mathcal{D}$ can be written in one of the forms commonly known as length, velocity and acceleration. The length form enhances the large distance contribution to the radial integral of Eq. (2). The variational ground state wave function is inaccurate at large distances which are negligible in terms of the energy. This puts certain limitation on the use of the length form of the electromagnetic interaction operator which enhances the area far from the origin. So, we restrict ourselves to the velocity and acceleration gauges only.

RESULTS

Double-to-single photoionization ratio

In recent years much of the attention has been focused on the ratio of the double to single photoionization cross sections $R = \sigma^{2+}/\sigma^{+}$ studied as a function of the photon energy ω. Although this ratio is now well established in the asymptotic region of large photon energies of several keV, there still exists uncertainty in this ratio at the low-to-intermediate photon energies from the double photoionization threshold up to several hundred eV.

The latest experimental undertakings by Dörner *et al.* [7] and Levin *et al.* [8] were aimed at eliminating any uncertainty in the value of $R(\omega)$. However, these most recent, and presumably most accurate, measurements produced conflicting results. Dörner *et al.* [7] reported systematically lower $R(\omega)$ with $R_{max} = 3.5\%$ as opposed to the value of 4.0% by Levin *et al.* [8]. Latest theoretical data [9–13] also vary greatly depending on the method used by the authors and the gauge (length, velocity, or acceleration) chosen for the electromagnetic interaction.

The results of our calculations for $R(\omega)$ are presented in Fig. 1 along with the latest experimental data due to Dörner *et al.* [7] and Levin *et al.* [8]. For clarity of presentation we compare with other theories which use a similar foundation for these calculations. These are the eigenchannel R-matrix method of Meyer and Greene[12] and the hyperspherical close-coupling approach of Tang and Shimamura [10]. Comparison with other experiments and theories may be found in these references. Convergence in our calculations was achieved by performing 52-state calculations consisting of 13 states for each He^+ target-space orbital angular momentum $\ell = 0, \ldots, 3$. Some unphysical oscillation is visible in our results, which are due to pseudoresonances associated with the pseudothresholds.

Our results, both in the velocity and acceleration forms, are in between the two sets of most recent experimental data with the difference of around 10% to either set. The calculation of Meyer and Greene[12] in the velocity gauge is close to our velocity gauge result. However, their acceleration form result is substantially lower than ours. This R-matrix method has also demonstrated an ability to obtain accurate ionization cross sections (a few eV above threshold) in the e-He^+ problem [14], and so we suspect that the source of the discrepancy may be the choice of the ground He state. The results of Tang and Shimamura [10] give a different shape to that obtained from the other calculations, but also generally lie in between the two sets of measurements. Given that both of their length and acceleration forms yielded much the same results, and that we are in agreement at the higher energies, we suspect that the difference may be due to the solution of the e-He^+ part of the problem.

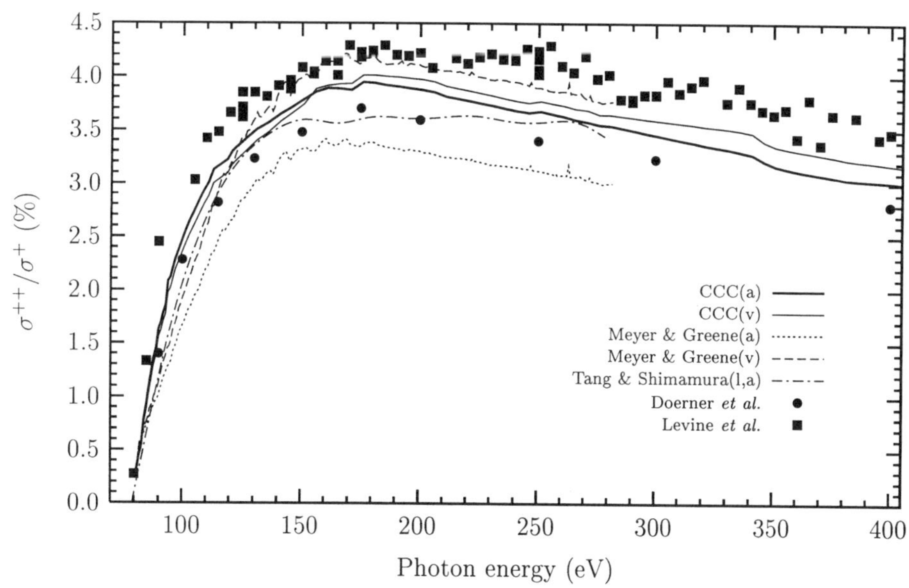

Figure 1: The ratio of double to single photoionization cross sections of the ground state of helium. The present calculations are denoted by CCC. The parenthesis indicate the gauge.

Single differential cross-section

In the CCC method the double photoionization cross section σ^{++} may be written as a sum of cross sections σ_n^N corresponding to excitation of the positive-energy states ϕ_n^N

$$\sigma^{++} = \sum_{n:0<\epsilon_n^N<E} \sigma_n^N. \tag{3}$$

We may put the σ_n, available at discrete positive energies ϵ_n^N, onto the continuum scale by defining [15]

$$\sigma^N(\epsilon_N^n) = |\langle \mathbf{q}(\epsilon_n^N)|\phi_n^N\rangle|^2 \sigma_n^N, \tag{4}$$

and then interpolating in the energy range $0 < \epsilon < E$. Here $\mathbf{q}(\epsilon)$ is the true continuum wave, eigenstate of the He^{2+} Hamiltonian with energy ϵ. We now write (3) as

$$\sigma^{++} = \int_0^E d\epsilon\, \sigma^N(\epsilon) = \int_0^{E/2} d\epsilon\, \Big(\sigma^N(\epsilon) + \sigma^N(E-\epsilon)\Big) = \int_0^{E/2} d\epsilon\, \frac{d\sigma}{d\epsilon}(\epsilon). \tag{5}$$

Thus obtained SDCS has worked very well at intermediate and high energies in the case of electron impact ionization of He [15,16] , but shows unphysical oscillations at lower energies [17] . These oscillations relate to fundamental problems with using the CCC method for calculating ionization, and are still being investigated. We find that as N is increased the $\sigma^N(\epsilon)$ for $0 < \epsilon < E/2$ show increasing oscillation, while for $E/2 < \epsilon < E$ they are very small and continue to diminish.

Suppose that in the limit $N \to \infty$ $\sigma^N(\epsilon)$ should be 0 in the range $E/2 < \epsilon < E$. We can mimick this with finite N calculations by choosing the states ϕ_n^N in such a way so that the integration rule ended at $E/2$ rather than E. With this integration rule we are able to produce fairly smooth SDCS almost free of unphysical oscillations. As an example we show the SDCS at photon energy of 120 eV calculated using different integration rules with 41 pseudostates included in the CCC basis. For comparison we

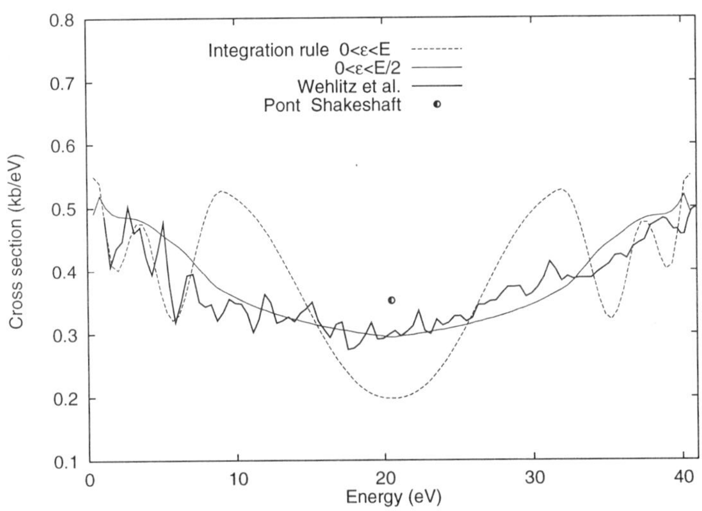

Figure 2: Single differential cross-section for the helium double photoionization at photon energy of 120 eV. Velocity form calculation. Experiment from Wehlitz *et al.* [18] is renormalized to the correct total double photoionization cross-section

also present the experimental SDCS from Wehlitz *et al.* [18] and calculation of Pont and Shakeshaft [11] at equal energy sharing. As it is seen from the figure the change of the integration rule has dramatic effect on the form of the SDCS making it much closer to the experiment.

REFERENCES

1. A. S. Kheifets and I. Bray, Calculation of double photoionization of helium using the convergent close-coupling method, *Phys. Rev. A*, 54:R995, 1996.
2. I. Bray, I. E. McCarthy, J. Wigley, and A. T. Stelbovics, Calculation of electron scattering on the He^+ ion, *J. Phys. B*, 26:L831, 1993.
3. I. Bray, Convergent Close-Coupling method for the calculation of electron scattering on hydrogen-like targets, *Phys. Rev. A*, 49:1066, 1994.
4. A. S. Kheifets, I. Bray, I. E. McCarthy, and Bo Shang, Theoretical triple differential cross section of the helium atom ionization with excitation to the $n = 2$ ion state, *Phys. Rev. A*, 50:4700, 1994.
5. T. Åberg, Asymptotic double-photoexcitation cross sections of the helium atom, *Phys. Rev. A*, 2:1726, 1970.
6. L. Spielberger, O. Jagutzki, R. Dorner, J. Ullrich, U. Meyer, V. Mergel, M. Unverzagt, M. Damrau, T. Vogt, I. Ali, K. Khayyat, D. Bahr, H. G. Schmidt, R. Frahm, and H. Schmidtbocking, Separation of photoabsorption and Compton scattering contributions to He single and double ionization, *Phys. Rev. Lett.*, 74:4615, 1995.
7. R. Dörner, T. Vogt, V. Mergel, H. Khemliche, S. Kravis, and C. L. Cocke, Ratio of cross sections for double to single ionization of He by 85–400 eV photons, *Phys. Rev. Lett.*, 76:2654, 1996.
8. J. C. Levin, G. B. Armen, and I. A. Sellin, Photoionization and compton double ionization of helium from threshold to 20 kev, *Phys. Rev. Lett.*, 76:1220, 1996.
9. C. Pan and H. P. Kelly, Calculation of the double-photoionization cross section of helium, *J. Phys. B*, 28:5001, 1995.
10. J. Z. Tang and I. Shimamura, Double photoionization of helium at low photon energies, *Phys. Rev. A*, 52:R3413, 1995.
11. M. Pont and R. Shakeshaft, Absolute cross sections for double photoionization of helium at energies from 0 to 80 ev above threshold, *J. Phys. B*, 28:L571, 1995.
12. K. W. Meyer and C. H. Greene, Double photoionization of helium using r-matrix methods, *Phys. Rev. A*, 50:R3573, 1994.
13. M. A. Kornberg and J. E. Miraglia, Double photoionization of helium - use of a correlated 2 electron continuum wave function, *Phys. Rev. A*, 48:3714, 1993.
14. K. W. Meyer, C. H. Greene, and I. Bray, Simplified model of electron scattering using R-matrix methods, *Phys. Rev. A*, 52:1334, 1995.
15. I. Bray and D. V. Fursa, Calculation of singly-differential cross sections of electron-impact ionization of helium at 100 eV, *J. Phys. B*, 28:L435, 1995.
16. I. Bray and D. V. Fursa, Calculation of ionization within the close-coupling formalism, *Phys. Rev. A*, 54:2991, 1996.
17. J. Röder, H. Ehrhardt, I. Bray, and D. V. Fursa, Absolute double differential cross sections for electron impact ionization of helium, submitted to *J. Phys. B*, 1997.
18. R. Wehlitz, F. Heiser, O. Hemmers, B. Langer, A. Menzel, and U. Becker, Electron-energy and -angular distributions in the double photoionization of helium, *Phys. Rev. Lett.*, 67:3764, 1991.

FROM (γ,2e) TO (γ,eR): KINEMATICALLY COMPLETE EXPERIMENTS WITH COLTRIMS

T. Vogt,[1] R. Dörner,[1] O. Jagutzki,[1] C.L. Cocke,[2] J. Feagin,[3] M. Jung,[4] E.P. Kanter,[4] H. Khemliche,[5] S. Kravis,[2] V. Mergel,[1] L. Spielberger,[1] J.Ullrich,[6] M. Unverzagt,[1] H. Bräuning,[1] U. Meyer,[1] and H. Schmidt-Böcking[1]

[1] Institut für Kernphysik, Universität Frankfurt, August-Euler-Str. 6, D-60486 Frankfurt, FRG
[2] J.R. McDonald Laboratory, Kansas State University, Manhattan, KS 66506, USA
[3] Dept. of Physics, California State Univ.-Fullerton, Fullerton, CA 92634,USA
[4] Argonne National Laboratory, Argonne, Il 60439, USA
[5] Lawrence Berkeley National Laboratory, Berkeley, CA 94720, USA
[6] Gesellschaft für Schwerionenforschung, D-64291 Darmstadt, FRG

INTRODUCTION

The technique of Cold Target Recoil Ion Momentum Spectroscopy (COLTRIMS) has been developed to study the reaction dynamics in atomic collisions[1,2]. Recently, this method has been applied to collisions of electrons and photons with helium[3,4,5,6]. Combining this technique with position sensitive large solid angle electron time-of-flight detectors it became possible to perform experiments of the type (γ,2e) [7] and (e,2e) [8], with the recoil-ion momentum measurement replacing the spectroscopy of one of the electrons.

In this paper we report on such kinematically complete experiments of the helium double photo-ionization at energies between 80 eV and 160 eV that were performed using COLTRIMS. These data can be compared to and expressed in triple differential cross sections $d^3\sigma/d\Omega_1 d\Omega_2 dE$ (TDCS) as one obtains by the common (γ,2e)-technique with a pair of electrostatic electron spectrometers.

The first results on kinematically complete double photo-ionization measurements on helium target were reported by Schwarzkopf *et al.*[9] and have attracted considerable interest on theoretical[10,11,12,13] and experimental[14,15,16,17] side. Although the experimental set-ups have been improved since then, the (γ,2e) technique with standard electron spectrometers generally suffers from very small detection efficiency. This results in data with less statistical significance and only for selected kinematical conditions in the Ω_1-Ω_2-E hypersphere. However, the combined theoretical and experimental efforts already resulted in considerable progress in our understanding of electron-electron correlation.

The COLTRIMS-technique has the advantage that at threshold energy the combined geometrical detection solid angle can reach 4π for *all* emitted particles. Even for higher energy the electron detection solid angle is still about 1π while for the recoil ions always the whole momentum sphere (4π) is measured. Furthermore, there are almost no apparative restrictions (none at all in the Wannier regime) on energy and angular configurations to be investigated.

We present triple differential cross sections for double photo-ionization of helium acquired with this technique for photon energies of 1eV above threshold (Wannier regime) and for 120 eV and compare those with theory.

EXPERIMENTAL SET-UP

The use of cold target recoil ion momentum spectroscopy (COLTRIMS) in ion-He, electron-He and photon-He collisions is now well established[3,5,18]. The key feature of this method is that the He target is cooled in a supersonic expansion (gas jet) to an internal temperature well below 1 Kelvin, i.e. to an internal thermal momentum spread of below 0.05 a.u. which is much smaller than the momentum transfers of interests in typical atomic collisions (≈ 1 a.u.).

In the version of the apparatus that we have used for the present studies the cooled gas jet was crossed by the photon beams of HASYLAB at DESY in Hamburg (E_γ = 120 eV and

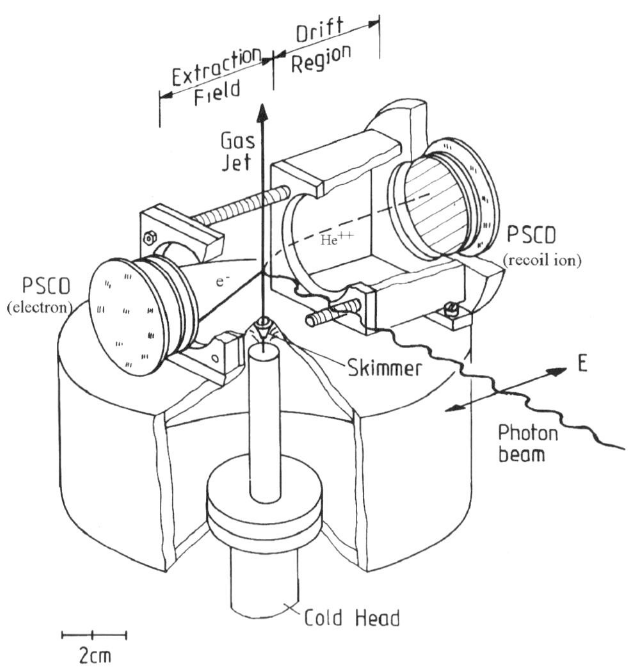

Figure 1. Sketch of the combined recoil-ion/electron momentum spectrometer

150 eV) and from the Advanced Light Source at the LBNL in Berkeley (E_γ = 80 eV, 85 eV, 100 eV, 160 eV), respectively. The interaction zone is located in a well defined weak (about 1 V/cm) electrostatic field region thus that all recoil ions produced in the collisions are extracted onto a two-dimensional position sensitive multi-channelplate detector (see fig. 1). The storage rings were operated in 2 or 5 bunch mode that allowed to measure the time-of-flight of the recoil ions with the precision of less than a nanosecond. The momentum vector of the helium ions was determined from the measured time-of-flight and the position of the ion impinging on the detector. Note, that the complete momentum sphere of the recoil ions is measured simultaneously (4π solid angle) with a resolution of 0.1 a.u. for each momentum component and that single and double ionization events are distinct in time-of-flight.

Figure 2 shows density plots of recoil-ion momentum distributions for He single (a) and double (b) photo-ionization at E_γ = 150 eV in the plane perpendicular to the photon beam direction (x-direction). Only events for a small x-momentum window of +/- 0.1 a.u. are projected onto the y-z-plane. For single ionization the distinct pattern of a dipole emission is visible. Since the recoil ion has to compensate the photo-electron momentum (the initial photon momentum can be neglected) the recoil ions have the same emission pattern in momentum space as the photo-electrons. The second structure in the centre of fig. 2a arises from simultaneous excitation of the residual bound electron, reducing the released kinetic energy in the process. For double ionization (fig. 2b) this pattern changes since the second electron also carries momentum in this three-body process. However, a dipole-like emission pattern of the recoil ions is maintained.

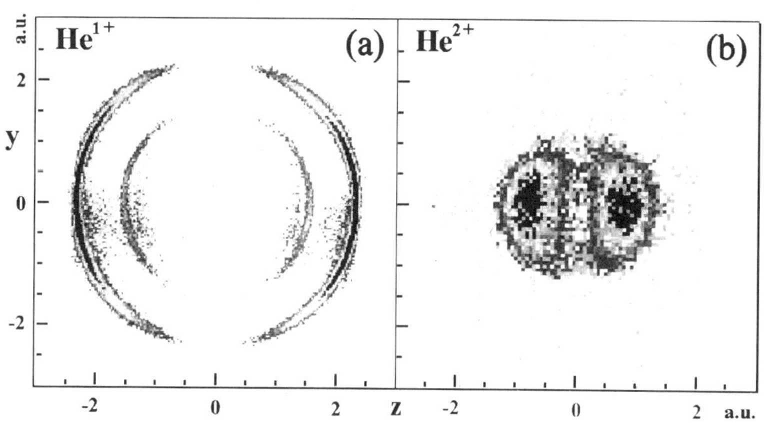

Figure 2. Density plots of the recoil-ion momentum distributions for helium single (a) and double (b) photo-ionization for 150 eV photon energy in the plane perpendicular to the photon direction x. Only events are plotted with $|p_x| \leq 0.1$ a.u.

A second position sensitive multi-channelplate detector facing the target region but located opposite to the ion detector detected one of the photo-electrons in coincidence with the ion. Events in which both electrons struck the detector could be rejected analyzing the pulse height distribution of the electron detector. The photo-electron momentum vector could be determined in the same way as for the recoiling ion. Up to 1 eV excess energy one of the photo-electrons could always be collected by the electrical field with 4π solid angle. For higher energies (as E_γ = 120 eV) only the geometrical detection solid angle (about 1π) was achieved so far since high energetic electrons not emitted into the direction of the active detector area escape from the weak electric field region in the spectrometer. However, the detection efficiency is still orders of magnitude bigger than for standard electron spectrometers and the orientation of the spectrometer was such that the polarization vector of the mainly linearly polarized light pointed in the direction of the electron detector. Due to the positive β-value of the electron emission that dominates even for double ionization events at these higher energies[19] most of the interesting emission pattern can still be recorded.

The investigation of single ionization give an excellent method to check the combined recoil ion/electron spectrometer system since electron and ion must show opposite momenta for each recorded single ionization event.

TRIPLE DIFFERENTIAL CROSS SECTIONS (TDCS) FOR DOUBLE PHOTO-IONIZATION OF HELIUM

By coincident detection of all three momentum components of both recoil ion *and* one photo-electron the momentum balance of the double photo-ionization is completely determined, even overdetermined by one coordinate. We directly measure cross sections* $d^2\sigma/d\mathbf{k}_{rec}d\mathbf{k}_e$ differential in the recoil ion momentum vector $\mathbf{k}_{rec}$ and one electron momentum vector $\mathbf{k}_e$, with only 5 of the 6 variables being independent. In order to compare the data with the usual TDCS notation in spherical coordinates as function of the electron emission angles and their energy sharing the measured data where transformed using simple algebra and momentum conservation

$$\mathbf{k}_{e1} + \mathbf{k}_{e2} + \mathbf{k}_{rec} = \mathbf{k}_\gamma \ (\cong \mathbf{0}) \qquad (1)$$

The energy conservation law

$$E_\gamma - E_{e1} - E_{e2} = E_{rec} \ (\cong 0) \qquad (2)$$

was not used in the transformation but gave another possibility to check the data due to the redundancy of the measuring technique.

Being restricted in momentum space only by the electron solid detection angle it is now possible to slice the momentum sphere in angular „cake pieces“ and energy-share bins to derive cross section images as was done in the work of Schwarzkopf *et al.*[9]. I.e., the five-dimensional cross sections are projected onto two-dimensional planes, e.g., for angle/energy combinations suggested by the cross section distribution´s symmetries.

The general shape of the TDCS in the next figures can be explained by simple selection rules[10]. The TDCS must be zero for $\mathbf{k}_{e1} = \mathbf{k}_{e2}$ due to the electron´s repulsive Coulomb force. Final states with $\mathbf{k}_{e1} = -\mathbf{k}_{e2}$ are forbidden because of parity violation. More exclusion principles exist for equal energy sharing and special angular configurations that result from parity and/or angular momentum conservation laws. This can be seen in fig. 3 where we present data obtained with our technique and transformed into TDCS for 41 eV excess energy and equal energy sharing. The angle α between the polarization vector ε and one electron was fixed at three different values (α_{e1}) and the TDCS as function of the second electron's angle

* Note that this cross section notation really represents a five-fold differential cross section like the „TDCS“.

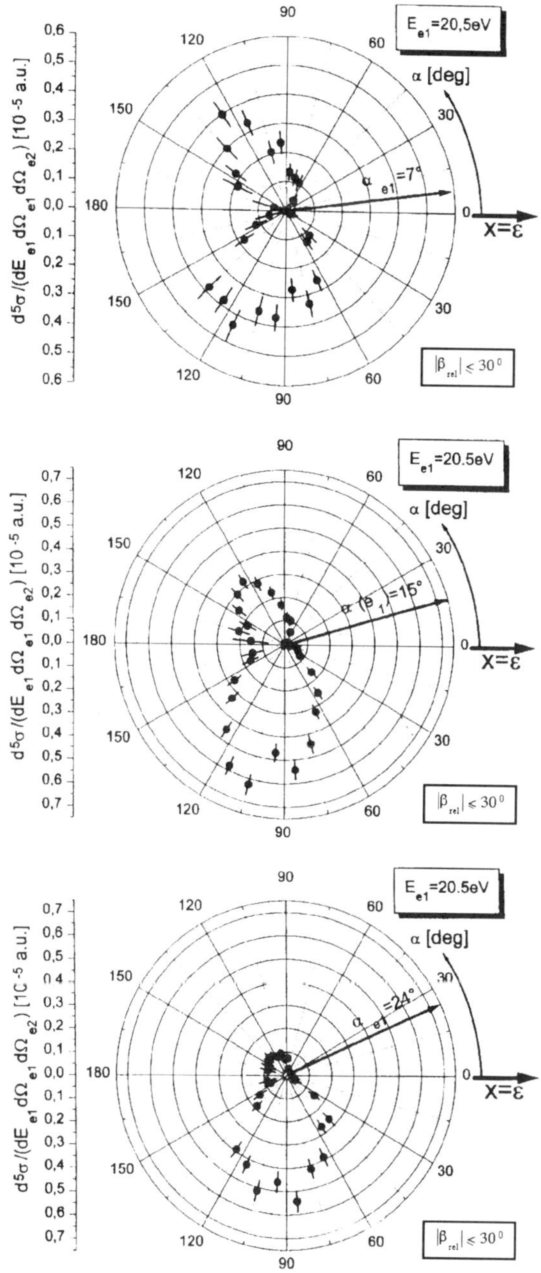

Figure 3. Experimental TDCS for planar geometry and equal energy sharing, for details see text.

α_{e2} was plotted for planar geometry, i.e., only electrons with relative azimuthal angles $\Delta\beta$ between +/- 15^0 are selected. This corresponds to a geometry that was chosen so far in the earlier (γ,2e)-measurements by setting the spectrometer's rotation axes and energy passes.

It is to note, that for obtaining these selected TDCS „views" from our data we had to introduce the respective kinematical conditions by software, i.e., we performed „projections" from our data set. Thus, all projections shown here are derived from only one single (and almost complete) five-dimensional data set, numerous other projection are possible.

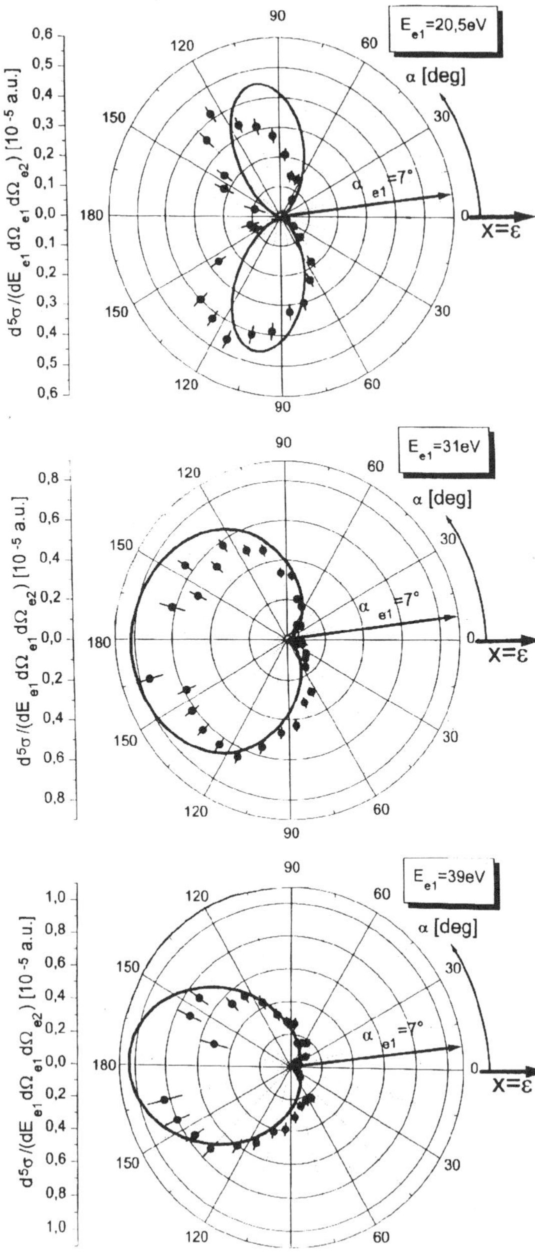

Figure 4. TDCS for unequal energy sharing. The solid lines indicate a 3C-calculation of Maulbetsch and Briggs[10] for slightly different energies and angles. Details see text.

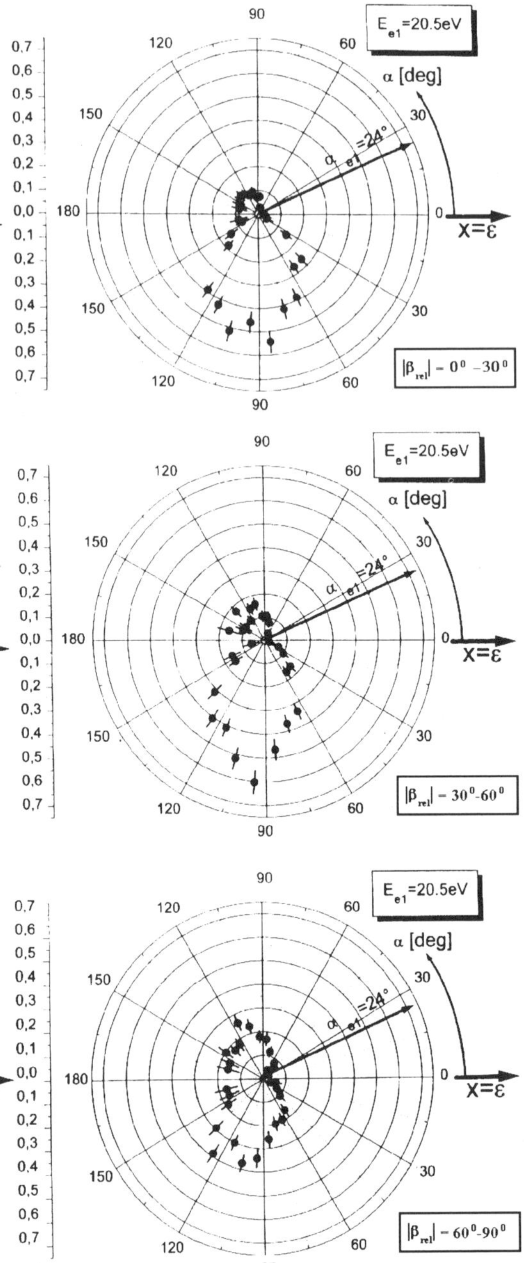

Figure 5. TDCS for equal energy sharing and „out-of-plane"-geometry. Details see text.

The shrinking of the upper loop in fig. 3 for increasing α_{e1} from 0 towards 30° is in qualitative agreement with findings of Maulbetsch and Briggs[10], however, there are so far no calculations for E_γ =120 eV and the experimental resolution and a non-perfect linear polarization of the beam do not yet allow stringent tests of theory.

Another „projection" obtained from our data shows in fig. 4 TDCS for fixed angle α_{e1} but varying energy sharing between both electrons. Again, theory[10] seems to describe the main feature of the TDCS correctly. This comparision is only for illustration since the energy and angle parameters in the calculation are not exactly the same as in the experiment.

As stated earlier, almost any 2D-projection, i.e., choice of some fixed E -, β -, α - combinations, for a TDCS presentation can be done, even out-of-plane geometries. In fig. 5 the „observation angle" β_{e2} for the second electron is tilted out of the plane normal to the beam direction which is in contrast to the earlier in-plane figures. Again, calculations for such geometries exist already[10], but for a detailed check of the theoretical exclusion rules and analysis for the *physics* behind all those TDCS loops, it still requires smaller angular and energy resolution bins with sufficient statistics than were achievable in our first experiment. Also „cleaner" polarization features of the photon beam are desirable.

CROSS SECTION REPRESENTATION IN JACOBY COORDINATES

For our experiments on the double photo-ionization at 1 eV excess energy we were able to measure the momenta of both the recoil ion and one electron with 4π solid angle. Covering the complete momentum space enables us to formulate the cross sections in a different coordinate system that intertwines the measured momenta of recoil ion and electron. We propose the use of Jacoby Coordinates as Feagin[21] and other suggested earlier for the Wannier process:

$$\mathbf{k}_r = (\mathbf{k}_1+\mathbf{k}_2) \qquad \text{and} \qquad \mathbf{k}_R = \tfrac{1}{2}\,(\mathbf{k}_1-\mathbf{k}_2) \tag{3}$$

$\mathbf{k}_r$ describes the motion of the centre-of-mass of the two electrons, $\mathbf{k}_R$ the relative motion of the electrons. The introduction of this coordinate system has several advantages. It is not only used in the theoretical description of the Wannier process[20] but it is also the „natural" coordinate system for our experimental method as the common TDCS-description suits the use of standard electron spectrometers. The value $\mathbf{k}_r$ is directly accessible in our measurement since $\mathbf{k}_r = -\,\mathbf{k}_{rec}$ (with $\mathbf{k}_\gamma \cong \mathbf{0}$). In this frame the kinematics of the Wannier process can directly be visualized[7].

Fig. 6 shows density plots of momentum distributions for the electron`s centre-of-mass (or recoil ion) motion (a), for a single electron (b) and for the relative electron motion (c) in the plane perpendicular to the photon beam direction (x), for details see figure caption. From this plot it is visible in a very intuitive way that the recoil ion emerges into the direction of the polarization vector of the linearly polarized light while the two electrons separate along a

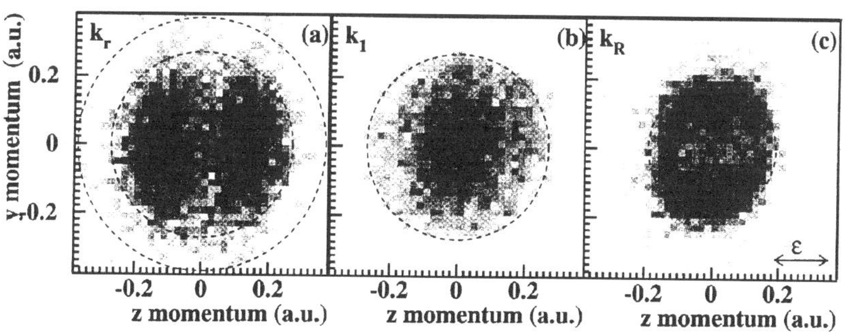

Figure 6. Density plots of projections of the momentum spectra for the centre-of-mass motion (a), the single electron motion (b) and the relative electron motion (c) in the plane perpendicular to the beam direction. Only events with $|k_{Rx}| < 0.1$ a.u. are plotted. The outer circles indicate the maximum possible values.[7]

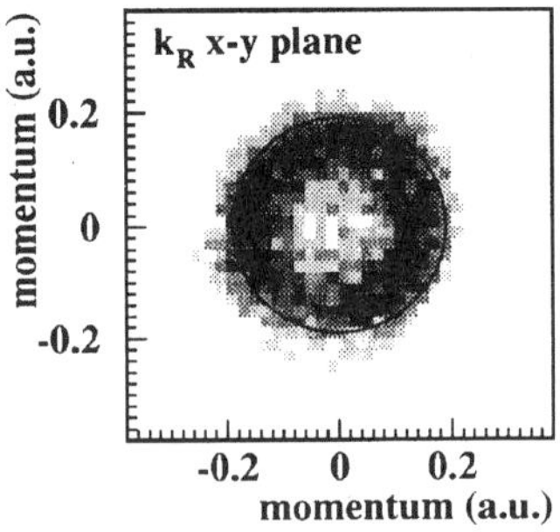

Figure 7. Density plots for the relative electron motion as in fig. 6c, but in the plane perpendicular to the polarization vector direction. Only events for $|k_{Rz}| < 0.1$ a.u. are plotted.

line perpendicular to that direction with their centre-of-mass moving opposite to the recoil ion. Fig. 7 shows another „view" on the break-up process. Here the plane of the figure is perpendicular to the polarization vector direction. The relative-motion distribution forms a doughnut around the polarization axis as the electrons separate perpendicular to the polarization axis but the distribution is symmetrical around this axis.

For energies higher than 1 eV above threshold a complete detection of all electron momenta was not ensured with the present experimental set-up, so no coordinate transformation in *both* $\mathbf{k}_r$ and $\mathbf{k}_R$ could be performed. However, since we always achieve 4π momentum detection for the recoil ion momentum (and such for $\mathbf{k}_r$!), we can still compare all our measured cross section distributions (for all photon energies investigated) as function of $\mathbf{k}_r$.

In fig. 8a the relative portion of the energy deposited into the electron´s centre-of-mass motion and into the relative motion of both electrons is shown. For 1 eV excess energy most kinetic energy is found in the relative electron motion which is in perfect agreement with the Wannier model. The line shows a calculation from Feagin in fourth-order Wannier theory[21] that fits the experimental results nicely[22].

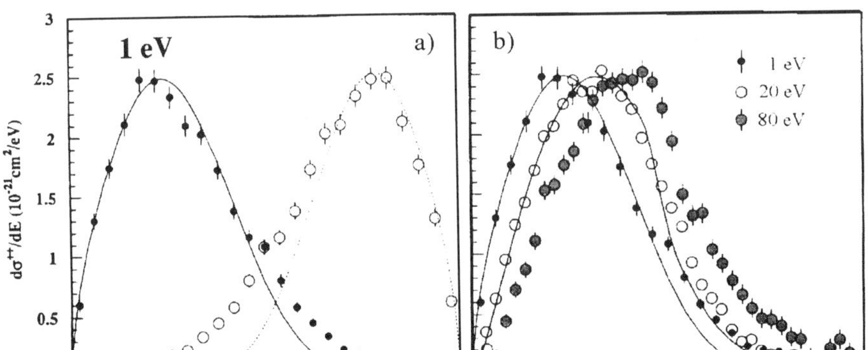

Figure 8a. Fractions of kinetic energy deposited in the electron´s centre-of-mass motion (solid dots, measured at 1 eV excess energy) and relative motion (open dots). The solid lines indicates the therory[21]
Figure 8b. Fractions of kinetic energy in the centre-of-mass motion for different photon energies. The second line indicates calculations of Pont and Shakeshaft[23].

At 20 eV excess energy the relative fraction of energy in the centre-of-mass motion increases (see fig. 8b) which is also in perfect agreement even on absolute scale with ab initio 2C calculations of Pont and Shakeshaft[23]. This energy portion in the centre-of-mass motion seems to increase further towards equilibrium in the high energy regime where one electron carries almost all the kinetic energy (shake-off limit).

SUMMARY AND OUTLOOK

We have measured the complete momentum balance after helium double photo-ionization at energies 1 eV above threshold and higher with the COLTRIMS technique. The data sets are complete in the sense that the momenta for all outgoing particles can be deduced for each event and almost the full detection solid angles for the particles could be achieved, at least for the Wannier regime. We can derive TDCS from our data without restrictions to any angular emission or energy sharing of the electrons. We propose the use of a cross section description in Jacoby Coordinates as the „natural frame“ of this atomic breakup process at low excess energies. The experimentally derived cross sections seem to be in good agreement with theory.

Further experiments and fine-tuned theoretical analysis might reveal a better understanding of the underlying physical processes, such as electron-electron correlation, for the different TDCS. Investigations at higher energies are desirable to probe electron-electron correlation in the initial state, i.e., in the helium atom itself. We hope to improve our spectrometer system in the electron detection device to enable the full solid electron detection angle also for higher photon (electron) energies in the next future by the use of a solenoid field to collect the electrons[24]. Such a system was recently succesfully tested in an experiment at the ALS in Berkeley, the data are currently analyzed[25]. With these data we also hope to achieve reliable cross section $d\sigma/dE_e$ for double photo-ionization of helium close to the threshold. Furthermore, we design (e,3e)-like experiments applying the COLTRIMS technique[26].

ACKNOWLEDGMENTS

The authors would like to thank the following supporting organisations:
Alexander von Humboldt Stiftung, Willkomm Stiftung, DFG, BMBF, Graduiertenförderung des Landes Hessens, Studienstiftung des Deutschen Volkes, DOE, Division of Chemical Sciences, Office of Basic Energy Sciences, Office of Energy Research

REFERENCES

1. J. Ullrich *et al.*, Comm. At. Mol. Phys. 30: 285 (1994)
2. L. Spielberger *et al.*, Nato ASI Series C414: 119 (1993)
3. O. Jagutzki *et al.*, Z. Phys. D36: 5 (1996)
4. L. Spielberger *et al.*, Phys. Rev. Lett. 74: 4615 (1995)
5. R. Dörner *et al.*, Phys. Rev. Lett. 76: 2654 (1996)
6. L. Spielberger *et al.*, Phys. Rev. Lett. 76: 4685 (1996)
7. R. Dörner *et al.*, Phys. Rev. Lett. 77: 1024 (1996)
8. A. Kounis, diploma thesis, private communication (1996)
9. O. Schwarzkopf *et al.*, Phys. Rev. Lett. 70: 3008 (1993)

10. F. Maulbetsch and J.S. Briggs, Phys. Rev. Lett. 68: 2004 (1992), J. Phys. B26: 1679 (1993), J. Phys. B26: L647 (1993), J. Phys. B27: 4095 (1994) and J. Phys. B28:551 (1995)
11. M. Pont *et al.*, Phys. Rev. A53: 3671 (1996)
12. S.C. Ceraulo *et al.*, Phys. Rev. A49: 1730 (1994)
13. G. Dawber *et al.*, J. Phys. B28: L271 (1995)
14. A. Huetz *et al.*, J. Phys. B27: L13 (1994)
15. O. Schwarzkopf *et al.*, J. Phys. B27: L347 (1994)
16. P. Lablanquie *et al.*, Phys. Rev. Lett. 74: 2192 (1995)
17. O. Schwarzkopf and V. Schmidt, J. Phys. B28: 2847 (1995)
18. R. Moshammer *et al.*, Phys. Rev. Lett. 73: 3371 (1994) and Phys. Rev. Lett.77: 1242 (1996)
19. R. Wehlitz *et al.*, Phys. Rev. Lett. 67: 3764 (1991)
20. G.H. Wannier, Phys. Rev. 90: 817 (1953)
21. J.M. Feagin, J. Phys. B28: 1495 (1995)
22. J.M. Feagin, J. Phys. B29: L1 (1996)
23. M. Pont and R. Shakeshaft, Phys. Rev. A, to be published
24. R. Moshammer *et al.*, Nucl. Instrum. Methods B108: 425 (1996)
25. H. Bräuning, private communication (1996)
26. M. Achler, dissertation, private communication (1996)

ANGLE RESOLVED ELECTRON ENERGY LOSS STUDY ON THE INNER SHELL EXCITED STATES OF NE

C.D. Schröter*, L. Avaldi*, R. Camilloni* and G. Stefani+

* IMAI del CNR, Area della Ricerca di Roma
CP10, 00016 Monterotondo Scalo, Italy
+ Dipartimento di Fisica "E. Amaldi", III Università di Roma e
Unità INFM di Roma III, Rome, Italy

INTRODUCTION

Electron-energy loss (EEL) is a successful tool to investigate inner-shell excitations of atoms and molecules[1,2]. In this technique a beam of energetic electrons is used to excite inner-shell transitions. The energy lost by the inelastically scattered electrons produces the excitation of the inner-shell state and the yield of the detected electrons provides information about the relative cross section for the excitation of the inner-shell state at that value of the incident energy. Then the dependence of the cross section from the momentum transfer in the collision can be studied by changing the scattering angle of the inelastically scattered electron. This latter characteristic together with the possibility to excite dipole forbidden transitions makes EEL a complementary technique to photoabsorption experiments. An important motivation for the study of inelastic collisions of electrons stems for the need of reliable cross section data in such diverse field as radiation physics, plasma physics, atmospheric physics, astrophysics and electron microscopy.

In this work we have studied the Ne $1s^{-1}nl$ excited states by EEL spectroscopy at about 2.5 keV incident energy and several scattering angles between 2.4° and 24.4°. While this energy region has been largely investigated by photoabsorption measurements[3-8], only one EEL measurement at the same incident energy as in the present work, but at the scattering angle of 1.15° has been previously reported[9]. That measurement has shown that the energy region below the Ne K edge is dominated by the $1s \rightarrow np$ ($n \geq 3$) dipole allowed transitions. However, also the $1s \rightarrow 3s$ dipole forbidden transition can be observed. In the present work we have extended the previous investigation measuring the EEL spectrum at several angles in order to study the dependence of the generalized oscillator strength (GOS)[10] versus the momentum transfer **K**. In addition we have performed a further EEL measurement at 5° scattering angle, but in the region above the Ne K edge up to an energy loss of 915 eV. In this region transitions which involve two electrons, namely the 1s and an outer shell electron, are expected to occur. These inner shell doubly excited states have been observed previously in photoabsorption[6,11,12] and threshold photoelectron spectroscopy[13], but so far not in EEL spectroscopy.

EXPERIMENTAL

The apparatus used for the present measurements is an electron-impact spectrometer specially designed for electron-electron coincidence experiments. It consists of a vacuum chamber, which contains an electron-beam source, two electrostatic analyzers and an effusive gaseous beam. A detailed description of the apparatus is reported elsewhere[14,15]. A home-made unselected gun provides an electron beam of typically 5 μA at E_0=2.5 keV. The inelastically scattered electrons are analyzed in energy by passing through an electron spectrometer, rotatable in the scattering plane from -15 to 150°. The angular acceptance of the analyzer was set to ±0.5°. The zero of the scattering angle scale was set by determining the symmetry of the scattered electron yield around 0°. The full width half maximum (FWHM) energy resolution of the analyzer, operated at a constant pass energy (E_p=80 eV) was $\Delta E \cong 1.0$ eV. The measurement procedure consisted in scanning E_0 with a step size of 0.1 eV, while the scattered electron energy E_a was kept fixed. The scanned range of the energy loss $\Delta E = E_0 - E_a$ was 11 eV for the region of the 1s→nl transitions and 15 eV for the region of the two electron transitions.

The variation of the field of view of the analyzer with the scattering angle θ has been checked by detecting Ne KLL (^{1}D) Auger electrons from 2.4° to 29.4°. Within the experimental uncertainty the Auger yield resulted to be constant. Furthermore the contribution of the background gas to the measured energy-loss yield has been determined by measuring an EEL spectrum after homogeneously filling the scattering chamber up to the same pressure as that one of the measurements, by a side inlet. At the scattering angle of 4.4° an almost constant contribution of 3 Hz has been measured.

The typical count rates in the EEL measurements were 135 and 20 Hz at θ=4.4° and 24.4°, respectively. Due to these low count rates accumulation times in the order of 15 and 80 hours were needed in the region of the singly and doubly inner shell excited states, respectively. During the measurements care has been taken to monitor the stability of the apparatus, by measuring simultaneously to the EEL spectra also the KLL Auger-electron yield with the second analyzer present in the scattering chamber. Finally the correct scaling with θ of the EEL spectra has been checked by measuring several times in sequence the yield of the scattered electron at a fixed ΔE value for all the measured angles. The ΔE values chosen correspond to the 1s→3p and 1s2p→3p^2 transitions.

Ten different measurements have been performed at 2.4°≤θ≤24.4°. The momentum transfer $\mathbf{K}=\mathbf{k}_0-\mathbf{k}_a$ at the ΔE of the 1s→3p transition varies from 2.67 to 5.74 a.u.. The minimum K achievable in an electron scattering experiment at the present incident energy is 2.62 a.u.

RESULTS AND DISCUSSION

A typical EEL spectrum in the ΔE region 863-874 eV, measured at θ=9.4°, is shown in Figure 1. Features due to the excitation of the 1s electron to the final 3s, 3p and 4p states as well as the continuum due to the 1s ionization can be seen clearly in the spectrum. A functional form composed by three gaussian functions, with the same FWHM, a gaussian function with a variable FWHM to simulate the higher np (n≥4) excited states, the arctan function[8]

$$f(E_0, IP) \propto \left[\tan^{-1}\left(\frac{E_0 - IP}{\Gamma}\right) + \frac{\pi}{2}\right]$$

where IP is the 1s ionization potential and Γ the natural linewidth of the 1s state, to represent the behaviour of the ionization threshold and finally a linear function to simulate the contribution to the EEL spectrum from the outer-shell ionization, have been fitted to the

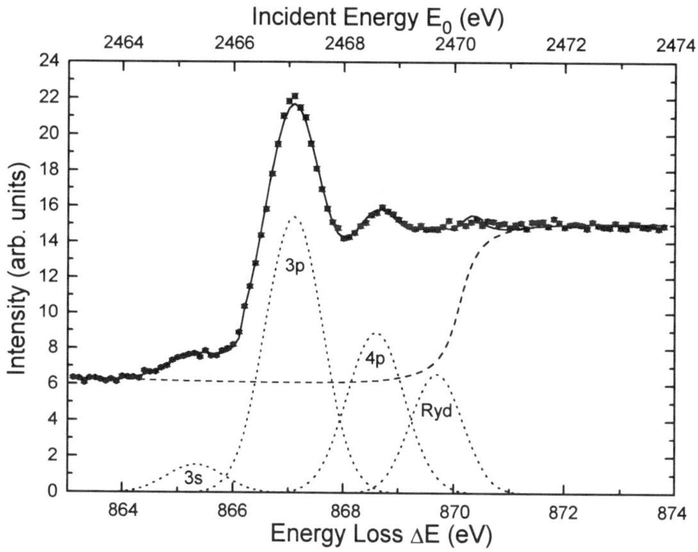

Figure 1. Typical EEL spectrum at a scattering angle of θ=9.4° and an accumulation time of 10 h. The full curve represents the fitted functional form. The dashed line is the contribution of the outer-shell ionization plus arctan function representing the behaviour of the ionization threshold. The dotted lines represent the gaussian functions for the 3s, 3p, 4p and higher lying unresolved Rydberg np (n ≥4) states.

experimental data. The energy positions of the 3s, 3p and 4p inner-shell excited states are in agreement with the previous experimental observations[3-9]. From the fit the relative intensities, Y_{nl}, of the three transitions are obtained and then the relative GOS, f_{nl}, is calculated[10]

$$f_{nl} \propto \frac{1}{2}\frac{k_o}{k_a}\Delta E_{nl} K^2 Y_{nl}$$

In Figure 2a the GOS's for the three transitions are shown. As expected for dipole allowed transitions f_{3p} and f_{4p} decrease as K increases, while f_{3s}, relative to a dipole forbidden transition, displays an opposite behaviour. This different behaviour is better displayed in Figure 2b where the f_{3s}/f_{3p} ratio is reported versus K, together with the previous determination by Hitchcock and Brion[9].

No calculations of the measured GOS's has been reported in literature, therefore only a qualitative discussion of the experimental results can be done at present. The transition 1s→3s leads to a final state with two possible J_f configurations, namely J_f=0 and 1. The J_f=0 configuration is reached via a monopole transition, while the J_f=1 one is due to a magnetic dipole transition. According to the Bethe theory[9] at high enough energy the GOS of electric dipole transitions scales as K^{-2}, while the GOS of monopole transitions scales with K^0. In Figure 2a f_{3p} and f_{4p} display a deviation from a pure K^{-2} dependence. This may be due to the fact that 2.5 keV is not an energy high enough to comply with the predictions of the first order theory for transitions with excitation energies of about 900 eV. Within the statistics of the present data it is also not possible to state whether f_{3s} is constant in the K range investigated. This might be due either again to the too low incident energy or to the contribution of the magnetic dipole transition.

In Figure 3a the EEL spectrum at θ=4.4° in the ΔE region 850-920 eV is shown, while the results of a scan with a smaller step size and longer accumulation time in the region 900-914 eV, is shown in Figure 3b. In the latter figure the contribution of the 1s ionization has been

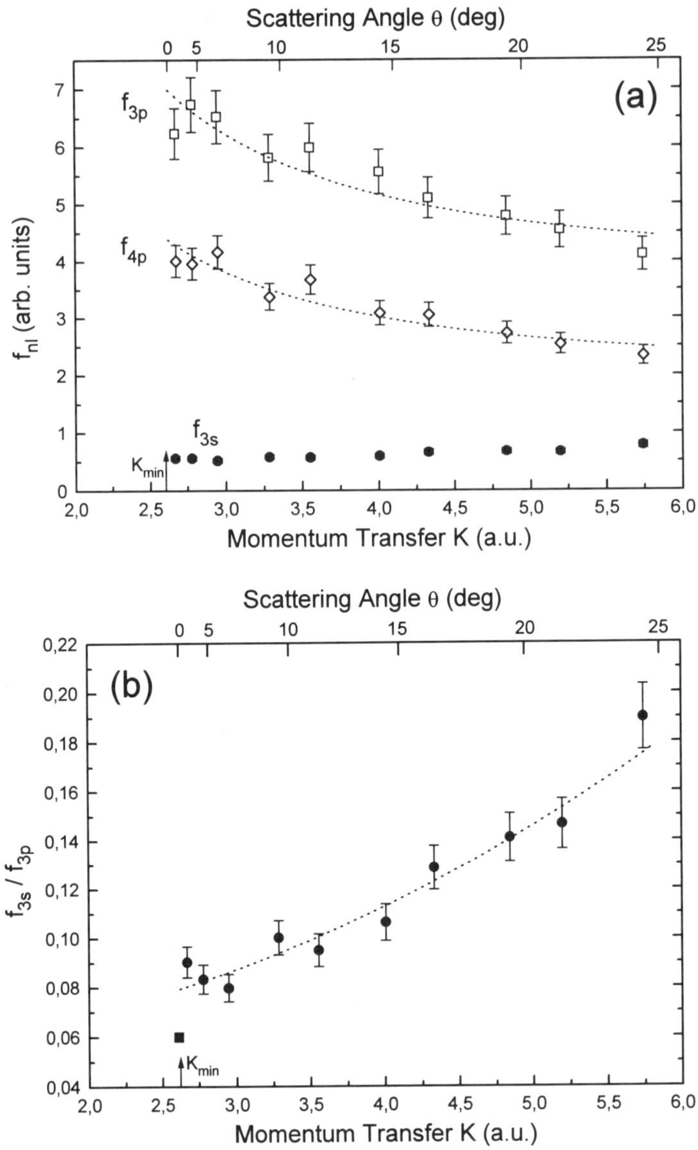

Figure 2. (a) GOS for the three transitions 1s→3s, 3p and 4p. The dotted lines are curves scaling as K^{-2}. (b) f_{3s}/f_{3p} versus momentum transfer K. The square point represents the measurement by Ref. 9. The dotted line is a curve with a K^2 dependence. The arrow on the bottom of both figures indicates the minimum K value achievable at this incident energy E_0.

subtracted. The intensity maximum of the features observed in Figure 3b is about 2.8% of the intensity of the 1s→3p transition. These features are interpreted as two electron excitations generated in the process

$$e\left(E_o\right) + Ne\left(1s^2 2s^2 2p^6(^1S)\right) \Rightarrow Ne^{**}\left(1s2s^2 2p^5(^{3,1}P)nln'l'\right) + e\left(E_a\right)$$

A zero order 2p-2h configuration interaction (CI) calculation[13] has predicted nine transitions, in the studied energy region, with an intensity that is at least 1% of the 1s→3p transition (Table 1). However, the $1s2p^5(^{3,1}P)3s3d$ states involve a monopole excitation of the inner electron and a dipole transition of the outer one and the $1s2p^5(^3P)4p^2$ state is the

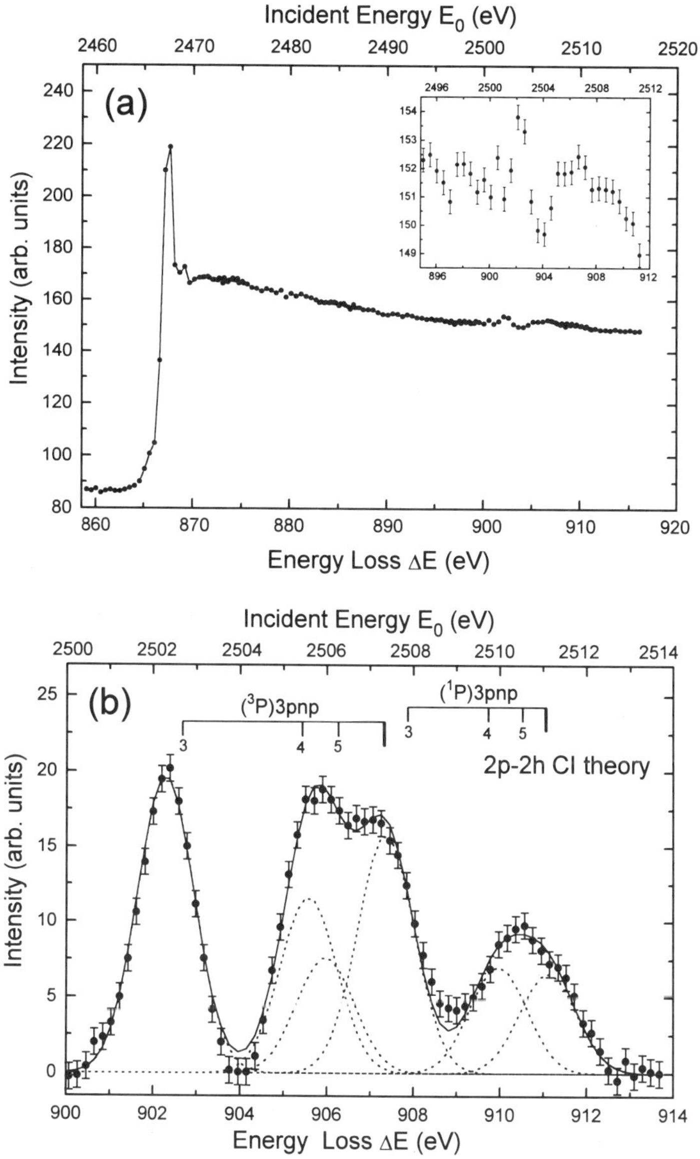

Figure 3. (a) EEL spectrum at a scattering angle of θ=4.4°, a point step size of 0.5 eV and an accumulation time of 72 h. The small inserted picture at the top shows the finer details in the region of the doubly excited states. (b) EEL spectrum at a scattering angle of θ=4.4°, a point step size of 0.2 eV and an accumulation time of 84 h. The contribution of the 1s ionization has been subtracted. The dotted lines represent 6 gausssian functions for the doubly excited states while the full line indicates the sum of this functions. The lines on top denote the energy positions of the (^{3}P)3pnp and (^{1}P)3pnp states given by a zero order 2p-2h CI calculation[13], converging to the (^{3}P)3p and (^{1}P)3p ionization potentials[16].

first one of a series converging to the higher up $1s2p^5(^3P)4p$ Ne^+ satellite state. Therefore they may be expected to give a minor contribution to the measured spectrum. Thus we have fitted only six gaussian functions to the experimental data. The results of the fit, shown in Figure 3b and reported in Table 1, confirm that the observed features can be assigned to the $1s2p^5(^{3,1}P)3pnp$ ($n\geq3$) series converging to the $1s2p^5(^{3,1}P)3p$ Ne^+ satellite states. While the assignment of the first two features in both series is not questionable, the third one of the first group might be due to the $1s2p^5(^{3,1}P)3p5p$ state with the rest of the series converging to the ionic state at 907.39 eV underneath the $1s2p^5(^1P)3p^2$ doubly excited state, or a combination of higher $(^3P)3pnp$ (n>4) states. On the other hand the third peak of the second series placed by the fit at 911.11 eV, practically at the Ne^+ $1s2p^5(^1P)3p$ ionization potential[16], seems to result from the overlap of the higher members (n>4) of the $1s2p^5(^1P)3pnp$ series.

Table 1. Measured energies for the observed features in the region of the Ne 1s inner-shell doubly excited states ([a]Ref. 13, [b]This work, [c]Ref. 12)

Theory[a]	Energy[a] (eV)	Energy[b] (eV)	Energy[c] (eV)
$1s2p^5(^3P)3p^2$	902.68	902.0 ± 0.3	902.42 ± 0.05
$1s2p^5(^3P)3p4p$	905.43	905.3	905.26
$1s2p^5(^3P)3s3d$	905.67		905.80
$1s2p^5(^3P)3p5p$	906.26	905.7	906.16
$1s2p^5(^3P)3p6p$			906.72
$1s2p^5(^1P)3s3d$	906.62		
$1s2p^5(^1P)3p^2$	907.87	907.1	
$1s2p^5(^1P)3p4p$	909.74	909.7	
$1s2p^5(^1P)4p^2$	909.91		
$1s2p^5(^1P)3p5p$	910.52	910.8	

The inner shell doubly excited states have been previously observed in photoabsorption mesurements[6,11,12] and threshold photoelectron spectroscopy[13]. Albeit the different energy resolution, the present EEL spectrum and the photoabsorption one[12] are consistent. The data analysis of the EEL spectrum and the assignment of the features at 907.36 eV to the $1s2p^5(^1P)3p^2$ state explain the sharp edge of the photoabsorption spectrum at 907.4 eV in contrast with the expected continuity of the GOS through an ionization threshold. On the other hand the higher resolution photoabsorption data enable the assignment of the peak at 905.7 eV in the EEL spectrum to the $1s2p^5(^3P)3p5p$ state and suggest that part of the yield of the $1s2p^5(^1P)3p^2$ state is due to the higher members (n>4) of the $1s2p^5(^1P)3pnp$ series.

CONCLUSIONS

The GOS for the 1s→3p, 4p dipole allowed and 1s→3s dipole forbidden transitions have been obtained by angular resolved EEL spectra at an incident energy of about 2.5 keV. We have shown that the EEL technique can be used successfully to study transitions with energy loss of about 900 eV and even experiments with a moderate resolution enable a detailed study of processes with low cross section as the dipole forbidden transitions.

The first observation of inner shell doubly excited states by EEL has been reported. The spectrum has been interpreted by comparison with photoabsorption data and a zero order 2p-2h CI calculation. Doubly excited states of neutral atoms offer ideal systems to investigate the departure from the single electron model and for the study of electron-electron correlations. Now that definite observations of inner shell doubly excited states of Ne have been reported, calculations at higher level of sophistication than the zero order 2p-2h one are highly desirable. Further experimental work to complete the characterization of the Ne $1s2p^5(^{3,1}P)3pnp$ states is in progress[17].

ACKNOWLEDGMENTS

This work was partially supported by the European Community Contract No. CHRX-CT93-0350.

REFERENCES

1. G.C. King and F.H. Read in: *Atomic Inner-Shell Physics,* B. Crasemann, ed., Plenum Publishing Corporation, New York, p.317 (1985)
2. A.P. Hitchcock and D. Mancini, *J. Electron Spectrosc. Rel. Phenom.* 67:1 (1994)
3. J. Backosky, Czech. *J. Phys.* 4:150 (1954)
4. R.J. Liefield, *Appl. Phys. Lett.* 7:276 (1965)
5. F. Wuilleumier, *C. R. Acad. Sci. Paris B* 270:825 (1970)
6. J.M. Esteva, B. Gauthe', P. Dhez and R.C. Karnatak, *J. Phys. B: At. Mol. Opt. Phys.* 16:L263 (1983)
7. M. Domke, T. Mandel, A. Puschmann, C. Xue, D.A. Shirley, G. Kaindl, H. Petersen and P. Kuske, *Rev. Sci. Instrum.* 63:80(1992)
8. C.M. Teodorescu, R.C. Karnatak, J.M. Esteva, A. ElAlif and J.P. Connerade, *J. Phys. B: At. Mol. Opt. Phys.* 26:4019 (1993)
9. A.P. Hitchcock and C.E. Brion, *J. Phys. B: At. Mol. Opt. Phys.* 13:3269 (1980)
10. M. Inokuti, *Rev. Mod. Phys.* 43:29 (1971)
11. F. Wuilleumier, *J. Physique* 32:88 (1971)
12. L. Avaldi, R. Camilloni, G. Stefani, C. Comicioli, M. Zacchigna, K.C. Prince, M. Zitnik, C. Quaresima, C. Ottaviani, C. Crotti and P. Perfetti , *J. Phys. B: At. Mol. Opt. Phys.* 29:L737 (1996)
13. L. Avaldi, G. Dawber, R. Camilloni, G.C. King, M. Roper, M.R.F. Siggel, G. Stefani, M. Zitnik, A. Lisini and P. Decleva , *Phys. Rev. A* 51:33 (1987)
14. E. Fainelli, R. Camilloni, G. Petrocelli and G. Stefani , *Nuovo Cimento D* 9:33 (1987)
15. L. Avaldi, R. Camilloni, R. Multari, G. Stefani, X. Zhang, H.R.J. Walters and C.T. Whelan, *Phys. Rev. A* 48:1195 (1993)
16. N. Martensson, S. Svensson and U. Gelius, *J. Phys. B: At. Mol. Opt. Phys.* 20:6243 (1987)
17. C.D. Schröter et al. to be submitted to *J. Phys. B: At. Mol. Opt. Phys.* (1996)

CIRCULAR DICHROISM IN DOUBLE-PHOTOIONIZATION OF HELIUM STUDIED BY ELECTRON TIME-OF-FLIGHT COINCIDENCE SPECTROSCOPY

J. Viefhaus[1], L. Avaldi[2], G. Snell[1,3], M.Wiedenhöft[1], R. Hentges[1], A. Rüdel[1], F. Schäfers[4], D. Menke[3], U. Heinzmann[3], A. Engelns[5], J. Berakdar[6], H. Klar[5] and U. Becker[1]

[1]Fritz-Haber-Institüt der Max-Planck-Gesellschaft, Faradayweg 4, D-14195 Berlin, Germany

[2]Istituto di Metodologie Avanzate Inorganiche del CNR, Area della Ricerca di Roma, CP10, 00016 Monterotondo Scalo, Italy

[3]Fakultät für Physik, Universität Bielefeld, Universitätsstraße 25, D-33615 Bielefeld, Germany

[4]BESSY GmbH , Lentzeallee 100, D-14195 Berlin, Germany

[5]Fakultät für Physik, Albert-Ludwig Universität Freiburg, Hermann-Herder-Straße 3, D-79104 Freiburg, Germany

[6]Atomic and Molecular Physics Laboratories, Research School of Physical Sciences and Engineering, Istitute of Advanced Studies, Australian National University, Canberra ACT 0200, Australia

INTRODUCTION

Double photoionization occurs when an energetic photon is absorbed by an atom and two electrons are promoted into the continuum

$$h\nu + A \Rightarrow A^{+} + e_1(E_1,\bar{k}_1) + e_2(E_2,\bar{k}_2) \quad (1)$$

The process is the more suited one to investigate electron-electron correlations. Indeed two-electron transitions are impossible within single-particle models, provided the one particle initial and final states are solutions of the same Hamiltonian[1]. Initial and final one-particle states, if eigenfunctions of different Hamiltonians, may allow for a finite cross-section, but such a description is generally poor. Reliable results can be expected only if correlated wavefunctions are used.

Experimental information on double photoionization can be provided by several techniques[2]. However the more direct way to study double photoionization is an experiment where the two photoelectrons are detected in coincidence after energy and angle selection. This experiment results in the measurement of the triply differential cross-section $d^3\sigma/d\Omega_1 d\Omega_2 dE_1$ (TDCS), where Ω_1 and Ω_2 are the angles of emission of the two electrons and E_1 is the energy of one of them. The energy of the other electron is defined by energy conservation $h\nu - I^{2+} = E_1 + E_2$, where $h\nu$ is the photon energy and I^{2+} is the double ionization potential. After the first experimental investigations in 1993[3] of the archetypal system for double photoionization, the helium atom, several coincidence experiments have been reported[4-9].

The year before the first experimental study on helium were reported, Berakdar and Klar[10] predicted that circular dichroism should be observed in double photoionization of helium with circularly polarized radiation. Chirality in a system with a fully symmetric state, the helium ground state is $^1S_0$, might surprise. According the definition by Lord Kelvin back in 1983[11] chirality implies the impossibility to overlap an object with its mirror image (fig.1a). In nature there are several examples of chiral molecules . However chirality can be also induced in photoemission experiments of linear molecules either adsorbed on surfaces[12] or fixed in space[13] (fig. 1b). In this latter case the quantal axis of the molecule, the direction of the incident radiation and the momentum of the photoelectron form a right handed reference frame which is turned into a left handed one in its mirror image. The same occurs in double photoionization, when we consider the direction of the incident radiation and the momenta of the two photoelectrons (fig. 1c).

An experiment, which aims to study dichroism effects in double-photoionization of helium, needs a source of circularly polarized radiation in the VUV or soft X-ray energy region and an efficient coincidence spectrometer to overcome the low value of the cross section. In this work we have achieved these goals using a trasmission multilayer, which acts as a quarter-wave plate for the linearly polarized incident light, together with time-of-flight spectrometers, which enable simultaneous detection of all the electron pairs.

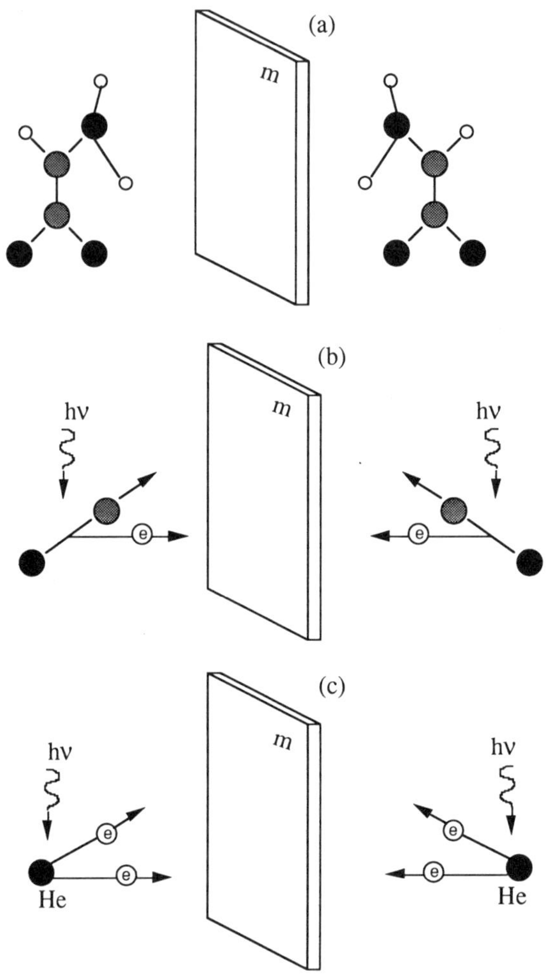

Figure 1 : Example of a chiral structure (a) and chirality induced in photoemission of a linear molecule(b) and in double-photoionization of helium (c)

THEORETICAL PREDICTIONS

The TDCS for the pair of photoelectrons, when they are detected in coincidence after energy and angular analysis, is given by [1,10,14]

$$\frac{d^3\sigma}{d\Omega_1 d\Omega_2 dE_1} = C\sum_{M_f}\frac{1}{2J_i+1}\sum_{M_i}\left|\left\langle \Psi^-_{k_1k_2}\left|\hat{\varepsilon}\cdot\overline{D}\right|\Phi_i\right\rangle\right|^2 \quad (2)$$

where $C=4\pi^2\alpha a_o^2(h\nu)k_1k_2$, α and a_o are the fine structure constant and the Bohr radius, respectively. $\overline{D}=\sum_n \overline{r}_n$ is the dipole operator in length form and $\hat{\varepsilon}$ is the polarization vector. Equation (2) averages over the initial magnetic sublevels M_i and sums over the magnetic sublevels M_f of the photoion.

Klar and Berakdar[10] and Berakdar et al.[14] showed that the dipole matrix element

$$M^{\pm}(\overline{k}_1,\overline{k}_2)=\left\langle \Psi^-{}_{k_1k_2}\left|\hat{\varepsilon}\cdot\overline{D}\right|\phi_i\right\rangle \quad (3)$$

where +(-) stems for right (left) circularly polarised radiation, is symmetric with respect to the exchange of the momenta of the photoelectrons

$$M^{\pm}(\overline{k}_1,\overline{k}_2)=M^{\pm}(\overline{k}_2,\overline{k}_1) \quad (4)$$

However this symmetric matrix element produces a TDCS with a portion antisymmetric with respect to :

- the angular exchange $\hat{k}_1 \Leftrightarrow \hat{k}_2$
- the energy exchange $E_1 \Leftrightarrow E_2$

of the two photoelectrons. This portion is the source of the dichroism. Necessary conditions to observe a non-vanishing dichroism are that (i) the two electrons unevenly share the excess energy and (ii) the direction of the incident radiation and the directions of the two ejected electrons are not lying in one plane.

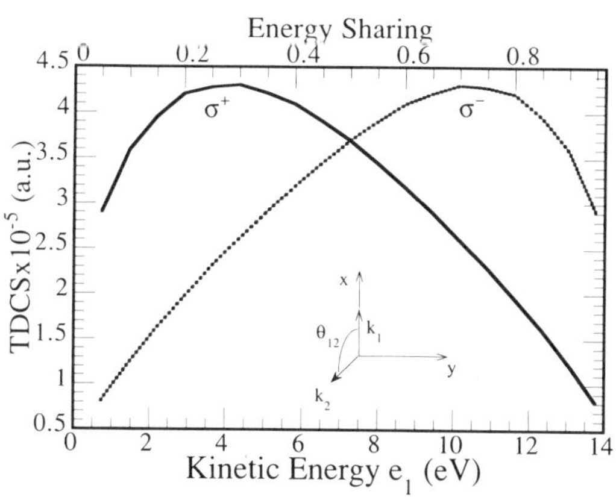

Figure 2: The TDCS of He at hν=93.5 eV as a function of the energy E_1 of one of the two electrons. The two electrons are ejected at $\vartheta_{12}=150°$ in a plane perpendicular to the direction (y) of the incident radiation.

In the present case we have calculated the TDCS at hν=93.5 eV using a Hylleraas-type wave function[15] as an approximation for the helium ground state and an approximated three-body scattering wave function[16] for the final state. In this approximation of the final state the electron-electron and electron-residual ion interactions are treated on equal footing. These approximations have been chosen because the extensive set of calculations performed by Maulbetsch and Briggs[17-20] has shown that such a model represents satisfactorily the shape of the TDCS of He measured in several kinematics. The results of the calculations for the energy sharing condition at ϑ_{12} and an angular pattern corresponding to $E_1/E_2=1/9$ are shown in figures 2 and 3, respectively.

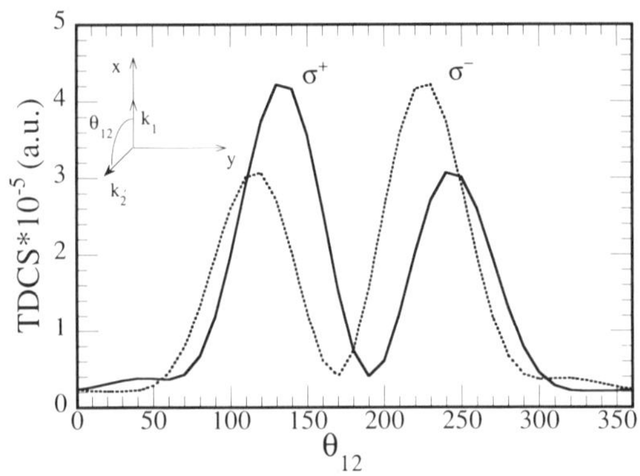

Figure 3 : Angular dependence of the TDCS of He at hv=93.5 eV. The energy sharing $E_1/E_2=1/9$, one electron is detected in the x direction, while the other one is detected at an angle ϑ_{12} in the x-z plane.

The results clearly show that the TDCS of helium with circularly polarised display dichroism effects. However the sign and size of the circular dichroism may change depending on the approximated wavefunctions used as shown for other experimental conditions by Berakdar et al.[14]. Thus experiments have to prove firstly the existence of the dichroism effects and then to disentangle among the predictions of models using different approximated wavefunctions.

EXPERIMENTAL

The experiment was performed at the Berliner Elektronenspeicherring-Gesellschaft für Syncrotronstrahlung (BESSY) under single bunch operation mode of the torage ring. The linearly polarized radiation from the undulator U1, monochromatized by the toroidal grating monochromator TGM5[21], was converted into circularly polarized radiation by a Mo/Si trasmission multilayer. The multilayer, composed by 50 bilayers each one 9 nm thick[22], has a working range of ±5 eV around 93 eV. For an incoming monochromatic photon flux of approximately 10^{13} photons per second per 100 mA ring current (gap setting 38.2 mm, no pinhole and resolving power $E/\Delta E \approx 300$) a flux higher than 10^{11} photons/s was achieved in the interaction region.

The characterization of circularly polarised radiation has been achieved in the following way. The polarization state of the radiation was monitored by a Rabinovitch-type[23] linear polarization analyzer consisting of a gold mirror (incident angle 42°) and a GaAs photodiode. The analyzing power of the polarimeter was higher than 0.97. First the degree of linear polarization was measured without the multilayer ($P_{lin}=0.95\pm0.02$). Afterwards the optimum incident angle of the multilayer for the given photon energy was determined by minimizing the intensity modulation measured by rotating the polarimeter (<2%). The polarimeter alone cannot distinguish between unpolarized and circularly polarized radiation. However, by successively changing the polar angles of the multilayer and rotating the polarimeter around the light axis, the changing ellipse of light polarization, being circles and lines in the extreme cases, could be observed. At given polar angles the incoming linear polarization was completely coverted into circular polarization ($P_{circ}=P_{lin}$)[22]. During the experiment the helicity of the radiation was changed every 1000 s by switching between two appropriate polr angles of the multilayer. The absence of linear polarized radiation after the multilayer was verified after each change.

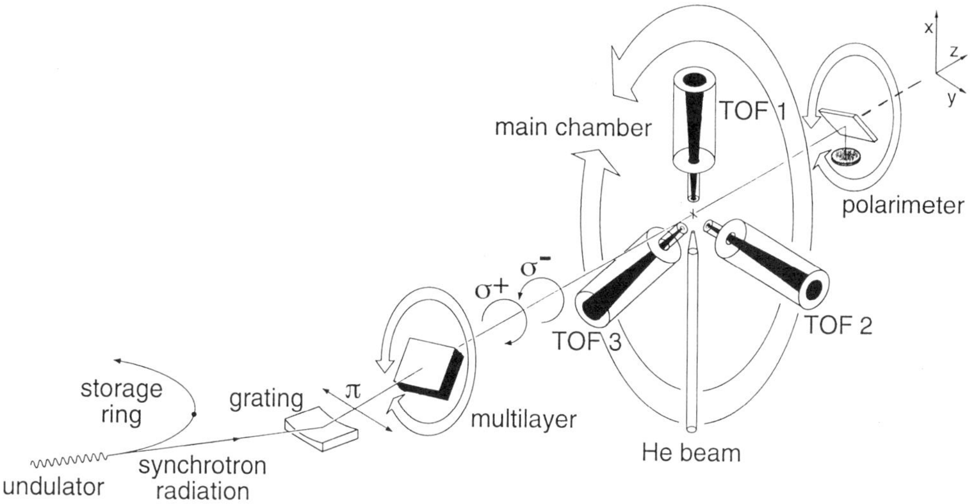

Figure 4 : Schematic drawing of the experimental setup. Three TOF electron analyzers are mounted in a plane perpendicular to the incoming photon beam. The linear polarized radiation is transformed into circular polarized radiation by a trasmission multilayer positioned at the entrance of the vacuum chamber. The helicity of the radiation can be switched using appropriate polar angles for the multilayers.

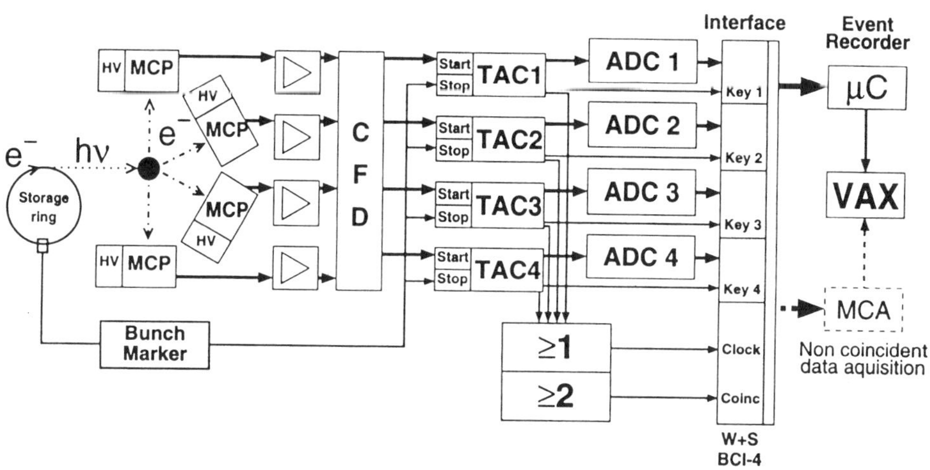

Figure 5: Electronic setup for TOF coincidence spectroscopy using a specially designed interface in addition to standard NIM modules.

The circularly polarized radiation crosses an effusive beam of helium atoms at the center of scattering chamber containing three time-of-fligh (TOF) spectrometers in a plane perpendicular to the incoming radiation (fig. 4). Two longer TOF spectrometers, with flight paths of 460 mm and acceptance angles of ±3° are mounted on the rotatable vacuum chamber, while the third short analyzer (140 mm length,±5° acceptance angle) is mounted inside the chamber and is independently rotatable around the light axis. An acceleration voltage applied to the drift tube of the TOF analyzers enables us to detect electrons with kinetic energies as low as 0.5 eV. The energy resolution of the analyzers was in the range of 2%-5% of the kinetic energy of the photoelectrons. The low energy Auger spectrum of xenon[24] has been used to determine the relative transmission of the analyzers. The three analyzers allow simultaneous coincidence measurements at three different relative emission angles ϑ_{12} of the two photoelectrons.

Each event detected from one of the TOF spectrometers is recorded by standard NIM modules and analyzed in a custom-made interface (fig.5)[25]. The non-coincidence signal is sent to a multichannel analyzer. The coincidence signal is sent to a custom-made event recorder, which pre-analyzes the coincidence events by a local processor and then stores them in a 4096x4096 channel map. This electronic set-up ensures that coincidence events as well as non-coincidence events are analyzed by the same electronic modules[25]. One of the time-scale maps is shown in figure 6(a). True coincidences lies on a curved line, that is generated by the fact that the sum of both kinetic energies of the electrons has to be constant. The same map after the random coincidence subtraction and the time to energy conversion is shown in figure 6(b). In this latter figure the locus of true coincidences is represented by a straight line. The sum of the coincidence counts along the diagonal lines, which correspond to $E=E_1+E_2$=constant, results in the photoelectron spectrum of He^{2+}, figure 6(c). The true-to-random ratio, being much higher than one, allows an unambiguous identification of the single accessible He^{2+} final state. Finally the TDCS was obtained by adding the coincidence counts over a certain energy interval (≈3eV, marked as rectangles in figure 6(d)) to achieve a sufficient statistical accuracy.

The experiment has been performed at 93.5 eV, a value which results as a compromise between the multilayer working range, photon flux of the undulator, and helium double photoionization cross section[26]. This resulted in a total true coincidence count rate of about 10 mHz.

The TDCS has been put on a relative scale using the simultaneously collected non coincident spectra of the three analyzers because all analyzers lay in a plane perpendicular to the direction of the propagation of the circularly polarized radiation.

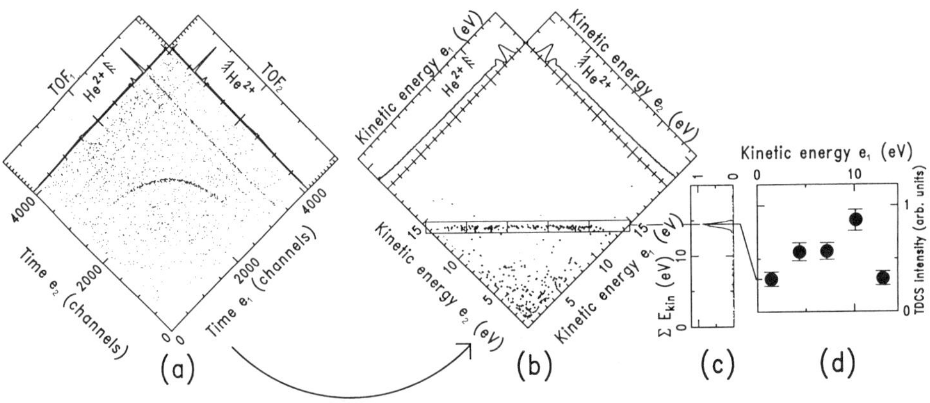

Figure 6: Overview of the data analysis of the helium double-photoionization electron spectra.(a) The two dimensional raw coincidence spectrum with the simultaneously recorded non-coincident electron spectra. (b) The same spectra after time-to-energy conversion and subtraction of random coincidences. (c)The double ionization photoelectron spectrum, generated by plotting all coincidence events versus E_1+E_2.(d) The TDCS intensity versus energy sharing

RESULTS AND CONCLUSIONS

The measured TDCS with positive (σ^+) and negative (σ^-) helicity of the circularly polarized radiation for three different relative emission angles ϑ_{12} are shown in figures 7(a). The experimental data have been then used to calculate the normalized circular dichroism Δ_n (fig. 7(b)) given by

$$\Delta_n = \frac{1}{P_{circ}} \frac{TDCS\,(\sigma^+) - TDCS\,(\sigma^-)}{TDCS\,(\sigma^+) + TDCS\,(\sigma^-)} \qquad (5).$$

In figure 7(a) we observe that the circular dichroism vanishes for equal and almost equal energy sharing. This is due to the fact that the observer is not able to distinguish between the two electrons under this conditions. Moreover the TDCS curves, measured with the two

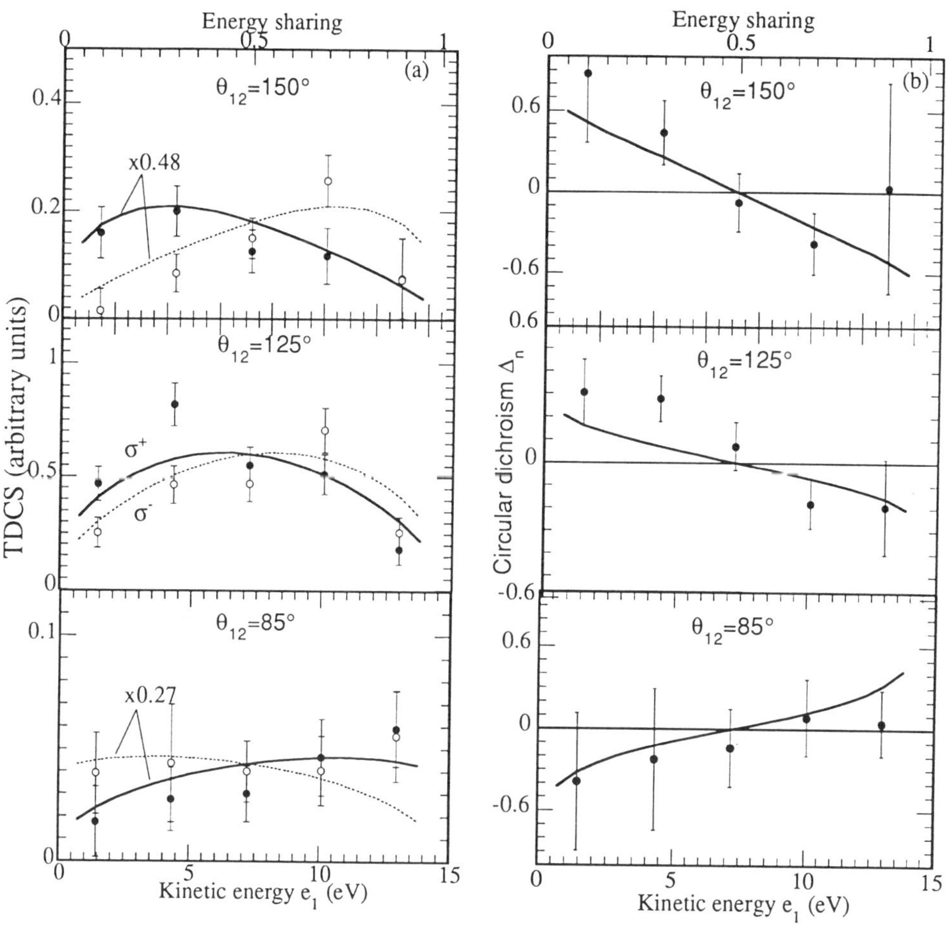

Figure 7 : (a) The He TDCS versus kinetic energy sharing at hν=93.5 eV for three different emission angles of the two electrons. The results with the positive(open circles) and negative (full circles) helicities of the circularly polarized radiation are compared with the theoretical results arbitrarily scaled (as quoted) at each angle. (b) Normalized circular dichroism for the three corresponding emission angles. The full curves represent the theoretical results.

helicities of the circularly polarized radiation, are mirror images with respect to reflection at the equal sharing line. This is a consequence of the fact that changing the helicity of the radiation as well as changing the "labels" ("fast", "slow") of the two electrons leads to the same geometry of the experiment. This latter finding is also a prove of the correct operation of the experimental set-up, i.e. the measured shape of the TDCS is not due to a transmission effect of the TOF's. Furthermore the mirror symmetry tells that the integration over all the possible energy sharings leads to a vanishing difference between cross sections measured with negative and positive helicity of the circularly polarized radiation. This clearly indicates that only coincidence measurements with a suitable energy and angular resolution enable the observation of circular dichroism.

The measurements at ϑ_{12}=125° have the strongest intensity and clearly show the circular dichroism. Despite the large error bars the data at ϑ_{12}=150° display a larger circular dichroism, whereas the results at ϑ_{12}=85° show, if any, an effect with the opposite sign. As far as the comparison with the theory is concerned, the numerical calculations show a qualitative agreement with the experiment. However to match the magnitude of the measured TDCS we had to rescale the theory for each angular settings. The scaling factors are indicated in figure 7(a). Nevertheless the theory is consistent with the experiments in assigning the largest intensity to the ϑ_{12}=125° geometry and the smallest intensity to the ϑ_{12}=85° one. On the other hand when the measured circular dichroism is compared with the theory a very good agreement is observed. The sign change as well as the general shape of the theory are consistent with the experiment, although at the ϑ_{12}=125° geometry a slightly larger dichroism is found in the experiment. The present comparison between theory and experiment with respect to the relative size of the TDCS suffers from the statistical uncertainties of the experiment, which hamper a definite judgment of the discrepancies observed. However the differences in the TDCS seem to indicate that further theoretical work has to be performed.

In summary we have shown that double-photoionization of helium with circularly polarized radiation displays dichroism effects. These effects strongly depend on the relative angle and energy sharing of the two emitted electrons. Satisfactory agreement is found with the results of the calculation. Even with the large efficiency of the TOF coincidence spectrometer the low value of the TDCS hampers the systematic study needed to distinguish among the predictions based on different wave functions. The advent of new undulator beam lines producing circularly XUV radiation will enable the accurate characterization of this effect, which adds a new facet to the three-body Coulomb system. Circularly polarized radiation and the observation of dichroism effects in double-photoionization will enable the investigation of differences between simultaneous and sequential double photoionization such as Auger decay following photoelectron emission. A deeper understanding of such a process would constitute a fundamental step towards a complete experiment in atomic double-photoionization[27,28].

This work was supported in part by the Deutsche Forschungsgemeinschaft and the Bundesminister für Bildung, Wissenschaft, Forschung und Technologie. The support of the BESSY staff is gratefully acknowledged.

REFERENCES

1. F.W. Byron and C.J. Joachain, *Phys. Rev.* 164:1 (1967).
2. J.H.D.Eland in *Vacuum Ultraviolet Photoionization and Photodissociation of Molecules and Clusters* , C.Y. Ng, Ed.,World Scientific, Singapore, (1991).
3. O. Schwarzkopf, B. Krässig, J. Elminger and V. Schmidt, *Phys. Rev. Lett.* 70: 3008 (1993).
4. A. Huetz, P.Lablanquie, L. Andric, P. Selles and J. Mazeau, *J. Phys. B :At. Mol. Opt. Phys.* 27:L13 (1994).
5. O. Schwarzkopf, B. Krässig, V. Schmidt , F. Maulbetsch and J. S. Briggs, *J. Phys. B :At. Mol. Opt. Phys.* 27:L347 (1994).
6. G. Dawber , L. Avaldi , A.G. McConkey, H. Rojas, M.A. MacDonald and G.C. King, *J. Phys. B :At. Mol. Opt. Phys.* 28:L271 (1995).
7. P. Lablanquie, J. Mazeau, L. Andric, P. Selles and A. Huetz, *Phys. Rev. Lett.* 74:2192 (1995).
8. J. Viefhaus, L. Avaldi , F. Heiser , R. Hentges , O. Gessner, A. Rüdel , M. Wiedenhöft , K. Wieliczek and U. Becker, *J. Phys. B :At. Mol. Opt. Phys.* 29: L729 (1995).
9. O. Schwarzkopf and V. Schmidt , *J. Phys. B :At. Mol. Opt. Phys.* 28:2847 (1995)
10. J. Berakdar and H. Klar, *Phys. Rev. Lett.* 70:3008 (1993).
11. R.T. Morrison and R.N. Boyd *Organic Chemistry* , Allyn and Bacon Inc., Boston (USA), 1968
12. C. Westphal, J. Bansmann , M. Getzlaff, and G. Schönhense, *Phys. Rev. Lett.* 63:151 (1989)

13. F. Heiser, O. Geßner, U. Hergenhahn , J. Viefhaus, K. Wieliczek, N. Saito and U. Becker, *J. Elect. Spectr. Rel. Phenom.* 79:415 (1996).
14. J. Berakdar , H. Klar , A. Huetz and P. Selles *J. Phys. B :At. Mol. Opt. Phys.* 26:1463 (1993).
15. E. A. Hylleraas *Z. Phys.* 54:347 (1929).
16. M. Brauner, J.S. Briggs and H. Klar *J. Phys. B :At. Mol. Opt. Phys.* 22:2265 (1989).
17. F. Maulbetsch and J.S. Briggs, *J. Phys. B :At. Mol. Opt. Phys.* 26:1679 (1993).
18. F. Maulbetsch and J.S. Briggs, *J. Phys. B :At. Mol. Opt. Phys.* 26: L647 (1993).
19. F. Maulbetsch and J.S. Briggs, *J. Phys. B :At. Mol. Opt. Phys.* 27: 4095 (1994).
20. F. Maulbetsch and J.S. Briggs, *J. Phys. B :At. Mol. Opt. Phys.* 28: 551 (1995).
21. W. Peatman, C. Carbone, W. Gudat, W. Heinen, P. Kuske, J. Pflüger, F. Schäfers and T. Schroeter, *Rev. Sci. Instrum.* 60:1445 (1989).
22. K. Rabinovitch, L. R. Canfield and R. P. Madden, *Appl. Opt.* 4: 1005 (1965)
23. D. Menke, G. Snell, M. Drescher, H.-J. Stock, U. Heinzmann and F. Schäfers, *BESSY Jahresbericht 1994*, p.472 (1994).
24. U. Becker, D. Szostak, H.G. Kerkhoff, M. Kupsch, B. Langer, R. Wehlitz, A. Yagishita and T. Hayaishi, *Phys. Rev.* A39: 3902 (1995)
25. R. Wehlitz, J. Viefhaus, K. Wieliczek, B. Langer, S.B. Whitfield and U. Becker, *Nucl. Instrum. Meth.* B99: 257 (1995)
26. J. M. Bizau and F.J. Wuilleumier, *J. Electron Spectrosc. Relat. Phenom.* 71: 205 (1995)
27. B. Kämmerling and V. Schmidt *Phys. Rev. Lett.* 67:1848 (1991)
28. K. Soejima, M. Shimbo, A. Danjo, K. Okuno, E. Shigemasa and A. Yagishita , *J. Phys. B :At. Mol. Opt. Phys.* 29: L367 (1996)

INTERFERENCE EFFECTS IN ARGON AUGER COINCIDENCE SPECTRA

D.K. Waterhouse and J.F. Williams

Centre for Atomic, Molecular and Surface Physics
Physics Dept., The University of Western Australia
Nedlands, Perth, Australia, 6907.

INTRODUCTION

The occurrence and significance of interference effects[1] observed in electron-impact (e,2e) coincidence measurements of Auger transitions[2,3], has aroused considerable interest. We report the first detailed observations of the strong angular behaviour of the interference phenomena, and the quantified behaviour of the interference and post-collision-interaction (PCI) effects on the argon $L_{2,3}M_{2,3}M_{2,3}$ Auger lines. Excess energies up to 18 eV above the $L_{2,3}$ inner-shell ionization threshold were used. The interference is attributed to interaction between the outer-shell satellite-state ionization processes, identified in coincidence[4], and the Auger processes.

Interference effects in the argon $L_3M_{2,3}M_{2,3}$ $^1D_2$ line were first observed by Sandner and Völkel[2]. Their experiment was designed so that the satellite state cross-section was small and the excess energy (the energy of the 'slow' electron ejected in the $L_{2,3}$ ionization) resolution was ± 4.5 eV. The observed interference effects were small and were difficult to quantify, but did exhibit some angular dependence. A later measurement by Lohmann[3] of the same transition for an excess energy of 2.0 ± 0.9 eV also suggested interference effects. Due to the small coincidence signal rates and the difficulty of these experiments, the interference effects were not quantified, and the selected scattering dynamics meant the angular dependence was not clearly observable.

The argon atom is particularly suited to (e,2e) coincidence observations of the interference effects because of the relative energies of the Auger $L_{2,3}M_{2,3}M_{2,3}$ transitions and the inner- and outer-shell ionization processes. The (e,2e) scattering dynamics can be selected so that the scattered and Auger electron energies in the Auger process are equal to the scattered and ejected electron energies in the satellite-state ionization process. An (e,2e) coincidence experiment cannot distinguish between different processes if the detected electrons have identical energies, and interference effects will be observed.

A brief overview of the models used in the data analysis of our observations is given. A short description of the (e,2e) spectrometer follows before the observed PCI and interference effects are discussed.

LINE SHAPES

The first Auger coincidence experiments were designed to observe the effects of PCI on Auger lineshapes, and several measurements were reported[5,6] for low excess energy. The effects of PCI have been modelled for electron impact[7] and observed in many non-coincidence experiments[8]. PCI effects are observed as a shift in the energy distribution of the Auger electron and asymmetric broadening of the lineshape. The magnitude of both effects increases as the excess energy decreases. The triple differential cross-section for an Auger transition that includes PCI effects[7] is given by

$$\frac{d^6\sigma}{dE_s dE_A d\Omega_s d\Omega_A} = \frac{\Gamma/2\pi}{\varepsilon_a{}^2 + \Gamma^2/4}\,\sigma(E_s,\Omega_s)K(\varepsilon_a,\bar{\xi})$$

Here $\varepsilon_a = (E - E_A)$ is the difference between the observed electron energy E and the nominal Auger transition energy E_A, and Γ is the Auger transition linewidth. The $L_{2,3}$ ionization cross-section is given by $\sigma(E_s, \Omega_s)$. The function $K(\varepsilon_a, \bar{\xi})$ accounts for the particle interactions, and $\bar{\xi}$ quantifies the Coulomb interaction between different final state electrons and is velocity-dependent. $K(\varepsilon, \bar{\xi})$ varies rapidly across an Auger linewidth due to its dependence on the ratio ε_a/Γ. When combined with an exponential dependence on the velocity term $\bar{\xi}$[7], the usual Lorentz profile is strongly distorted for low excess energies.

While PCI makes the Auger lineshape asymmetric, the model could not match the lineshape observed by Lohmann[3] for the argon $L_3M_{2,3}M_{2,3}$ $^1D_2$ line. At 2 ± 0.9 eV excess energy and an Auger electron detection angle of 70°, the data were fitted better by a Fano lineshape given by[9]

$$\sigma_{obs} = \sigma_1\,(q + \varepsilon)^2\,(\varepsilon^2 + 1)^{-1} + \sigma_2$$

This form of the observed cross-section σ_{obs} indicates the contributions from the cross section σ_1, arising from that part of the continuum that interferes with the resonant transition, and σ_2, the cross section of the non-interacting continuum. The relative energy parameter ε differs from ε_A above by a normalisation factor of $1/\Gamma$. The resonant states are the Auger transitions arising from electron-impact ionization of the argon $L_{2,3}$ inner-shells. The satellite states observed in coincidence are much broader than the Auger transitions and are treated as the continuum. The ratio of the symmetric and asymmetric parts of the resonance amplitude is given by the 'q' parameter. This parameter can be determined from a Fano lineshape fitted to the data by comparing the height of the peak, h, and the depth of the dip, d, relative to the continuum background using $q = \sqrt{h/d}$.

EXPERIMENT

The angular dependence of the interference was tested by selecting (e,2e) kinematics that corresponded to a maximum and a minimum in the satellite-state cross-section, since the Auger cross-section is approximately isotropic. A 3μA incident electron beam with an energy of 1000 eV and an energy spread of 0.75 eV was used. The scattered electrons were detected in about a 30 eV range centred on 750 eV with ± 1.3 eV resolution. The scattered electron energy determined the excess energy of the slow ejected electron using energy conservation and knowledge of the inner-shell ionization potentials (L_3 = 248.6 eV, L_2 = 250.8 eV). The scattered electron detection angle of 21° determined the kinematics for which the (e,2e) satellite-state coincidence intensity at 60° was a maximum. Interference effects are expected to be greatest here. At an Auger electron detection angle of 90° the satellite-state continuum background is about a factor of 20 smaller, and the interference is expected to be correspondingly reduced. The Auger detector resolution was 420 meV at 60° and 500 meV at 90°.

Apparatus

The essential feature of the apparatus that enables this coincidence experiment is a multiparameter coplanar (e,2e) coincidence spectrometer[10]. Microchannel-plate detectors on the output of each of two hemispherical electrostatic energy analysers allow the simultaneous collection of data for a large range of excess energies with good energy resolution for more than one entire Auger transition lineshape. In that way the excess energy dependence of PCI and interference effects can be observed for a number of Auger transitions simultaneously, as shown in figure 1, which contains coincidence data recorded with the Auger detector at 60°.

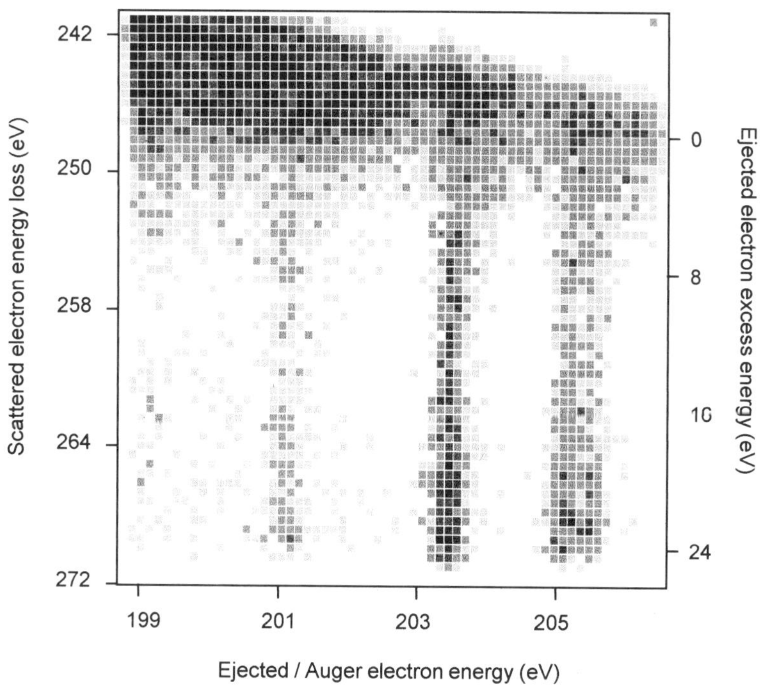

Figure 1. The (e,2e) coincidence energy grid containing data recorded for an Auger electron detection angle of 60°. Darker colours represent more counts. The incident electron energy was 1000eV. Scattered and Auger electron energies are indicated. See text for further details.

Each cell in the coincidence grid contains a 256-channel time spectrum, in which 'true' coincidence data form a peak on a uniform background. A 'window' is selected that contains the 'true' data and after subtraction of the background signal estimated from areas outside this window, the 'true' coincidence data are displayed. The vertical lines in figure 1 correspond to the $L_3\ {}^1S_0$ (at 201.07 eV), $L_2\ {}^1S_0$ (203.18 eV), $L_3\ {}^1D_2$ (203.46 eV), $L_3\ {}^3P_{0,1,2}$

(205.18 eV) and the L_2 $^1D_2$ (205.60 eV) Auger lines. The nearly horizontal lines at the top of the grid are the satellite-state lines, belonging predominantly to the $3s^23p^4(^1D)nd\,^2S$ (n = 3,4,5) manifold[4]. By selecting particular rows in the coincidence energy grid, one can observe the coincidence Auger lineshapes for a particular excess energy.

The non-coincidence Auger lineshapes measured simultaneously with the coincidence signal are shown in figure 2(a). In the absence of PCI effects they are Lorentzian with a linewidth of about 120 meV. The coincidence data forms a true subset of the non-coincidence data, and any changes in experimental conditions should have comparable influences on both data sets. Figure 2(b) shows the non-coincidence scattered-electron energy loss spectrum that includes the $L_{2,3}$ ionization threshold. Figure 2(c) shows the argon 3s peak at about 29 eV and the satellite-state 'continuum', which were measured in a separate experiment for similar conditions with a coincidence energy resolution of 1.8 eV.

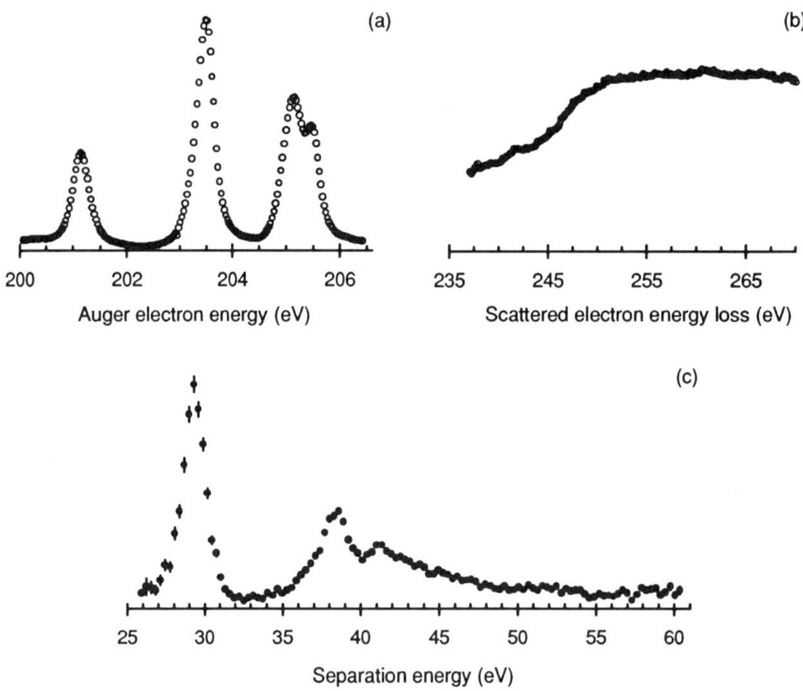

Figure 2. (a) The non-coincidence argon $L_{2,3}M_{2,3}M_{2,3}$ Auger electron spectrum, (b) the scattered electron energy-loss spectrum near the $L_{2,3}$ ionization threshold, and (c) the argon 3s and satellite-state separation energy spectrum measured in coincidence. The satellite states start at about 34eV.

For this type of experiment, where the energy dependencies of the interference and PCI processes are being measured, the spectrometer requires a similar amount of time as a conventional apparatus to obtain statistically significant data for a particular excess energy or Auger transition energy. However, simultaneous acquisition over a large range of excess and Auger electron energies means similar data are acquired concurrently for a number of whole Auger lineshapes for a number of excess energies, both with good energy resolution.

RESULTS

The data acquired in the energy grid shown in figure 1 were analysed by rows corresponding to particular excess energies. A least-squares fitting routine was used to fit the models to the data. In this way the excess-energy dependence of the PCI and interference effects could be observed and the parameters could be quantified. The relative intensities of the Auger lines were determined from the non-coincidence data, and the lines were grouped according to the initial inner-shell vacancy (L_2 or L_3) involved in their production. The excess energies were determined from the calibrated coincidence grid energy scale, and the values quoted in the text are for the $L_3M_{2,3}M_{2,3}$ Auger transitions. The excess energies of the L_2 Auger transitions are always 2.2 eV lower.

Separate models were used to analyse the data acquired in two separate measurements with Auger detection angles of 60° and 90°. The model lineshapes were convoluted with a suitable Gaussian spectrometer function in each model. At 90° a preliminary analysis showed that interference effects were small and the PCI model described earlier could be used without needing to include interference effects. At 60° it was clear that PCI was not the only effect, and a Fano lineshape model was used. The data recorded at 90° are described first since the results were used in the analysis of the 60° data.

Post-Collision Interaction

The PCI effects could be observed clearly in coincidence data acquired with an Auger electron detection angle of 90°. The PCI lineshape model was used in the fitting procedure. The only adjustable parameters were the continuum background intensity and an overall scaling factor for the Auger transition amplitudes.

No interference effects were observed in this data, even for very low excess energies, which is consistent with the small satellite-state continuum background observed at this angle. The continuum background was uniformly flat and of the same intensity over the entire coincidence energy grid.

Four examples of the PCI model fit to the observed data are shown in figure 3. The Auger PCI lineshape that was fitted to each of the L_3 transitions is shown separately to the left of each set, and is normalised to accentuate the change in shape rather than the relative intensity. The L_3 excess energy is indicated in the top right corner of each plot. The width and asymmetry of the Auger PCI lineshape clearly increase as the excess energy decreases in the data of figure 3. In all cases the PCI model predicted the observed Auger lineshapes quite well. For the data at about 1 eV the finite excess energy resolution is taken into account and the L_2 transitions can still be seen.

The data were analysed to determine the observed lineshifts relative to the non-coincidence lines. The magnitudes of the observed shifts were generally under-estimated using the formula[7] $\Delta\varepsilon = \bar{\xi}\,\Gamma/2$, where $\bar{\xi}$ is the velocity-dependent parameter in the PCI model and Γ is the nominal Auger transition linewidth as before. The difference arises because of the finite energy resolution of the Auger electron detector, which tends to enhance the observed lineshift[11]. A uniform additive factor of about 24 meV, corresponding to the residual shift[12] of the non-coincidence Auger lines, was also required before agreement was reached between the modelled and observed lineshifts. The model did reproduce the observed shift quite well if the nominal transition linewidth was replaced by the observed linewidth, which included the broadening caused by the Auger spectrometer resolution.

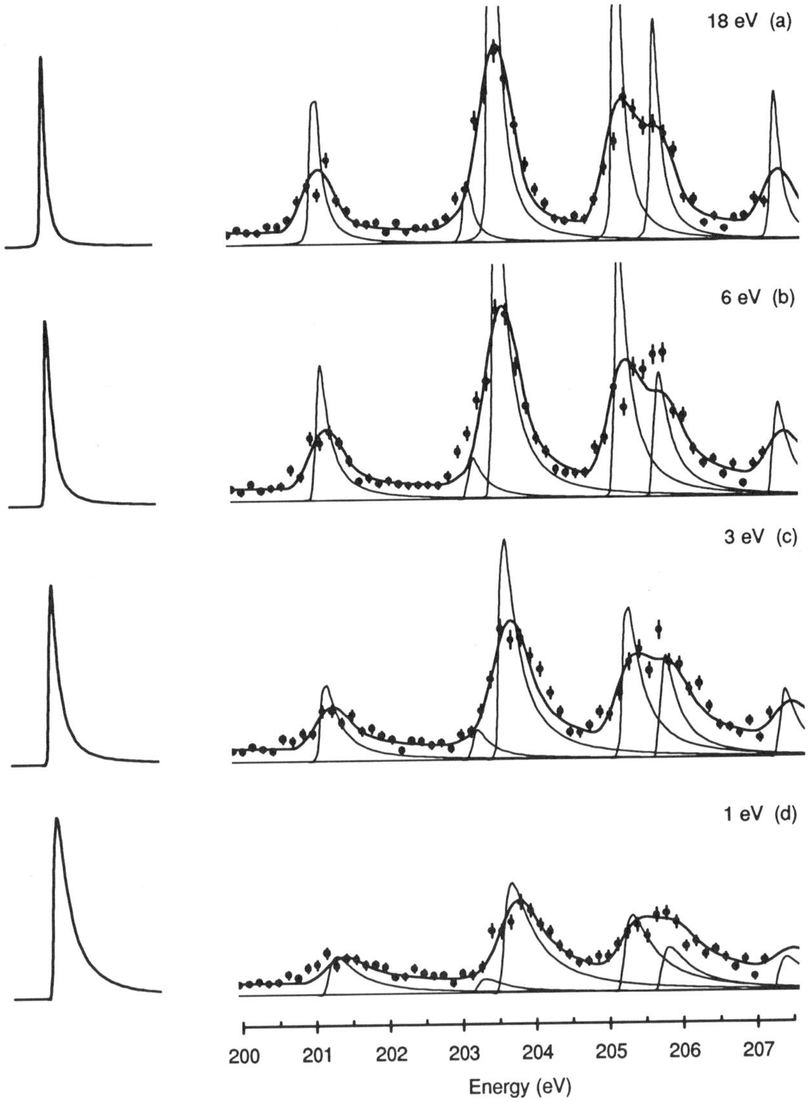

Figure 3. Selected rows from the coincidence grid recorded at an Auger electron detection angle of 90°. The ejected electron excess energies are (a) 18 eV, (b) 6 eV, (c) 3 eV and (d) 1 eV. The solid line is the best fit to the data.

Interference

The data for an Auger detection angle of 60° show interference effects which are large when the satellite-state cross section is also large. The data for low excess energies showed considerable oscillations that extended below the average background intensities. This feature could not be reproduced by the PCI model. Instead the Fano model was used to fit the data. The adjustable parameters were the continuum background level, the 'q' parameter and a scaling factor for each of the L_2 and L_3 initial vacancy states. Since the PCI effects are expected to be isotropic, a shift in the nominal Auger transition energy of $\Delta\varepsilon_a = \xi\,\Gamma/2$, as well as a Fano linewidth consistent with the full-width-at-half-maximum of the predicted PCI-broadened lineshape[7], were included. The combined model gives a better indication of the amount of interference and usually resulted in a better fit. Figure 4 shows a selection of the data and corresponding model fits. The fitted lineshape is shown on the left and the excess energy on the top right of each plot. The excess energies are equal to those in the PCI data of figure 3 to emphasise the angle dependence of the interference.

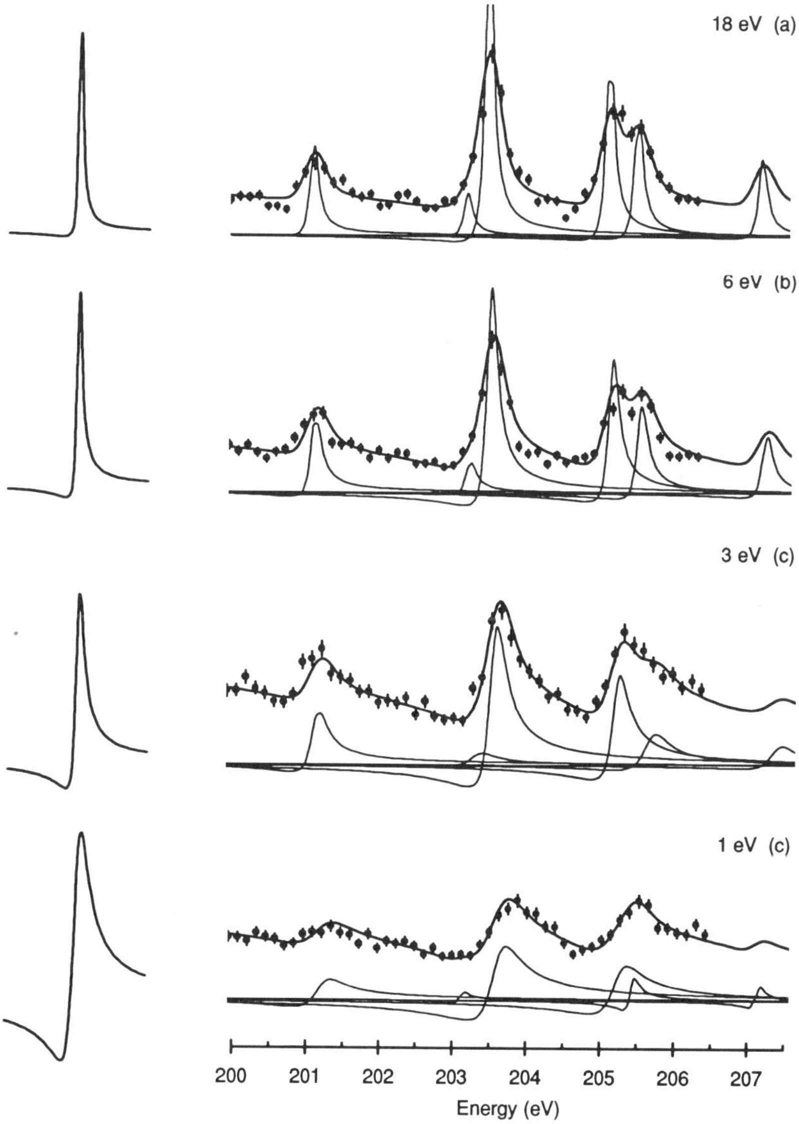

Figure 4. As for figure 3 but the Auger electron detection angle is 60°. The Fano lineshape shown to the left of each plot includes the predicted PCI energy shift and broadening.

Figure 4 shows the interference increases significantly as the excess energy decreases. An excellent fit was obtained using the Fano lineshape model. For an excess energy of 1 eV for the L_3 transitions, the data are fitted best if the L_2 lines are treated as if they did not experience any PCI. This is reasonable, since the excess energy of the slow electron associated with the $L_2M_{2,3}M_{2,3}$ transitions would be 'negative', and it would not be ejected. The Auger electron would therefore not experience any PCI.

Above the $L_{2,3}$ ionization threshold the interference is attributed to recapture of the slow ejected electron, especially since the slow electron loses energy in the PCI process. An $L_{2,3}M_{2,3}M_{2,3}$ Auger process leaves a fast scattered electron, an Auger electron, a slow ejected electron and a doubly-ionized ion with two vacancies in the 3p orbital. Following recapture of the slow ejected electron, the ion is only singly ionized, but has two 3p vacancies and a loosely-bound electron in an outer orbital. The Auger electron has a slightly different energy due to the PCI. In a satellite-state ionization a fast scattered electron and an ejected electron (with energy that may be the same as the Auger electron energy) are produced, and the argon ion is left singly ionized but with two vacancies in the 3p orbital and an electron loosely bound in an outer orbital. By choosing the (e,2e) kinematics appropriately, the Auger recapture and satellite-state processes will result in identical final configurations and interference will occur.

CONCLUSION

Using a multiparameter (e,2e) coincidence spectrometer, we have observed PCI effects for a range of excess energies in the argon $L_{2,3}M_{2,3}M_{2,3}$ Auger transitions. The data are in excellent agreement with the PCI model[7]. Interference between the argon Auger and satellite-state ionization processes was shown to be angle-dependent, and increased as the excess energy decreased. The interference was quantified by fitting a Fano lineshape[9] to the data. The model included the expected PCI shift and broadening effects.

ACKNOWLEDGEMENTS

We are grateful to Dr Peter Hayes and Mr John Flexman for helpful discussions. This work was supported by the Australian Research Council and The University of Western Australia. DKW was supported by a University of Western Australia Postgraduate scholarship and a research assistantship.

REFERENCES

1. U. Fano, Phys. Rev. 124:1866 (1961).
2. W. Sandner and M. Volkel, J. Phys. B 17:L597 (1984).
3. B. Lohmann, J. Phys. B 24:L249 (1991).
4. I.E. McCarthy, R. Pascual, P. Storer and E. Weigold, Phys. Rev. A 40:3041 (1989).
5. E.C. Sewell and A. Crowe, J. Phys. B 17:2913 (1984).
6. B. Lohmann, Aust J. Phys. 49:365 (1996), and references therein.
7. S.A. Sheinerman, W. Kuhn and W. Mehlhorn, J. Phys. B 27:5681 (1994).
8. H. Ishii, Y. Iketaki, T Watabe, T. Takayanagi, K. Wakiya, H Suzuki and F. Koike, Phys Rev. A 43:134 (1991), and references therein.
9. B.W. Shore, J. Opt. Soc. Am. 57:881 (1967).
10. D.K. Waterhouse, J.F. Williams and P.A. Smith, Rev. Sci. Instrum. 67:1769 (1996), and references therein.
11. G.B. Armen, Phys. Rev. A 37:995 (1988).
12. W. Sandner and M.Völkel, Phys. Rev. Lett. 62:885 (1989).

TRIPLE DIFFERENTIAL CROSS SECTIONS FOR ELECTRON IMPACT IONIZATION OF ARGON BETWEEN 50 EV AND 215.8 EV

Stéphane Rioual, Bruno Rouvellou, Jochen Röder*, and Amédée Pochat

Laboratoire des Collisions Electroniques et Atomiques, UFR Sciences et Techniques; 6, avenue V. Le Gorgeu, 29285 Brest Cédex, France

*Permanent adress: Fachbereich Physik, Universität Kaiserslautern, Erwin Schrödinger Strasse, D6750, Kaiserslautern, Germany

INTRODUCTION

The most detailed information about electron impact ionization of atoms is available from (e,2e) experiments in which both electrons following an ionizing event are detected in coincidence. The kinematics of the collision is fully determined and provides the measurement of triple differential cross sections (TDCS). Numerous experiments on helium and on atomic hydrogen performed from threshold to high energy have shown the extreme sensitivity of the coplanar energy sharing symmetric geometry to short- and long-range interactions. At intermediate energy, the TDCS present two maxima interpreted in terms of binary and backscattering processes. Then, the standard DWBA calculations of Whelan et al.[1] reproduce correctly the measurements. When decreasing the incident energy, the two maxima merge into a unique structure; the inclusion of post-collisionnal interaction (PCI) and polarization effects in the standard DWBA calculations is needed to give a correct agreement with the experiments.[1,2]

For heavier targets, (e,2e) experiments have been performed in coplanar symmetric geometry between 500 and 65 eV on neon,[3,4] at 215.8 and 115.8 eV on argon[5] and at 200 eV on xenon.[6] Although the experimental TDCS on neon is completely reproduced at 500 eV by the standard DWBA calculations, at intermediate energy the agreement is not so good as for helium. Below 100 eV, the inclusion of PCI and polarisation effects in the DWBA model provides a better agreement with the measurements on neon, nevertheless important disagreements remain.[4] Thus, further measurements at low energies seem to be necessary to get a better understanding of p electron ionization and to test the validity of the DWBA model for heavier targets than helium or hydrogen. We present here TDCS measurements carried out on argon in coplanar geometry between 215.8 and 50 eV and over an extended angular range from 25 to 140 degrees. The measurements at the two highest energies are compared with those performed by Bell et al.[5] over a smaller angular range. At 50 eV, additional measurements have been performed in coplanar θ_{ab} constant geometry. The DWBA model represents the only available model to describe ionization of heavy targets, so the data are compared with DWBA calculations including PCI effects with the Gamow factor.

EXPERIMENT

The experiments have been performed with the coincidence spectrometer previously used by Rioual et al.[4] for the study of neon. Briefly, an electron beam of few μA produced by an electron gun crosses at right angles a gas jet emerging from a 1.5 mm diameter capillary tube; the residual pressure in the vacuum chamber is about $5x10^{-6}$ torr. The two outgoing electrons are detected in coincidence using two identical 127 degrees electrostatic analyzers which can be independently rotated around the scattering region. The angular acceptance angle of the analyzers is about +/-1 degree. The statistical error on the measurements is small; this is due to the good values of the coincidence count rate and of the true-to-accidental coincidences ratio. For example at 60 eV, the maximum coincidence rate is about 2 counts/sec and the true-to-accidental coincidences ratio about 1.

Measurements have been made in two kinds of kinematical arrangement in which the energy is equally shared between the two outgoing electrons. In coplanar symmetric geometry the momenta of both outgoing electrons make the same angle with the direction of the incident electron; in the coplanar θ_{ab} constant geometry, the mutual θ_{ab} angle between both outgoing electrons is kept constant.

RESULTS IN COPLANAR SYMMETRIC GEOMETRY

The measurements made at 215.8 and 115.8 eV, presented in figure 1, are compared with the previous data of Bell et al.[5]. The two sets of data are normalized to each other near 45° where the error bars are very small; they present an overall good agreement. However, our new measurements allow to observe a maximum near 37° and also probably another one near 135°. Near 45°, the TDCS are explained by a direct binary collision and near 135° by a backscattering process followed by an ionizing event. The slight minimum observed at 45° is explained by the momentum density of the p orbital. When decreasing the incident energy, like on neon, this minimum seems to disappear and the first maximum near 37° is shifted to low angles.

The measured TDCS are also compared with the DWBA calculations[7] with and without the semiclassical Gamow factor which takes into account the electron-electron repulsion in the final state. It should be noted that the effect of this factor is to give the dominant angular behavior of the TDCS and a complete theoretical treatment of the PCI effect is not made.[1,2] The measurements are normalized to the theory at 90°. At these energies and particularly at 215.8 eV, the effects of polarization must be very weak and the standard DWBA model provides a correct description of the collision. At 115.8 eV, including PCI effects improves the behavior of the TDCS at low angles. Our measurements have been performed with a better angular resolution than Bell et al.[5], nevertheless the strong minimum predicted by the calculation near 45° is not observed. This discrepancy can be partially due to the finite angular resolution of the experiment. At 215.8 eV, near 120°, the theory predicts a slight minimum whereas the experimental TDCS presents a flattening. At 200 eV, a minimum has been also observed at large angles on xenon[6] but not on neon. Whelan and Walters[8] have suggested that there exist a rough mirror effect about 90°, and then, the minimum at large angles could be attributed to the image of the split at 45°. The ratio of the intensity of the main maximum near 37° to the large angles peak is correctly reproduced by calculations. With respect to the ionization potential, the contribution of the large peak to the TDCS is more important for argon than for helium; but it is lower than for neon.

The measurements performed at 85.8 and 50 eV are compared with the DWBA calculations in figure 2. As above, the theory is normalized to the measurements at 90°. The experimental TDCS present a completely different shape like for neon or helium. With decreasing the incident energy, the minimum near 90° is rapidly shifted towards large angles and the TDCS are increased near 65°. Measurements made at 50 eV with the apparatus previously used at Kaiserslautern and currently installed at Brest are also presented. There is a good agreement between the two sets of data.

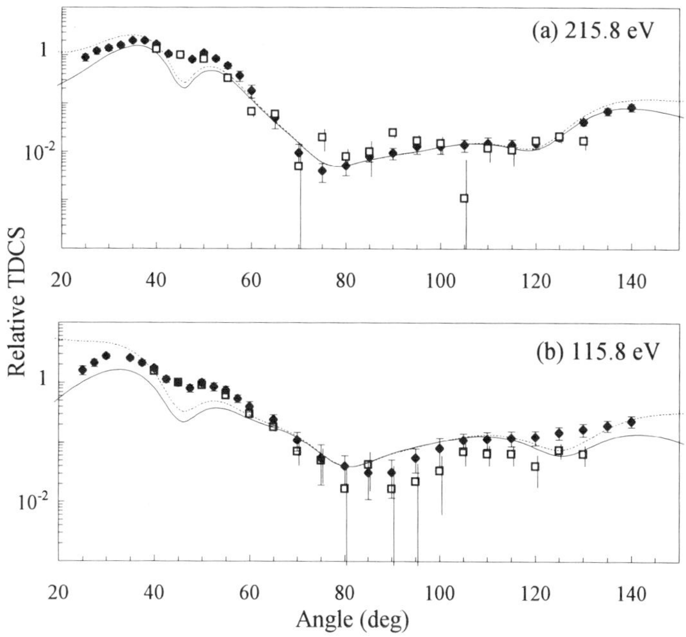

Figure 1. Coplanar symmetric TDCS for Ar(3p), (a) at 215.8 eV and (b) at 115.8 eV. (▫) measurements of Bell et al. (1995); (•), this work. DWBA calculations with (——) and without (········) the Gamow factor.

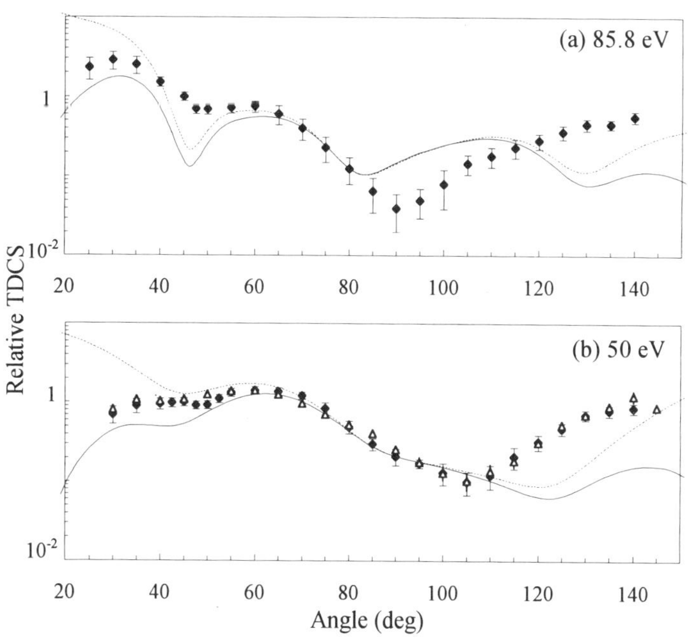

Figure 2. Coplanar symmetric TDCS for Ar(3p), (a) at 85.8 eV and (b) at 50 eV. Measurements performed with (•) the Brest and (▵) the Kaiserslautern apparatus. DWBA calculations with (——) and without (········) the Gamow factor.

Below 100 eV, the simple DWBA calculations do not correctly describe the process. Including polarization effects in the DWBA model could provide a better agreement in this energy region, however as for neon, this would probably not reproduce completely the measurements.

RESULTS IN COPLANAR θ_{ab} CONSTANT GEOMETRY

To get a better understanding of the effects of the different interactions at 50 eV, additional measurements have been made in coplanar θ_{ab} constant geometry where the PCI effects are kept constant. Figure 3 presents the data obtained for the mutual θ_{ab} angle between both outgoing electrons equal to 180, 150 and 120° together with the DWBA calculations. A very good agreement is observed between the measurements performed with the Brest and Kaiserslautern apparatus. The normalization between theory and experiments is again made at $\theta_a = \theta_b = 90°$. For $\theta_{ab} = 180°$, the DWBA calculations give a quite good description of the collision; as shown on hydrogen,[9] the effect of the polarization should be very weak in this case. In this particular kinematical arrangement, the PCI is radial and can be treated with success, even at very low energies, by effective charges.[10,11] When decreasing the θ_{ab} angle, the main structures are still well represented, however some discrepancies appear. They concern essentially the description of the ratio of the TDCS between low and large angles.

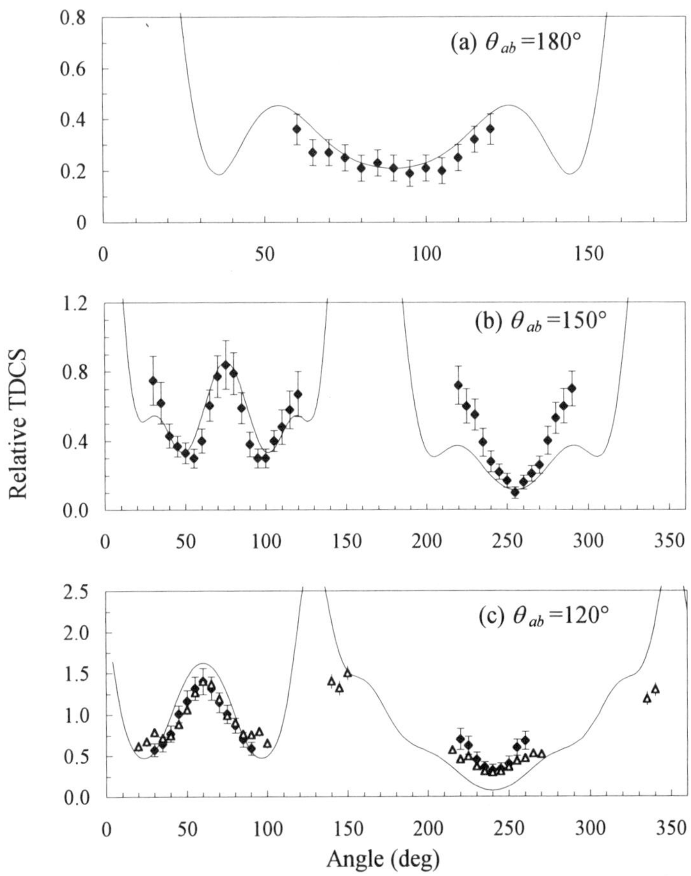

Figure 3. TDCS in coplanar θ_{ab} contant geometry at 50 eV for Ar(3p). The angle θ_{ab} between the two outgoing electrons is fixed at angles of (a) 180°, (b) 150° and (c) 120°. Measurements performed with the Brest (•) and the Kaiserslautern (▲) apparatus. DWBA calculations including the Gamow factor.

CONCLUSION

This paper confirms that above 100 eV the measured TDCS in coplanar symmetric geometry are quite well reproduced by the DWBA model. When the incident energy decreases, the agreement between experiments and calculations is improved by including the Gamow factor into the theoretical model. However, it would be useful to get a better understanding of the collisionnal effects to take also into account polarization effects. This should also be crucial to describe correctly the measurements made currently at lower energies and near threshold.

ACKNOWLEDGMENTS

The authors would like to thank Jens Rasch and Colm T. Whelan for providing the distorted wave Born approximation calculations. This work is supported by the EEC contract ERB-CHRX-CI93-0350.

REFERENCES

1. C.T. Whelan, H.R.J. Walters, R.J. Allan, and X. Zhang, *(e,2e) and Related Processes*, Dortrecht: Kluwer, p1-32 (1993).
2. C.T. Whelan, R.J. Allan, J. Rasch, H.R.J. Walters, X. Zhang, J. Röder, K. Jung, and H. Ehrhardt, Coulomb three-body effects in (e,2e) collisions: The ionization of H in coplanar symmetric geometry, *Phys. Rev. A* 50:4394 (1994).
3. T. Rösel, C. Dupré, J. Röder, A. Duguet, K. Jung, A. Lahmam-Bennani, and H. Ehrhardt, Coplanar symmetric cross sections on helium and neon, *J. Phys. B: At. Mol. Opt. Phys.* 24:3059 (1991).
4. S. Rioual, A. Pochat, F. Gélébart, R.J. Allan, C.T. Whelan, and H.R.J. Walters, Low and intermediate energy coplanar symmetric (e,2e) cross sections for ionization of neon, *J. Phys. B: At. Mol. Opt. Phys.* 28:5317 (1995).
5. S. Bell, C.T. Gibson, and B. Lohmann, Triple-differential cross section for electron-impact ionization of argon in coplanar symmetric geometry at intermediate energies, *Phys. Rev. A* 51:2623 (1995).
6. T. Rösel, K. Jung, H. Ehrhardt, X. Zhang, C.T. Whelan, and H.R.J. Walters, The electron impact ionization of Xe in coplanar symmetric geometry, *J. Phys. B: At. Mol. Opt. Phys.* 23:L649 (1990).
7. J. Rasch and C.T. Whelan, Private communication (1996).
8. C.T. Whelan and H.R.J. Walters, A new version of the distorted-wave impulse approximation-application to coplanar symmetric ionization, *J. Phys. B: At. Mol. Opt. Phys.* 23:2989 (1990).
9. J. Roder, J. Rasch, K. Jung, C.T. Whelan, H. Ehrhardt, R.J. Allan, and H.R.J. Walters, Coulomb three-body effects in low-energy impact ionization of H(1s), *Phys. Rev. A* 53:225 (1996).
10. S. Jones, D.H. Madison, and M.K. Srivastava, Near-threshold ionization of hydrogen and helium by electron impact, *J. Phys. B: At. Mol. Opt. Phys.* 25:1899 (1992).
11. C. Pan and A.F. Starace, Angular distributions for near-threshold (e,2e) processes for H, He, and other rare-gas targets, *Phys. Rev. A* 45:4588 (1992).

MULTIPLE SCATTERING EFFECTS IN HIGH-ENERGY ELECTRON MOLECULE IONIZATION OF H_2

R.J. Allan[1], J. Hanssen and B. Joulakian

Laboratoire de Physique Moleculaire et des Collisions, Institut de Physique et d'Electronique, Universite de Metz, 1 rue Arago, 57078 Metz Cedex 3, France

[1]usual address: CCLRC Daresbury Laboratory, Daresbury, Warrington, WA4 4AD, U.K.

INTRODUCTION

The electron impact ionization of molecules is of growing interest. Most previous work has considered molecular targets, such as H_2, which are not oriented in space, see e.g. Weigold[1], Cherid et al.[2] and references therein. Experiments are now possible using both surfaces and fields to orient molecules, e.g. Palmer et al.[3], Volkmer et al.[4], and our knowledge of the ionization process will be further tested in new geometries.

In this paper we consider orientation effects which can be seen in the (e,2e) cross section for certain special geometrical arrangements. We extend previous work by including multiple collision processes which are demonstrated to be important. H_2 is chosen as an example as it has been studied previously and its electronic structure is not too complicated enabling a straightforward discussion of the features observed.

THEORY

Molecular Wavefunction

We will in general write the molecular wavefunction as a Slater determinant

$$\Phi = \frac{1}{\sqrt{N!}} \left| O_A^1 O_B^1 O_A^2 O_B^2 \ldots\ldots\ldots \right|$$

in which O_i^j represents the jth orbital on atom i. The orbitals are orthonormal and Slater's integration rules[5] are applied so that only 1-e and 2-e excitations are required in evaluation of transition matrix elements. We further assume that the system following ionization shall have the same determinental form except that one orbital (from which the electron is ionized) is replaced by an orthogonalised 1-e continuum wavefunction.

The static potential experienced by an electron approaching the molecule (neglecting exchange) is

$$V(r_1) = \frac{1}{N}\langle\Phi|\sum_j \frac{1}{r_{1j}}|\Phi\rangle = \frac{1}{N}\sum_j \langle O_A^j|\frac{1}{|r_1 - r_j - r_a|}|O_A^j\rangle + \frac{1}{N}\sum_j \langle O_B^j|\frac{1}{|r_1 - r_j - r_b|}|O_B^j\rangle$$

$$= V_A(r_{1a}) + V_B(r_{1b})$$

plus nuclear terms.

Within this formalism the method therefore resembles the "atoms in molecules" method which has previously been used in elastic scattering and photoionization[6,7].

Multiple Scattering Method

In previous work[8] on H_2^+ we used the lowest order of multiple scattering theory[9]. In an improved theory[10] the ejected electron in the ionization of H_2^+ was treated to all orders using a two-centre continuum wavefunction. This showed considerable differences in the (e,2e) cross section which depend on molecular orientation and directions of final electron momenta.

It is known that a multiple scattering expansion may be used as an approximate treatment of effects due to a potential which may be written as the sum of two or more potentials, $V = V_A + V_B$. As a well known example of such effects we cite the usual interpretation of (e,2e) cross sections which includes both direct ejection of the atomic electron (binary peak) and scattering from the nucleus followed by ejection (recoil peak). These effects are included to all orders in the Distorted Wave Born Approximation (DWBA) as discussed by Whelan et al.[11] In this paper we include similar effects in multi-centre scattering using the Multiple Scattering Method (MSM) to avoid explicit calculation of multi-centre wavefunctions. This approach has been successfully used for many years for instance in the interpretation of Low Energy Electron Diffaction (LEED) experiments with solid targets[12].

We briefly revise the first order MSM theory fundamental to the interpretation of our results. We consider only the incident electron channel for clarity. The transition operator for a scattering process in potential V due to two atoms A and B may be written as

$$T = (T_A + T_B) + (D_A G_0 T_B + D_B G_0 T_A) + \text{higher terms}$$

where we are to evaluate the first Born direct scattering transition amplitudes

$$T_A = \langle k_s k_e|\ V_A\ |k_i 0_A\rangle \text{ and } T_B = \langle k_s k_e|\ V_B |k_i 0_B\rangle \text{ etc.}$$

Here k_s, k_e and k_i represent continuum wavefunctions defined with respect to a common origin (e.g. the centre of mass of the molecule) and V_A and V_B are potentials due to each atom. The first two terms represent independent scattering from the two atomic centres and their linear combination includes interference due to the relative shift of the origin in the coordinate system in each term. This yields a simple optical interpretation[8] and may be generalised trivially for many scatterers. The next two terms include the free-particle Green's function G_0. We evaluate these using a Fourier transform procedure similar to that of Messiah.

$$f_A = \frac{1}{(2p)^3}\int e^{i\vec{k}'\vec{r}'}\langle \vec{k}'0 \; |V_A| \vec{k}_i 0\rangle d\vec{k}'$$

$$f_B = \frac{1}{(2\pi)^3}\int e^{i\vec{k}''\vec{r}''}\langle \vec{k}_s \vec{k}_e \; |V_B| \vec{k}''0\rangle d\vec{k}''$$

$$G_0 = -\frac{1}{2\pi}\frac{e^{ik|r''-r'|}}{|r''-r'|}$$

$$D_A G_0 T_B = -\frac{1}{2\pi}\int dr' \int dr'' \frac{e^{ik|r''-r'|}}{|r''-r'|} f_A(r') f_B(r'')$$

In applying the MSM to the (e,2e) process T_i is therefore interpreted as the ionization from atom i and D_j as elastic scattering by a different atomic centre j. G_0 connects atom j to atom i.

The integrals are simplified if a "peaking" approximation is used as follows. We suppose that $f_B(r'')$ is localised and has its maximum value in the region of r''=b. Because the Green's function varies more slowly than f_B we replace 1/|r''-r'| by 1/|b-r'| and obtain

$$D_A G_0 T_B = -\frac{1}{2\pi}\int dr' \frac{f_A(r')}{|b-r'|} e^{ikr'} \int dr'' f_B(r'') e^{ikr''}$$

Similarly if f_B is maximal in a region r'=a we obtain

$$D_A G_0 T_B = -\frac{1}{2\pi}\frac{1}{|b-a|}\int dr' f_A(r') e^{ikr'} \int dr'' f_B(r'') e^{ikr''} = -\frac{1}{2\pi|b-a|} D_A T_B$$

where the operators are now to be evaluated (in the first Born approximation) as

$$D_A = \langle k_\rho 0_A| V_A |k_i 0_A\rangle \text{ and } T_B = \langle k_s k_e| V_B |k_\rho 0_B\rangle,$$

k_ρ is in the direction b-a and |b-a| is just the internuclear distance between atom B and atom A. This is just the result given by Messiah for the case of elastic scattering.

Elastic scattering must be included for both initial, scattered and ejected electrons so a total of six terms of O(1) needs to be evaluated, in addition to two terms of O(0). We also evaluate terms of O(2). This is done explicitly so that all observable effects can be identified. Exchange contributions may be evaluated by antisymmetrising the wavefunction, but are not considered in this paper. We note that the Two Centre Continuum (TCC) method[10] is expected to be equivalent to using MSM to all orders for the ejected electron channel only.

RESULTS

We restrict our discussion to the high-energy region of electron coincidence spectroscopy. In previous work we took the energy of incident and ejected electrons to be 2keV and 50eV respectively fixing the ejected electron in the momentum transfer direction $\underline{k}_e = \underline{k}_i - \underline{k}_s$. We will consider this case first. Cherid et al. considered even higher initial energy in the asymmetric geometry with k_i and k_s fixed but varying the direction of k_e. They gave experimental results for the triple differential cross section averaged over all molecular orientations. We consider this case second.

In this high-energy regime the incident and scattered electron wavefunctions in the field of the neutral molecule may effectively be replaced by plane waves. This allows some simplification in the theory and integrals over coordinates for this fast electron are carried out analytically with the usual Bethe integral. In evaluating the (e,2e) transition elements, T, integrals over the bound and ejected electron coordinate are computed using numerical quadrature and an expansion in partial waves. The ejected electron sees the asymptotic charge of H_2^+ and a Coulomb wavefunction is used which is evaluated numerically[13]. The elastic scattering elements, D, are evaluated analytically (see Joachain[14]) assuming scattering by a modified atomic hydrogen bound state. This is not strictly correct for the ejected electron, but is unlikely to cause significant errors as will be shown below.

The H_2 bound-state wavefunction is taken to be a Slater determinant composed of two modified H(1s) orbitals with exponent 1.19. Accurate values are taken for the first ionization energy and internuclear distance of 16.4eV and $1.4a_o$ respectively.

Momentum transfer geometry

Results comparable to our previous results[10] for H_2^+ are shown in Figures 1 and 2. In this case the incident and ejected electron energies are E_i=2000eV and E_e=50eV respectively and the ejected electron is taken in the direction of momentum transfer. The angle of the scattered electron is varied, and molecular orientations of θ_ρ=90° and 0° with respect to the incident beam and in the plane defined by the incident and scattered electron are considered in Figs. 1 and 2 respectively. These show results of calculations in O(0), O(1) and O(2) of the MSM expansion. The relatively small difference between the O(1) and O(2) results indicates that convergence has been achieved.

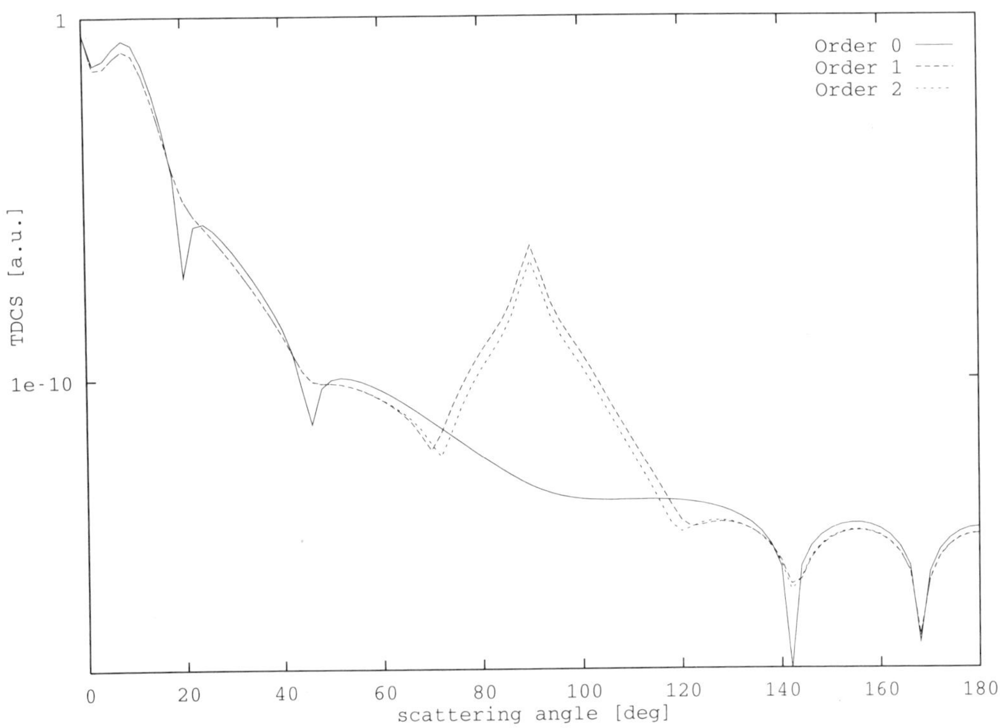

Figure 1. Triple differential cross section for (e,2e) on H_2 in coplanar momentum-transfer geometry. E_i=2000eV, E_e=50eV. H_2 molecule perpendicular to incident electron direction.

Consider first Fig. 1. Using MSM a sharp peak appears near θ_ρ=90°, which is just the molecular orientation. This feature is due to elastic scattering in the incident channel from atom A followed by ionization from atom B. At an angle $\theta_s=\theta_\rho$ the (e,2e) transition element corresponds to the case of forward scattering and is very large. Scattering from atom B followed by ionization from atom A will lead to backward scattering and is negligible, and direct ionization is also small, so there is little interference at this angle.

The case of θ_s=0° is much more interesting, see Fig. 2. For angles θ_e>18° a plateau appears in the first and second order calculations. To investigate this effect we selectively suppressed the elastic scattering in the three channels. It was found that with only the scattered electron channel results were indistinguishable from the full calculation, but without scattering in this channel results were almost identical to the O(0) ones. The only event which plays a role is ionization from atom A followed by diffusion from atom B, so no interference is seen. This is again a case involving a dominant (e,2e) transition in a forward scattering direction.

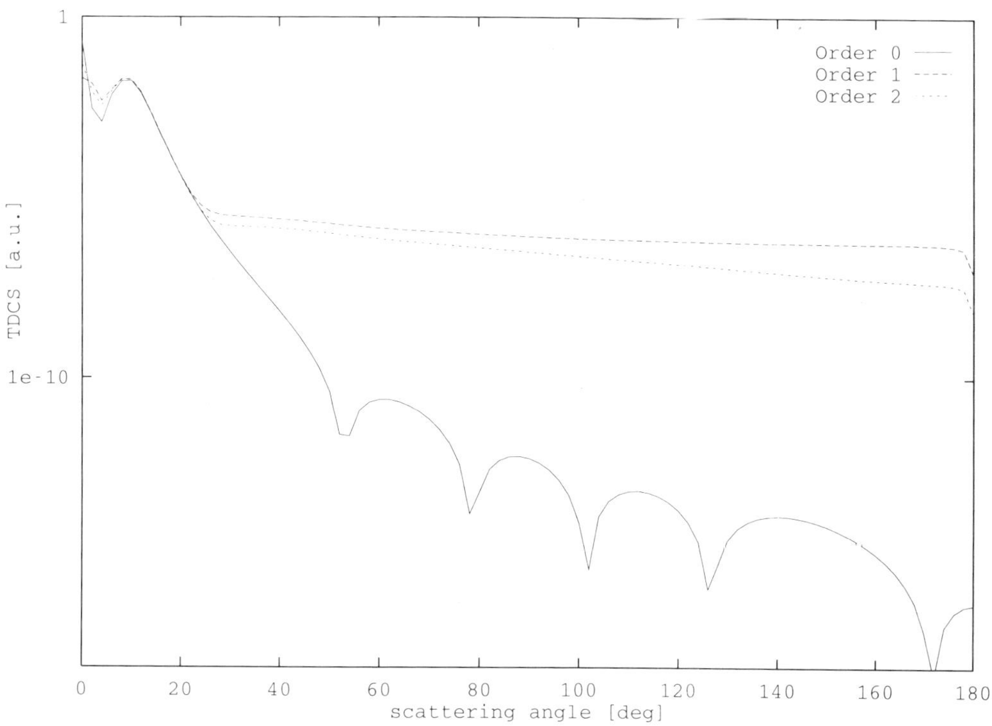

Figure 2. Same as Fig. 1 but H_2 molecule parallel to incident electron direction.

The shape of this curve can also be explained. θ_e, which is in the momentum transfer direction, rises rapidly as θ_s increases, to almost a value of 90° in the "near Bethe ridge" condition, θ_s=10°. θ_e then decreases steadily while θ_s rises. The product TD is approximately constant because T increases while D decreases leading to a flat plateau.

We note that in previous work[8] on H_2^+ matrix elements were calculated with a potential which included some effect from the second nucleus (see e.g. eqn. 16 in that paper). Some elastic scattering was thus implicitly included, and a similar nearly constant behaviour was found.

Finally we compare the values of the cross section for the scattering angle θ_s=90°. For this angle, following the above mechanism, the (e,2e) transition element is always the one for the forward direction. In each case the elastic transition element is for 90° scattering. In Figure 1(a) the scattering angles are 90° followed by 0° whereas in Fig. 2 it is 0° followed by 90°. These two cases are shown schematically in Fig. 3. Note that the magnitue $k_i \approx k_s$ so the values of the matrix elements are almost the same in the two cases, except that the direction of k_e remains fixed in space relative to the incoming electron direction. This means that in case 1(a) the (e,2e) element has θ_e=135° whereas in case 1(b) the (e,2e) element has θ_e=305°. Comparing the values of these two matrix elements yields a ratio of 0.7 which is exactly as seen in the computed values of the cross sections, confirming the above analysis.

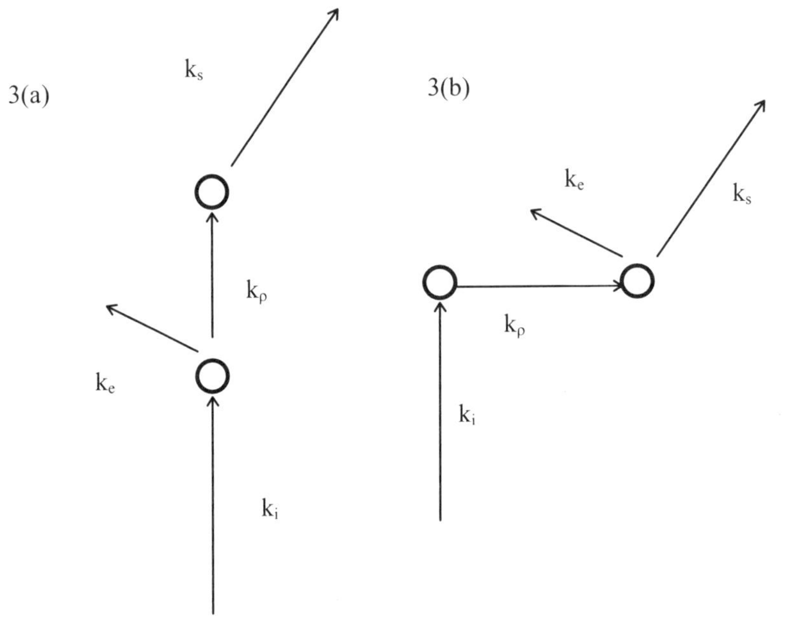

Figure 3. (a) Primary two-step event for molecule parallel to incident electron direction k_i, yields a large cross section if k_s parallel to k_i. (b) primary two-step mechanism for perpendicular molecule, yields a large cross section if k_s perpendicular to k_i.

Unfortunately these effects are not likely to be experimentally observable in the near future, so we turn now to other geometries.

Asymmetric geometry

In this geometry, which has been thoroughly studied experimentally in the case of (e,2e) from atoms, the scattered electron angle is held fixed and the ejected electron angle is varied. In previous work on H_2 Cherid et al.[2] measured for E_i=4087eV, E_e=20eV and for E_i=4168eV, E_e=100eV. They reported absolute experimental values at a variety of angles. We will consider only the most complete set for the first energy at θ_s=1°. Cherid et al. did not use oriented molecules, but theoretically we are able to calculate cross sections for various molecular orientations. We again restrict the molecule to be in the scattering plane.

The usual shape of the atomic (e,2e) cross section in this geometry with the first Born approximation is well understood[11] and for S states consists of a two-peaked structure with a binary peak (direct electron-electron impulsive ionization) centred about the direction of momentum transfer which satisfies $\underline{k}_e=\underline{k}_i-\underline{k}_s$ and recoil peak (elastic scattering from nucleus followed by ionization) in the opposite direction. The binary angle is θ_e=284° (recoil θ_e=10°) for the case of Figs. 4 and 5. This two-peaked structure is retained in a molecular first Born calculation at O(0) as seen in Fig. 4, however there is a shift in the position of the peak values, in this case giving a binary peak at θ_e=298° and a recoil peak at θ_e=64°. This effect is due to interference between the two atomic centres. Introduction of MSM terms changes the picture completely.

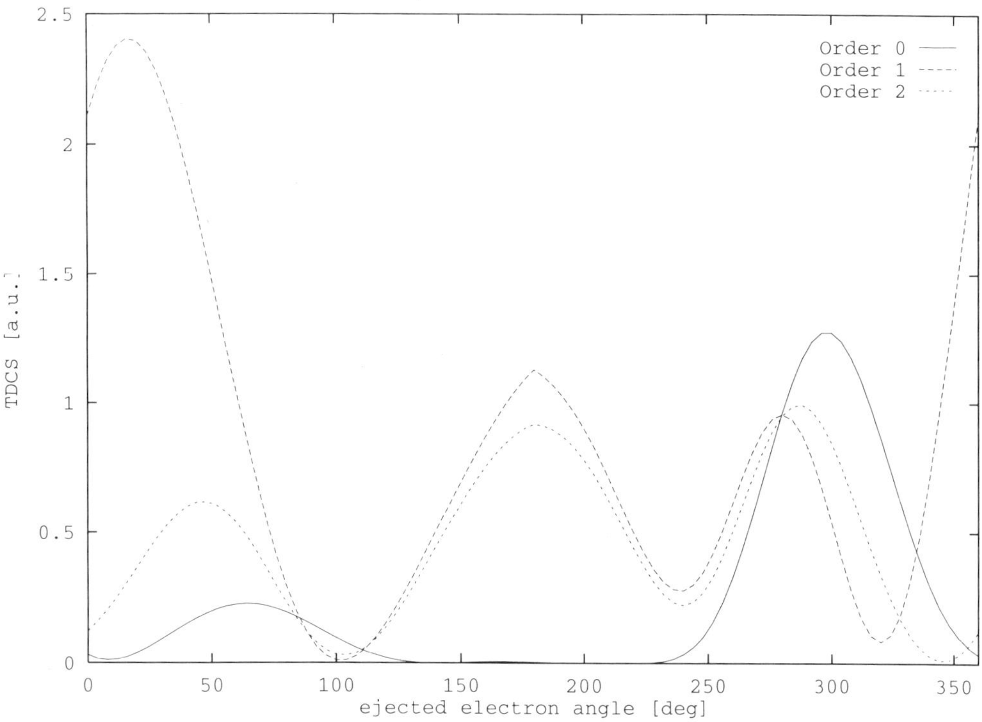

Figure 4. Triple differential cross section for (e,2e) on H_2 in coplanar asymmetric geometry. E_i=4087eV, E_e=20eV, scattered electron deflected by 1° in opposite direction to ejected electron. H_2 molecule parallel to incident electron direction

Consider Fig. 4 in which θ_ρ=0°. A forward-scattering (e,2e) event is possible at atom A followed by elastic scattering by 1° at atom B (see Fig. 3(a)). The binary and recoil peaks for the forward scattering event are large and at θ_e=0° and θ_e=180° respectively. These are almost as seen in Fig. 4 except that the binary peak is shifted due to interference effects with terms of O(0). The forward peak is reduced in the O(2) calculation. To summarise, the molecular (e,2e) matrix element is formed from a combination of the atomic matrix elements with θ_s=1° and θ_s=0° with interference leading to a three-peaked structure in this case.

Consider next Fig. 5 in which θ_ρ=90°. In this case there is no possible forward scattering (e,2e) event and the shape of the curve is subtantially unaltered by adding higher order terms. Convergence is moreover very rapid. The binary peak is reduced in the O(1) calculation. This is probably an effect of interference between the O(0) and O(1) scattering since the momentum transfer direction (peak of cross section) is roughly in line with the molecule. It is less rapid, in fact poor, if the molecule is aligned in the direction of the scattered electron θ_s=θ_ρ. Clearly similar effects to the ones described above occur, but are greatly exaggerated. It should be possible to observe these effects experimentally.

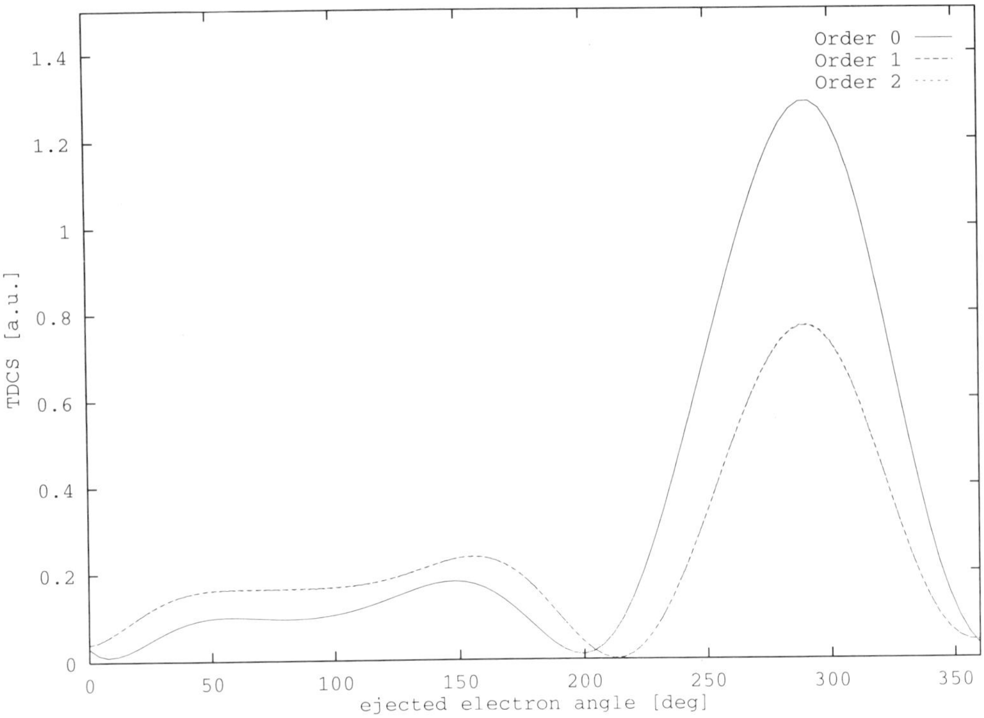

Figure 5. Same as Fig. 4 but H_2 molecule perpendicular to incident electron direction.

Average over molecular orientations

The cross section has been computed averaged over orientations of the molecular axis. This can be done analytically for the O(0) result[8] since phases factorise. This is however no longer the case in O(1) and higher MSM calculations and a numerical average has been computed. 81 samples were used over a half sphere in (θ_ρ,ϕ_ρ). To improve accuracy of the integral the difference between numerical O(2) and O(0) results was integrated and added to the analytical O(0) result. This is shown in Figure 6. The cross section is reduced in the region of the binary maximum mainly because of the reduced cross sections for orientations of $\theta_\rho \approx 90°$ and $\phi_\rho \approx 0°$ as discussed above.

Whilst it would seem from the figure that the experimental results of Cherid et al. are in better agreement with the O(0) calculation, this may be due to the poor averaging procedure used here. We would welcome other experimental work to enable a more conclusive investigation.

CONCLUSIONS

We have investigated some effects in the application of an explicit form of multiple scattering theory to electron-impact ionization of the H_2 molecule. We predict that large effects, due to both multiple scattering and interference, should be seen at certain geometrical arrangements if the molecule can be oriented in space. We also observe a noticable effect in the binary maximum of the cross section averaged over all orientations which appears to disagree slightly with the only available experimental data. We would welcome further in-depth investigations, both experimental and theoretical of this subject.

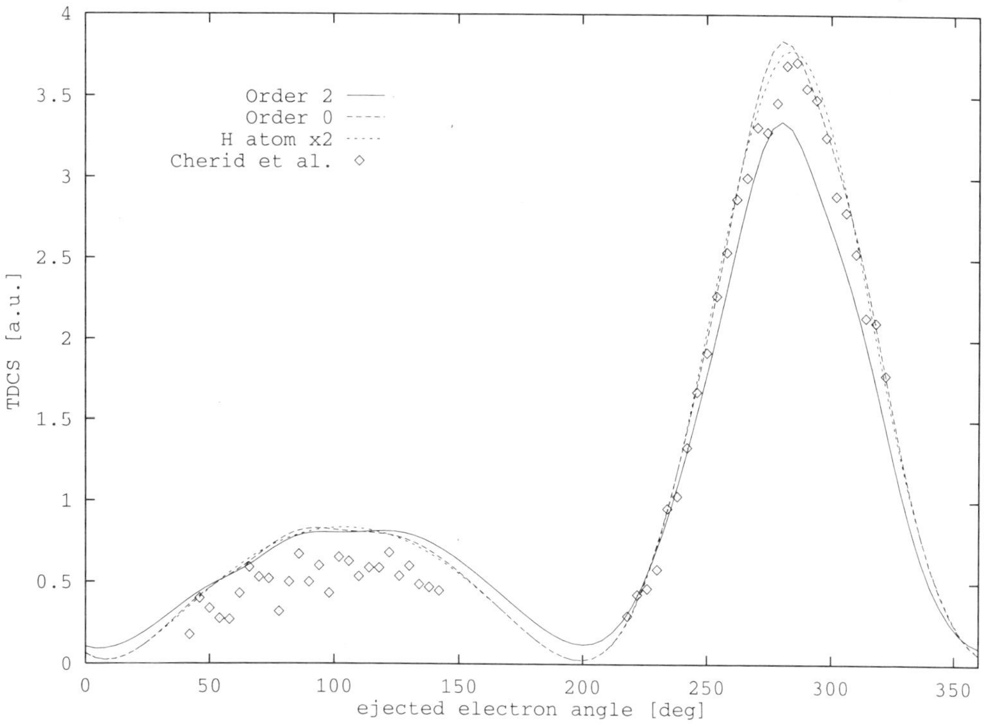

Figure 6. Same as Fig. 4 but averaged over all molecular orientations. Results of order 0 and order 2 calculations are compared with H atom (x2) and experimental points of Cherid et al.[2]

ACKNOWLEDGEMENTS

We acknowledge support from an EC HC&M Network grant CHRX-CT93-0350. RJA thanks the University of Metz for funding to enable this work. We also acknowledge use of IBM SP2 parallel computing facilities at CNUSC, Montpellier and Daresbury Laboratory.

REFERENCES

1. E. Weigold, Electron momentum spectroscopy of molecules: a review of recent developments, in: *Electron-Molecule Scattering and Photoionization,* P.G. Burke and J.B. West eds., Plenum (1987) ISBN 0-306-42998-5.
2. M. Cherid, A. Lahmam-Bennani, A. Duguet, R.W. Zurales, R.R. Lucchese, M.C. Dal Cappello and C. Dal Cappello, Triple differential cross sections for molecular hydrogen, both under Bethe ridge conditions and in the dipolar regime. Experiments and theory, *J. Phys. B.* 22:3483-99 (1989).
3. R.E. Palmer, P.J. Rous and R.F. Willis, Negative ion resonance electron scattering from oriented physisorbed O_2, in: *Electron-Molecule Scattering and Photoionization,* P.G. Burke and J.B. West eds., Plenum (1987) ISBN 0-306-42998-5.
4. M. Volkmer, Ch. Meier, J. Lieschke, R. Drier, M. Fink and N. Bowering, Azimuthal scattering dependencies in electron diffraction from free oriented molecules, in: *Contributed Papers to 5th EPS Conference on Atomic and Molecular Physics, Edinburgh, 3-7/5/95.*
5. J.C. Slater, *Quantum Theory of Atomic Structure,* McGraw-Hill (1960).
6. D.A. Kohl and E.J. Shipsey, Elastic electron scattering from state-selected molecules. I. intensities, *Z. Phys. D* 24:33-8 (1992).

7. D.A.L. Kilcoyne, C.M. McCarthy, S. Nordholm, N.S. Hush and P.R. Hilton, An atomic diffraction theory of molecular photoionisation cross sections, *Journal of Electron Spectroscopy and Related Phenomena* 36:153-85 (1985).
8. J. Hanssen, B. Joulakian, R.D. Rivarola and R.J. Allan, Diffferential cross section of the dissociative ionization of H_2^+ by fast electron impact, *Physica Scripta* 53:41-45 (1996).
9. A. Messiah, *Mecanique Quantique,* Dunod (1964).
10. B. Joulakian, J. Hanssen, R.D. Rivarola and A. Motassim, *Phys. Rev. A* (1996) in press.
11. C.T. Whelan, H.R.J. Walters, X. Zhang and R.J. Allan, e-2e, effective charges, distorted waves and all that!, in: *(e,2e) and Related Processes Proc. NATO ARW Cambridge 27/9-1/10/92,* C.T. Whelan, H.R.J. Walters, A. Lahmam-Bennani and H. Erhardt eds., Kluwer Academic Press (1993) p1-32 ISBN 0-7923-2458-7.
12. J.B. Pendry, *Low Energy Electron Diffraction,* Academic Press (1974)
13. A.R. Barnett, COULFG: Coulomb and Bessel functions and their derivatives, for real arguments, by Steed's method, *Comp. Phys. Comm.* 27:147-66 (1982).
14. C.J. Joachain, *Quantum Collision Theory,* 3rd edn., North-Holland (1983) ISBN 0-444-8677-32.

PARAMETRISATION OF TDCS IN A PLANE

Matjaž Žitnik

J. Stefan Institute and University of Ljubljana
Jamova 39, 1000 Ljubljana, Slovenia

INTRODUCTION

In $(e, 2e)$ experiments, the electrons with momentum $\boldsymbol{k}_0$ impinge onto the gas target with an energy large enough to ionize the atoms. After the collision, the pairs of electrons leaving the target with momenta $\boldsymbol{k}_a$ and $\boldsymbol{k}_b$ are counted in coincidence. Since 1969, when $(e, 2e)$ technique was introduced[1,2] into the atomic physics, a large number of experiments were performed on various targets and under various kinematical conditions. One of the main reasons for the continuous interest in the field is due to the fact that triple differential cross section (TDCS) measured in the reaction, represents the most sensitive test for the theoretical treatment of electron impact ionisation.

In order to investigate the quality of the ionisation model in a greater detail, the measured TDCS may be deconvoluted into a set of angular functions, suggested either by the model itself or derived from the symmetry properties of the experimental set-up. The model analysis is performed by comparing weights of angular functions, usually called asymmetry parameters, with the calculated ones. The advantage of symmetry based deconvolution is its universality, namely, it is justified for all possible tunings of the experimental set-up. As such it can provide a link between the different theoretical approaches which themselves are valid for a specific set of experimental conditions only.

The first general parametrisation of TDCS for the electron impact ionisation[3] was presented in 1983. In atomic physics the subject became more widely known with the work of Klar and Fehr[4]. Their parametrisation was build on the rotational invariance as an exact symmetry of the $(e, 2e)$ experiment. Indeed, since in the experiment no information is obtained on the target polarisation nor on the polarisation of incoming and outgoing electrons, TDCS must be a scalar which depends on three vectors only, $\boldsymbol{k}_0$, $\boldsymbol{k}_a$ and $\boldsymbol{k}_b$. The cross section is parametrized as a linear combination of all possible scalars formed from the three vectors

$$\sigma(\boldsymbol{k}_b, \boldsymbol{k}_a, \boldsymbol{k}_0) = \sum_{L,P,T} B_{LPT}(k_b, k_a, k_0) I_{LPT}(\hat{k}_b, \hat{k}_a, \hat{k}_0), \tag{1}$$

The angular-independent part of the cross section is represented by the parameters

B_{LPT} while the angular behaviour is determined by the set of scalar invariants

$$I_{LPT} = [\mathbf{C}_L(\hat{k}_b) \wedge \mathbf{C}_P(\hat{k}_a)]_T \cdot \mathbf{C}_T(\hat{k}_0) = \sum_{m_T} B^{LP}_{Tm_T}(\hat{k}_b, \hat{k}_a) C^*_{Tm_T}(\hat{k}_0). \tag{2}$$

The components of the spherical tensor $\mathbf{C}_T$ are denoted by C_{Tm_T}. The components of the tensor product are bipolar harmonics $B^{LP}_{Tm_T}$, in agreement with definition[5].

The six angles are needed in (1) to describe the angular dependence of the cross section in an arbitrary coordinate system. For physics, however, only the angles between three pairs of directions $\hat{k}_b, \hat{k}_a$ and $\hat{k}_0$ are relevant. To exclude the three Eulerian degrees of freedom from further treatment, a specific choice of the coordinate system is made, with respect to the directions of the three momenta. If Z-axis is chosen along $\hat{k}_0$, the scalar invariants have a particulary simple form. In case $L+P+T$ is even, they are given by

$$B^{LP}_{T0}(\hat{k}_b, \hat{k}_a) = \sum_{m=0}^{\min(L,P)} \epsilon_m (-)^{T+m} [T]^{\frac{1}{2}} \begin{pmatrix} L & P & T \\ m & -m & 0 \end{pmatrix} d^L_{m0}(\theta_b) d^P_{m0}(\theta_a) \cos m(\varphi_b - \varphi_a). \tag{3}$$

Above, ϵ_m is the Newmann factor, the 2×3 matrix embraced by round brackets is the Wigner 3-j symbol, and

$$d^j_{m0}(\theta) = (-)^m \left[\frac{(j-m)!}{(j+m)!}\right]^{1/2} P^m_j(\cos\theta)$$

is the reduced rotation matrix[5]. The same choice of the coordinate system was made by Huang[3] although later on he used the parametrisation form

$$\sum_{L,P,m} b_{LPm}(k_b, k_a, k_0) d^L_{m0}(\theta_b) d^P_{m0}(\theta_a) \cos m(\varphi_b - \varphi_a) \tag{4}$$

to deconvolute the experimental data. However, the two sets of parameters, (1) and (4), are completely equivalent. They are connected by the linear transformation

$$b_{LPm} = \epsilon_m \sum_{T=|L-P|}^{L+P} (-)^{m+T} [T]^{\frac{1}{2}} \begin{pmatrix} L & P & T \\ m & -m & 0 \end{pmatrix} B_{LPT}. \tag{5}$$

The values of parameters B_{LPT}, which determine angular distribution, depend on the physics of the $(e,2e)$ process. Due to orthogonality of bipolar harmonics, the parameter values might be obtained via integration of TDCS over all angles of outgoing electrons, but this would imply to measure the cross section over a wide range of angles. The internally normalised sets of experimental angular distributions are available almost exclusively for specific types of $(e,2e)$ kinematics, eg. the coplanar or perpendicular plane, and even then relatively large angular regions remain inaccessible for measurement due to the specific design of the experimetal set-up. Such situation makes the integration approach questionable. The other way to obtain parameter values is to fit the functional form (1) to the experimental data. When the nontrivial form of (1) is used, the set of parameters B_{LPT} is too large to be determined by the fitting procedure which is restricted to the coplanar or perpendicular plane[3]. The work of Murray et al[6] is the only case up to now, where the measured set of TDCS was large enough to permit extraction of the angular independent parameter values. Without referring to any specific theory, the parametrisation form (1) was used to deconvolute a series of fully symmetric $(e,2e)$ experiments going from the coplanar to the perpendicular plane kinematics. The authors did not give a clear physical meaning of the parameters which entered the fit. They concluded that a variety of parameters appears due to the interactions involving angular momenta of all three particles in the final state. Here we suggest to simplify the form (1) in order to be suitable for parametrisation of TDCS measured in the coplanar $(e,2e)$ kinematics. Besides the fact that a relatively smaller number of parameters is involved, the advantage of the kinematicaly restricted parametrisation would be that both, integration and fitting approach may be possible. In the following, the idea is developed further, starting with a short discussion of the general parametrisation form (1).

FROM A GENERAL TO THE COPLANAR CASE

At least for the purpose of book-keeping, a simplified view of the $(e,2e)$ reaction is used below, essentially the one on which the Distorted Wave Born Approximation (DWBA) is based[7]. In order to specify the number of active asymmetry parameters, it is assumed that the wave functions of free electrons are not coupled and can be written as

$$\chi(\boldsymbol{k},\boldsymbol{r}) = \sum_n i^n [n] \exp(i\eta_n) R_n(k,r) P_n(\hat{k}\cdot\hat{r}) \tag{6}$$

at all energies. The partial wave expansion approach (6) is not justified in general. At low energies of outgoing electrons, the three body effects become important and they can not be easily incorporated into the two-body distorting potentials. Although they are making use of (6), most of the results given below follow a more general line and are not strictly bound to the given physical interpretation.

In the two-body formalism, the estimates for the maximum values of L,P and T which are used to describe the angular distribution are *twice* the highest partial wave components $\mathcal{L}$, $\mathcal{P}$ and $\mathcal{T}$, needed to represent the wavefunction of ejected, scattered and incoming electron, respectively. In other words, inclusion of terms in (6) with angular momentum larger than $\mathcal{L}$, $\mathcal{P}$ and $\mathcal{T}$ would not alter significantly the cross section since their contribution to the already existing or newly formed parameters B_{LPT} is too small. This seems a bit misleading since at high energies more and more partial waves contribute to (1), but on the other hand, it is known[8] that relatively small number of angular functions is needed there to deconvolute TDCS. Indeed, in such a case it is advantageous to introduce the momentum transfer $\boldsymbol{q}$ and use the parametrisation form

$$\sigma(\boldsymbol{k}_b,\boldsymbol{q}) = \sum_L A_L(k_b,q) P_L(\hat{k}_b\cdot\hat{q}), \tag{7}$$

instead of (1) since the plane wave character of the incoming and scattered projectile wave allows to replace the extended summation in (1) by a simpler form (7). The way the partial wave approach is used here is most closely connected with the reality in the intermediate energy range, where the waves are distorted by the atomic (ionic) potential.

There are two general constraints on the scalar invariants in (1) to reduce their number, i.) indices have to fulfil the triangular condition, ii.) the sum $L+P+T$ must be even. The later restriction follows from the fact, that the cross section is to be invariant with respect to the simultaneous reflection of $\boldsymbol{k}_b$ and $\boldsymbol{k}_a$ on the ZX plane or alternatively, to the change of sign of $\varphi_b - \varphi_a$. This is true only if the bipolar harmonics with odd $L+P+T$ have their weights equal to zero. Further constraints may be found for the special case of symmetric $(e,2e)$ experiments[7]. If $\mathcal{T} \geq \mathcal{L}+\mathcal{P}$, the number of parameters B_{LPT} does not depend on $\mathcal{T}$ and is given by

$$N(\mathcal{L},\mathcal{P}) = \frac{1}{3}(\min+1)(2\min+1)(6\max-2\min+3), \tag{8}$$

where $\min = \min(\mathcal{L},\mathcal{P})$ and $\max = \max(\mathcal{L},\mathcal{P})$. When the energies of outgoing electrons are small, only a few partial waves are needed in (6) to calculate TDCS. Table 1 gives the total number of parameters for various pairs $(\mathcal{L},\mathcal{P})$.

Table 1. The number N of parameters B_{LPT} and the number N_p of parameters $\mathcal{A}_{m_1m_2}$ as a function of the maximum angular momenta of the outgoing electrons. The table is valid if $\mathcal{P} \geq \mathcal{L}$ and $\mathcal{T} \geq \mathcal{L}+\mathcal{P}$.

$\mathcal{L}$	$\mathcal{P}$	$N(\mathcal{L},\mathcal{P})$	$N(\mathcal{L},\mathcal{L})$	$N_p(\mathcal{L},\mathcal{L})$
0	n	$2n+1$	1	1
1	n	$2(6n+1)$	14	13
2	n	$5(6n-1)$	55	41
3	n	$28(2n-1)$	140	85

As one notes, the number of asymmetry parameters rises fast with the angular momenta of outgoing electrons. For example, to calculate the cross section at arbitrary directions of two outgoing electrons, 55 asymmetry parameters have to be calculated when only the s, p, d partial waves of outgoing electrons are involved in the reaction. Alternatively, if a more or less complete experimental angular distribution at certain electron energies is available, one would already need 55 free parameters to fit the set of angular functions (10) to the experimental data. If TDCS is measured in the coplanar kinematics only, the number of fitting parameters should be smaller, since the set of scalar invariants (10) is not linearly independent anymore. Using the notation of (8), the number of linearly independent angular functions in the coplanar case is

$$N_p(\mathcal{L},\mathcal{P}) = 8\min\max+2(\min+\max)+1, \tag{9}$$

which is always smaller than $N(\mathcal{L},\mathcal{P})$ except in the trivial case $\mathcal{L}=\mathcal{P}=0$ (Table 1). In the ZX plane the invariants can be written

$$B_{T0}^{LP}(\hat{k}_b,\hat{k}_a) \rightarrow \sum_{i,j}\alpha_{ij}(L,P,T)P_i(\cos\theta_b)P_j(\cos\theta_a)+\sum_{i,j}\beta ij(L,P,T)P_i^1(\cos\theta_b)P_j^1(\cos\theta_a),$$

if $L, P > 2$. Indices $i = L, L-2, \dots$ and $j = P, P-2, \dots$ are non-negative integers in the first, and positive integers in the second summation on the right, respectively. The coeficients α_{ij} and β_{ij} do not depend on angles. The coplanar TDCS can be therefore parametrized as

$$\sigma_p = \sum_{i=0}^{2\mathcal{L}}\sum_{j=0}^{2\mathcal{P}} A_{ij}P_i(\cos\theta_b)P_j(\cos\theta_a) + \sum_{i=1}^{2\mathcal{L}}\sum_{j=1}^{2\mathcal{P}} B_{ij}P_i^1(\cos\theta_b)P_j^1(\cos\theta_a), \tag{10}$$

The parameters A_{ij}, and by analogy also B_{ij}, can be expressed as linear combinations of the terms $\sum_T \alpha_{ij}(L,P,T)B_{LPT}$ with $L = i, i+2, \dots$ and $P = j, j+2, \dots$ where L and P are not greater than $2\mathcal{L}$ and $2\mathcal{P}$, respectively. Inversely, if $L, P > 2$, the same parameter B_{LPT} contributes to all A_{ij} and B_{ij} for which $i = L, L-2, \ldots$ and $j = P, P-2, \ldots$. It can be shown that not even one parameter B_{LPT} can be determined from the coplanar measurements when $\mathcal{L}, \mathcal{P} > 1$. From the fit one can only obtain the $N_p(\mathcal{L}, \mathcal{P})$ constraints for the $N(\mathcal{L}, \mathcal{P})$ parameters B_{LPT}. The situation here is different from the case of the complete double differential cross section (DDCS) measurement, where a specific subset of parameters B_{LPT} *can* be obtained from the fitting procedure[4]. Due to this fact and because of the lack of complete TDCS measurements it is worth to study the parametrisation in a plane alone. One possible way of doing it is (10), but other functional forms may be used as well. We can start from the begining as in the three-dimensional case (2) to write down the scalar invariants constructed from the three directions in the two-dimensional space. Selecting X-axis along $\hat{k}_0$, the following parametrisation form is obtained

$$\sigma_p = \sum_{m_1=1}^{2\mathcal{L}} \sum_{m_2=-2\mathcal{P}}^{2\mathcal{P}} \mathcal{A}_{m_1 m_2} \cos(m_1 \varphi_b + m_2 \varphi_a) + \sum_{m_2=0}^{2\mathcal{P}} \mathcal{A}_{0 m_2} \cos(m_2 \varphi_a). \tag{11}$$

For a given pair $(\mathcal{L}, \mathcal{P})$ the number of angular independent parameters $\mathcal{A}_{m_1 m_2}$ is exactly the same as in (10). It can be shown easily that (11) is obtained also directly from the three-dimensional parametrisation (2) when the X-axis points into the projectile direction and the Z-axis is perpendicular to the plane, specified by $\hat{k}_0$ and $\hat{k}_a$. The coplanar parameters (11) are given in terms of general parameters B_{LPT} as

$$\mathcal{A}_{m_1 m_2} = \kappa_{m_1 m_2} \sum_{L=m_1}^{2\mathcal{L}} \sum_{P=|m_2|}^{2\mathcal{P}} \sum_{T} (-)^T [T]^{\frac{1}{2}} \begin{pmatrix} L & P & T \\ m_1 & m_2 & m_0 \end{pmatrix} d^L_{m_1 0}(0) d^P_{|m_2| 0}(0) d^T_{|m_0| 0}(0) B_{LPT}, \tag{12}$$

where $\kappa_{m_1 m_2} = 2\mu_{m_1}\mu_{m_2}\mu_{m_0}$ for $m_1, m_2 \neq 0$ and $\kappa_{00} = 1$. Also, $\mu_m = (-)^m$ if $m > 1$ and $\mu_m = 1$ if $m \leq 0$.

In conclusion, the parametrisation of TDCS in a plane was discussed and appropriate fitting forms were presented. The constraints for the parameters B_{LPT} were given which can be determined from the coplanar $(e, 2e)$ experiments. Although the two-body partial wave model of the reaction was used to count the parameters, the main results are valid generally.

References:

1. L. Amaldi, A. Egidi, R. Marconero and Pizzela, Use of a two channeltron coincidence in a new line of research in atomic physics, *Rev.Sci.Instrum.* 40:1001(1969).
2. H. Ehrhardt, M. Shultz, T. Tekaat and K Willmann, Ionisation of helium: angular correlation of the scattered and ejected electrons, *Phys.Rev.Lett.* 48:1807(1969).
3. K.N. Huang, Theoretical triply differential cross section of electron-impact ionisation of atoms, *Phys.Rev.* A28:1869(1983).
4. H. Klar and M. Fehr, Parametrization of multiply differential cross sections, *Z.Phys.* D23:295(1992).
5. D.M. Brink and G.R. Satchler. *Angular Momentum*, Clarendon Press, Oxford (1968).
6. A.J. Murray, F.H. Read and N.J. Bowring, Decomposition of experimentaly determined atomic $(e, 2e)$ ionisaton measurements, *Phys.Rev.* A49:R3162(1994).
7. D.H. Madison, R.V. Calhoun and W.N. Shelton, Triple differential cross sections for electron-impact ionisation of helium, *Phys.Rev.* A16:552(1977).
8. M. Žitnik, G.Stefani, L. Avaldi and R. Camilloni, TDCS for the electron impact ionisation approaching dipolar limit, *J.Phys.* B26:L851(1993).

AN EXPLANATION OF THE STRUCTURE OBSERVED IN OUT-OF-PLANE SYMMETRIC MEASUREMENTS ON HELIUM

J. Rasch,[1] Colm T. Whelan[1] R. J. Allan,[2] H. R. J. Walters[3]

[1]Department of Applied Mathematics and Theoretical Physics
University of Cambridge, Silver Street, Cambridge, CB3 9EW, England
[2]Daresbury Laboratory, Warrington, WA4 4AD, England
[3]Department of Applied Mathematics and Theoretical Physics
The Queen's University of Belfast, Belfast, BT7 1NN, England

INTRODUCTION

The study of (e,2e) processes on Helium at impact energies of 100eV and below has yielded a wealth of experimental data and has contributed greatly to the understanding of the Coulomb few body problem. The great advantage of coincidence measurements is that one can manipulate the geometry of the experiment to study delicate processes which yield striking structures in a particular arrangement, but are largely masked by stronger effects in virtually all other setups. Quite recently Murray[1] identified a very strong structure in the ionization of Helium at an impact energy of 64.6eV in a very specific out of plane geometry. In this paper we will show that this structure is well represented in a Distorted Wave Born Approximation (DWBA) calculation and we will interpret it as an interference, that is to say a pure quantum, effect. We predict that similar structures will be present in the Triple Differential Cross Section (TDCS) for some targets but not for others.

THE IONIZATION OF HELIUM

It will be helpful to fix our notation and to define the geometries used. In Figure 1 we show the general set up. We assume that an electron, wave vectors $\boldsymbol{k}_0$, is incident on a target and two electrons $\boldsymbol{k}_1$ and $\boldsymbol{k}_2$ are collected in coincidence. Let ψ define the angle of the incident electron with respect to the plane defined by $\boldsymbol{k}_1$ and $\boldsymbol{k}_2$ and ϑ_{12} be the angle between the electrons. We take our origin at the target and our z-axis to lie in the direction of the projection of the incident vector on the detectors plane. The two exiting electrons will be detected at angles ϑ_1 and ϑ_2 right and left of this z-axis, i.e. $\vartheta_1 + \vartheta_2 = \vartheta_{12}$. If $\psi = 0°$ we will describe the arrangement as being coplanar, $\psi = 90°$, perpendicular, and out of plane otherwise. We describe it as a symmetric measurement if $\vartheta_1 = \vartheta_2 = \vartheta_{12}/2 =: \vartheta$ and $|\boldsymbol{k}_1| = |\boldsymbol{k}_2|$.

Murray et. al.[2] and Murray[1] reported results on Helium in symmetric perpendicular plane geometry at impact energies ranging from 44.6 to 104.6eV. Excellent agreement with

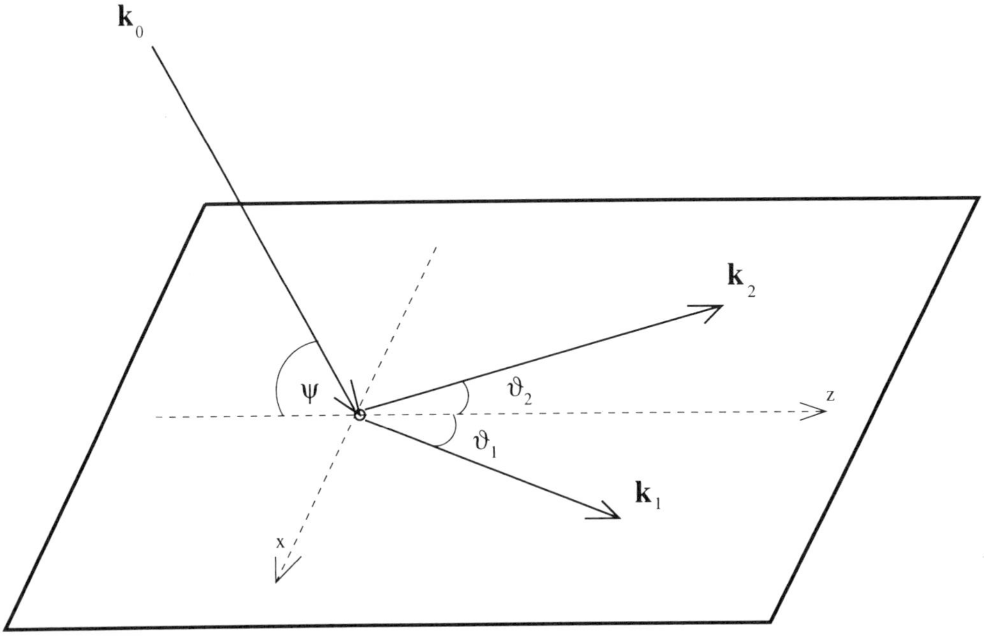

Figure 1. Shown is the general setup for an (e,2e) experiment. $\boldsymbol{k}_0$, $\boldsymbol{k}_1$ and $\boldsymbol{k}_2$ denote the incident and two outgoing electrons. ψ denotes the angle by which the incident electron gun is raised out of plane.

experiment was achieved by Whelan et. al.[3] who used a distorted wave Born approximation. The TDCS is given for an arbitrary atomic target by

$$\frac{d^3\sigma^{\mathrm{DWBA}}}{d\Omega_1 d\Omega_2 d\epsilon} = 2(2\pi)^4 \frac{k_1 k_2}{k_0} \left(|f|^2 + |g|^2 - Re(f^* g)\right) \tag{1}$$

where

$$f = \left\langle \chi^-(\boldsymbol{k}_1 \boldsymbol{r}_1)\chi^-(\boldsymbol{k}_2, \boldsymbol{r}_2) \left| \frac{1}{|\boldsymbol{r}_1 - \boldsymbol{r}_2|} \right| \chi^+(\boldsymbol{k}_0, \boldsymbol{r}_1)\phi(\boldsymbol{r}_2) \right\rangle \tag{2}$$

$$g = \left\langle \chi^-(\boldsymbol{k}_1 \boldsymbol{r}_2)\chi^-(\boldsymbol{k}_2, \boldsymbol{r}_1) \left| \frac{1}{|\boldsymbol{r}_1 - \boldsymbol{r}_2|} \right| \chi^+(\boldsymbol{k}_0, \boldsymbol{r}_1)\phi(\boldsymbol{r}_2) \right\rangle \tag{3}$$

The target wave function $\phi(\boldsymbol{r}_2)$ was taken from Clementi and Roetti[4]. The distorted wave χ^+ was generated in the static-exchange potential of the atom, the distorted waves χ^- were each generated in the static exchange potential of the ion and each orthogonalised to ϕ. Exchange was included using the spin singlet local exchange approximation of Furness and McCarthy[5] for the final states and the closed shell equivalent for the incident electron.

At energies above 100eV the DWBA as defined in (1) also gave excellent agreement in coplanar symmetric geometry[6,7,8], however for energies of 50eV and below agreement was very poor. This disagreement prompted Whelan et. al.[9] to suggest that it would be valuable to carry out a series of measurements relating the coplanar symmetric geometry of Rösel et. al.[8], with the perpendicular plane geometry of Murray et. al.[2]. This Murray[1] did and while rotating the plane came across a deep minimum in the TDCS for an impact energy of 64.6eV and $\psi = 67.5°$ which is the subject of this paper. A full discussion of the TDCS for all gun angles ψ will be given elsewhere[10]. Here we only remark that there is quite some evidence that the shape of the TDCS in the coplanar case is the result of a delicate interplay between 3 body effects in the incident and final channels[3,11,12,13,14] while in the perpendicular plane the

3 body effects tend to enhance existing structures and not to interfere. For the spherical case we are interested in here, the deep minimum observed has a different origin.

In order to see how such a structure can arise we made a number of calculations. Firstly we performed a DWBA calculation, as defined by equations (1)-(3), then we considered a number of model calculations i.e.:

- a plane wave in the incident channel but distorted waves for the two outgoing electrons;
- plane waves for the outgoing but a distorted wave for the incident electron.

The results of these two model calculations are shown in the figures for all atoms considered here. A sharp dip is only found when we have distorted waves in both the incident and final channels. The dip is thus clearly an interference effect and only exists when we allow for elastic scattering from the atom in the incident channel and from the ion in the final channel. It should be emphasised that we have not allowed for either the polarisation of the target or post collisional interaction in the final. We may to some extent allow for the latter by the inclusion of a multiplicative factor in the TDCS, i.e.

$$\frac{d^3\sigma}{d\Omega_2 d\Omega_2 dE} = M_{ee} \frac{d^3\sigma^{\mathrm{DWBA}}}{d\Omega_1 d\Omega_2 dE} \tag{4}$$

where $\frac{d^3\sigma^{\mathrm{DWBA}}}{d\Omega_1 d\Omega_2 dE}$ is given by (1) and the M_{ee} factor is that introduced by Ward and Macek[15], i.e.

$$M_{ee} = N_{ee} |{}_1F_1(-i\nu_3, 1, -2k_3 r_{3\mathrm{ave}})|^2 \tag{5}$$

where

$$N_{ee} = \frac{\gamma}{e^{\gamma} - 1} \qquad \text{with} \qquad \gamma = \frac{2\pi}{|\boldsymbol{k}_1 - \boldsymbol{k}_2|} \tag{6}$$

and

$$\nu_3 = -\frac{1}{|\boldsymbol{k}_1 - \boldsymbol{k}_2|}. \tag{7}$$

The parameter $r_{3\mathrm{ave}}$ is given by

$$r_{3\mathrm{ave}} = \frac{\pi^2}{16\epsilon} \left(1 + \frac{0.627}{\pi} \sqrt{\epsilon} \ln \epsilon \right)^2 \tag{8}$$

where ϵ is the total energy of the two exiting electrons.

The Gamov factor N_{ee}, defined in (6) above[16,3,11,14], was used to represent the final state electron-electron interaction. The Gamov factor was first used in the theory of alpha decay and is intimately related to the normalisation of the Coulomb function, it plays an interesting role in the BBK approximation of Brauner et. al.[17] where it corresponds to the Coulomb normalisation of the third wave function, i.e. the electron-electron wave function, in the zero energy limit. The problem with it is that it does not represent well the absolute size of the cross-section at very low energies. It predicts a yield which goes to zero exponentially as the impact energy approaches threshold in direct contradiction to the power law predicted by Wannier[18]. N_{ee} does do very well at predicting the shape of the TDCS over a wide range of geometries and kinematics[3,11,13,14]. The M_{ee} factor is chosen to give the Wannier behaviour near threshold. There is little difference between the shapes of the TDCS with the M_{ee} or N_{ee} factors, but we use the former rather than the latter since it gives a better normalisation.

In Figure 2 we compare calculations of DWBA + M_{ee} and the experimental data of Murray, together with the calculation of Berakdar and Briggs[19], which we will discuss later. We note that the DWBA, DWBA + M_{ee} and the Berakdar and Briggs[19] calculations all agree with respect to absolute size. Murray[1] normalised his data with respect to the absolute coplanar value of Gélébart and Tweed[20]. This value was later scaled down by the Brest group[21] and the latest results from the Brest-Kaiserslautern collaboration have suggested that it may be necessary to reduce it even further (Röder 1996 - personal communication). The experimental determination of absolute cross-sections is an extremely delicate and difficult task, and very slight differences in the setup can produce very large errors [22,23]. We have used a normalisation of the Murray data consistent with the new preliminary results of the Brest-Kaiserslautern experiments and which gives the best agreement with DWBA + M_{ee} over all geometries, i.e. not just the present out of plane case. We remark that many theories can predict reasonable shapes for the TDCS in a range of geometries (Brauner et. al.[17], Berakdar and Briggs[19], 19, our own work) but differ wildly in absolute size. It is now the case, in our view, that only absolute cross-sections can discriminate between the models especially when final state Coulomb interactions are important. We strongly encourage our experimental colleagues to focus their attention on this very difficult problem.

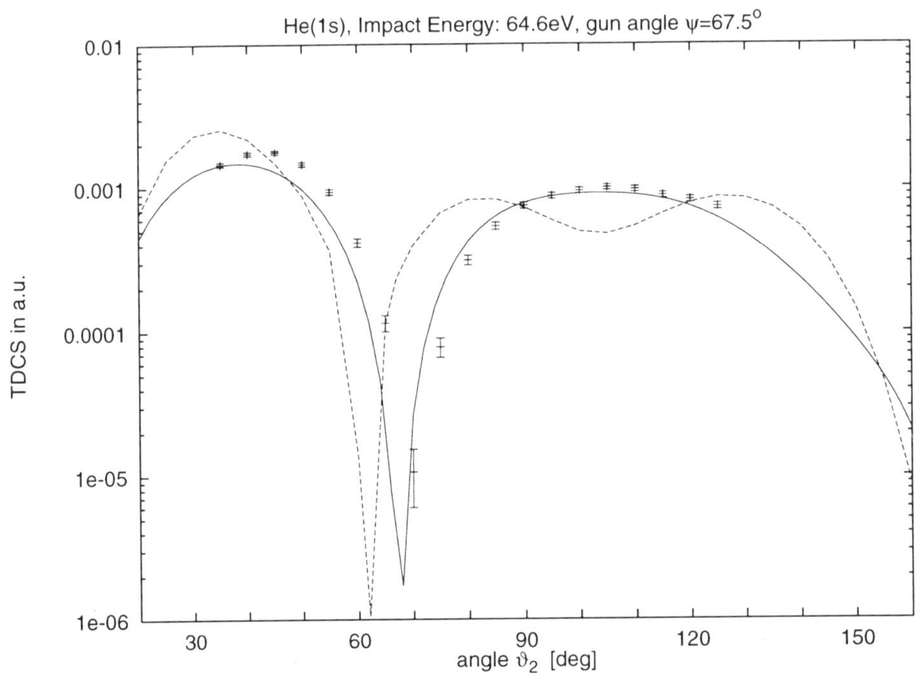

Figure 2. TDCS for Helium in symmetric equal energy sharing geometry an an impact energy of 64.6eV. Electron gun raised to $\psi = 67.5°$. Shown are pure DWBA using singlet potential (solid curve) compared with BBK results (dashed-dotted).

The correlated final state approach of Berakdar and Briggs[19] starts from the TDCS given by

$$\frac{d^3\sigma}{d\Omega_1 d\Omega_2 dE} = (2\pi)^4 \frac{k_1 k_2}{k_0} \left\langle \Psi_f^- \left| V_i \right| \Phi_i \right\rangle \tag{9}$$

where the initial asymptotic wave function is given by

$$\Phi_i(\boldsymbol{r}_1, \boldsymbol{r}_2, \boldsymbol{r}_3) = \frac{1}{(2\pi)^{3/2}} e^{i\boldsymbol{k}_0 \cdot \boldsymbol{r}_1} \phi(\boldsymbol{r}_2, \boldsymbol{r}_3) \tag{10}$$

and $\phi(\boldsymbol{r}_2, \boldsymbol{r}_3)$ represents the He(1s) bound state (see Berakdar and Briggs[19] for more details).

Berakdar and Briggs[19] then approximate the exact wave function Ψ_f^- by

$$\Psi_f^-(\boldsymbol{r}_1, \boldsymbol{r}_2, \boldsymbol{r}_3) \approx \frac{1}{(2\pi)^3} e^{i\boldsymbol{k}_1 \cdot \boldsymbol{r}_1} e^{i\boldsymbol{k}_2 \cdot \boldsymbol{r}_2} \phi^{+1}(\boldsymbol{r}_3) \times \prod_j N_j \, {}_1F_1 \left[i\alpha_j(\boldsymbol{k}_1, \boldsymbol{k}_2, \boldsymbol{k}_{12}), 1, -i(k_j r_j + \boldsymbol{k}_j \cdot \boldsymbol{r}_j) \right] \tag{11}$$

where $N_j = e^{\alpha_j \pi/2} \Gamma(1 + i\alpha_j)$. The momentum $\boldsymbol{k}_{12}$ is defined as $\boldsymbol{k}_{12} = (\boldsymbol{k}_1 - \boldsymbol{k}_2)/2$ and its conjugate co-ordinate as $\boldsymbol{r}_{12} = (\boldsymbol{r}_1 - \boldsymbol{r}_2)/2$. $\phi^{+1}(\Omega_3)$ is the hydrogenic wavefunction with charge $Z = 2$. The parameters α_j are Sommerfeld parameters[19].

The initial channel interaction potential in (9) is taken to be

$$V_i = \frac{1}{|\boldsymbol{r}_1 - \boldsymbol{r}_2|} - \frac{2}{\boldsymbol{r}_1} + \frac{1}{|\boldsymbol{r}_1 - \boldsymbol{r}_3|} \tag{12}$$

Due to the fact that this potential consists of 3 terms one has to add 3 complex amplitudes which need to cancel to produce the dip. Since the third term represents interaction with the remaining target "spectator" electron, Berakdar and Briggs[19] argued that such a dip could not be observed in a hydrogen target. In Figure 2 we compare the 19 calculation with the DWBA + M_{ee} calculation. We see that the DWBA calculation is in better agreement with the experiments in that the Berakdar and Briggs[19] calculation has the location of the dip shifted somewhat to the left of the experiment and predicts an additional structure at around 110° which is neither in the DWBA calculations nor the measurement.

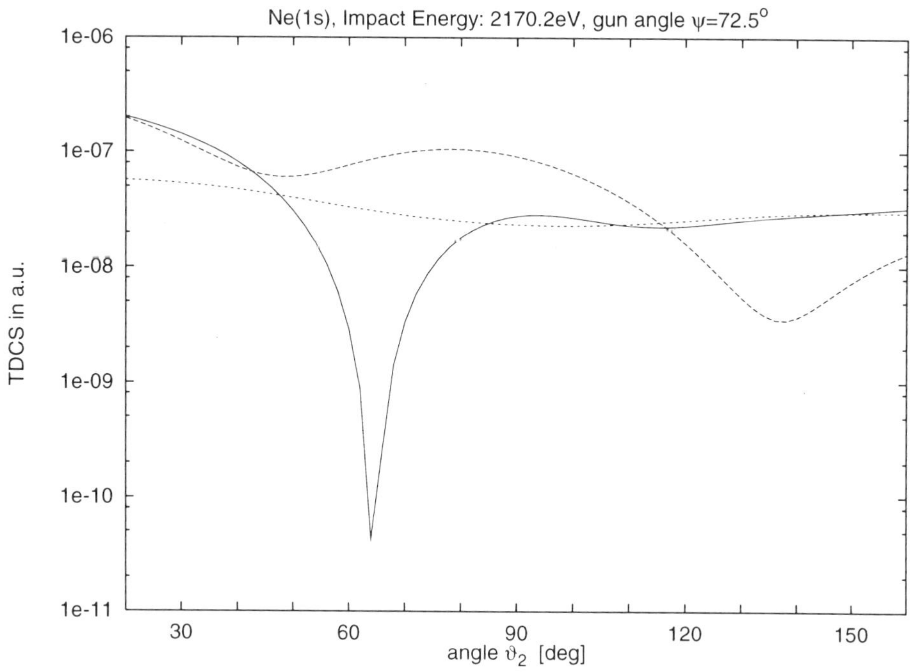

Figure 3. TDCS for Neon(1s) in symmetric equal energy sharing geometry at an impact energy of 2170.2eV. Electron gun raised to $\psi = 72.5°$. Shown are pure DWBA (solid curve), DWBA with incoming plane wave (long dashed), DWBA with outgoing plane waves (short dashed).

It is of interest to see if such sharp interference effects are to be observed in other systems. We have performed calculations on Ne(1s) at an impact energy of 2170.2eV with

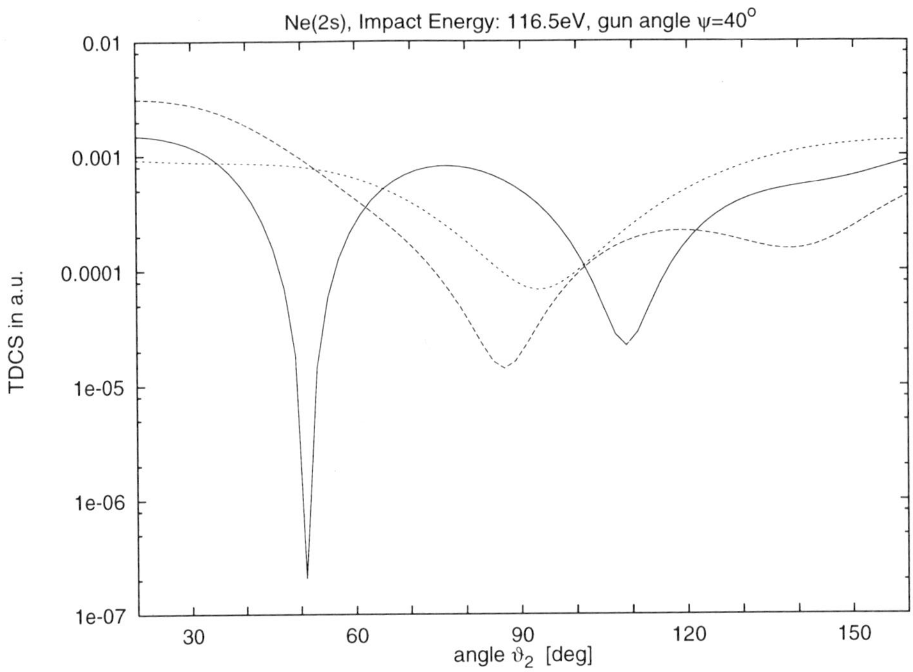

Figure 4. TDCS for Neon(2s) in symmetric equal energy sharing geometry at an impact energy of 116.5eV. Electron gun raised to $\psi = 40°$. Shown are pure DWBA (solid curve), DWBA with incoming plane wave (long dashed), DWBA with outgoing plane waves (short dashed).

$\psi = 72.5°$, Ne(2s) at an impact energy of 116.5eV with $\psi = 40°$ and on Ar(2s) at an impact energy of 826.3eV with $\psi = 50°$ and predict that very strong interference structures should be observable (Figures 3-5). Switching off any of the distorting potentials in either the initial or final channels causes the dip to disappear. There are a number of similarities between the atoms. This applies to the ratio of the impact to bound state energy, gun angle ψ and angle θ_2 at which the dips occur (see Table 1).

Table 1. Similarities in the occurrence of the interference structure in different closed shell atoms. First column: ratio of the impact (E_0) to bound state (E_b) energy, second column: gun-angle ψ, third column: angle at which the minimum of the dip occurs.

	$\frac{E_0}{E_b}$	ψ	ϑ_2
He(1s)	2.63	67.5°	68.0°
Ne(1s)	2.49	72.5°	64°
Ne(2s)	2.40	40°	51.0°
Ar(2s)	2.53	50°	58.0°

It is interesting to see if similar interference structures can be observed in open shell systems. Mathematically we are merely dealing with a 6-dimensional integral over a highly

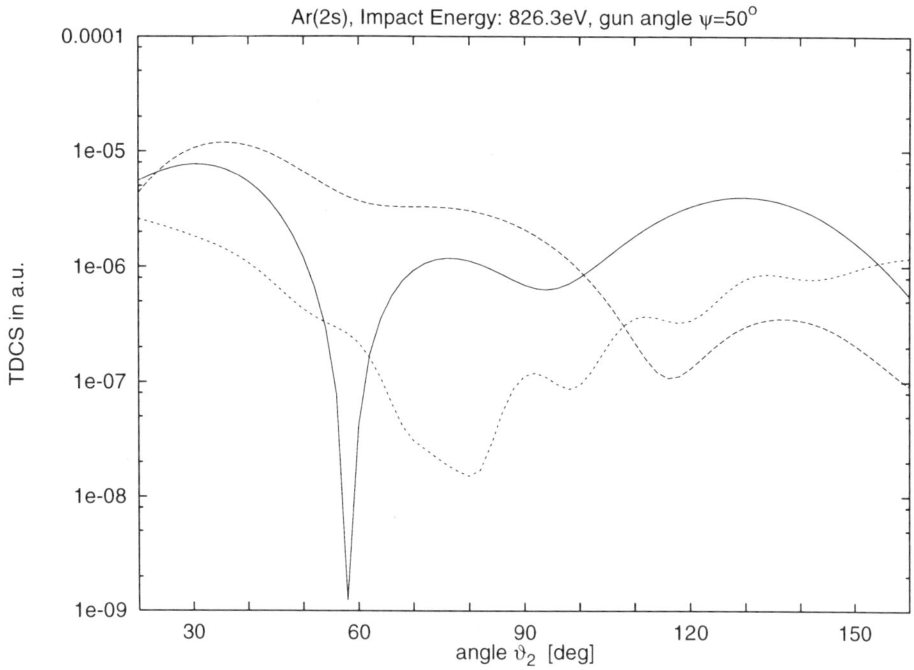

Figure 5. TDCS for Argon(2s) in symmetric equal energy sharing geometry at an impact energy of 826.3eV. Electron gun raised to $\psi = 50°$. Shown are pure DWBA (solid curve), DWBA with incoming plane wave (long dashed), DWBA with outgoing plane waves (short dashed).

oscillatory argument and as such destructive interference effects may yield very small values for certain cases. Indeed a pure DWBA calculation for Hydrogen reveals a sharp dip in symmetric coplanar geometry at 29eV. However, as we mentioned earlier, polarisation effects are very important in the coplanar case and if we include these in our calculation the dip disappears, in contrast to the out-of-plane He case discussed here where the sharp dip persists in the presence of the very strong polarisation potential of Whelan et. al.[3] (for more details see Rasch[23]). For Hydrogen we found no evidence of a dip out of plane, nor could we find such a structure in He^+.

CONCLUSION

We have demonstrated that the sharp dip observed by Murray[1] in symmetric out of plane geometry can be well represented in a distorted wave Born approximation. The dip exists even in the simplest calculation of this type where neither polarisation nor post-collisional electron-electron interaction is included. It is necessary to allow for elastic scattering of the incident electron from the atom and elastic scattering of the exiting electrons from the ion if this structure is to manifest itself. The DWBA calculations automatically include the effect of the spectator electron as well as the nucleus on these channels and it is impractical to try to separate them. Results of simpler calculations seem to indicate that both play a part and this is partially confirmed by the non-existence of such a dip in hydrogenic systems.

We have predicted where similar structures are likely to be found for other atoms. Interference effects of this kind can only be sensibly interpreted in quantum mechanical terms and arise when we integrate over wave functions and as such are important to identify and study.

ACKNOWLEDGEMENTS

Thanks to Jochen Röder and Amedé Pochat for communicating information of their absolute measurements prior to publication. We would also like to thank Joe Macek for discussion concerning the M_{ee} factor. We gratefully acknowledge financial support for this work from the British Council and DAAD (ARC), NATO through grant CRG 950407 and the European Union (HCM CHRX-CT93-0350). J.R. holds a Research Fellowship at Wolfson College, Cambridge.

REFERENCES

1. A. J. Murray, Coupling the coplanar (e, 2e) to the perpendicular plane geometry, in: *(e,2e) and Related Processes*, Colm T. Whelan, H. R. J. Walters, A. Lahmam-Bennani, and H. Ehrhardt, eds., Kluwer Academic Publisher, (1993).
2. A. J. Murray, M. B. J. Woolf, and F. H. Read, Results from symmetric and non-symmetric energy sharing (e,2e) experiments in the perpendicular plane. *J. Phys. B*, 25:3021, (1992).
3. Colm T. Whelan, R. J. Allan, H. R. J. Walters, and X. Zhang, (e,2e), effective charges, distorted waves and all that!, in: *(e,2e) and Related Processes*, Colm T. Whelan, H. R. J. Walters, A. Lahmam-Bennani, and H. Ehrhardt, eds., Kluwer Academic Publisher, (1993).
4. Clementi and Roetti, Roothan-hartree-fock atomic wavefunctions. *Atomic Data and Nuclear Data Tables*, 14:177, (1974).
5. J.B. Furness and I.E McCarthy, Semiphenomenological optical model for electron scattering on atoms. *J. Phys. B*, 6:2280, (1973).
6. X. Zhang, Colm T. Whelan, and H. R. J. Walters, (e,2e) cross sections for ionization of helium in coplanar symmetric geometry. *J. Phys. B*, 23:L509, (1990).
7. L. Frost, P. Freinstein, and M. Wagner, 200eV coplanar symmetric (e,2e) on helium: a sensitive test of reaction models. *J. Phys. B*, 23:L715, (1990).
8. T. Rösel, C. Dupré, J. Röder, A. Duguet, K. Jung, A. Lahmam-Bennani, and H. Ehrhardt, Coplanar symmetric (e,2e) cross sections on helium and neon. *J. Phys. B*, 23:3059, (1991).
9. Colm T. Whelan, X. Zhang, and H. R. J. Walters. in: *Proceedings of the European Conference on (e,2e) and related processes*, G. Stefani, ed., CNR Rome, (1990).
10. J. Rasch, Colm T. Whelan, and H. R. J. Walters. . *in preparation*, (1996).
11. H. R. J. Walters, X. Zhang, and Colm T. Whelan. Directions in (e,2e) and related processes, in: *(e,2e) and Related Processes*, Colm T. Whelan, H. R. J. Walters, A. Lahmam-Bennani, and H. Ehrhardt, eds., Kluwer Academic Publisher, (1993).
12. Colm T. Whelan, R. J. Allan, J. Rasch, H. R. J. Walters, X. Zhang, J. Röder, K. Jung, and H. Ehrhardt, Coulomb three-body effects in (e,2e) collisions: The ionization of H in coplanar symmetric geometry. *Phys. Rev. A*, 50:4394, (1994).
13. J. Röder, J. Rasch, K. Jung, Colm T. Whelan, H. Ehrhardt, R. J. Allan, and H. R. J. Walters, Coulomb three-body effects in low energy impact ionisation of H(1s). *Phys. Rev. A*, 53:225, (1996).
14. F. Rouet, R. J. Tweed, and J. Langlois, The effect of target atom polarisation and wavefunction distortion in (e,2e) ionisation of Hydrogen. *J. Phys. B*, 29:1767, (1996).
15. S.J. Ward and J.H. Macek, Wave functions for contiuum states of charged fragments. *Phys. Rev. A*, 49, (1994).
16. J. Botero and J.H. Macek, Threshold angular distributions of (e,2e) cross sections of helium atoms. *Phys. Rev. Lett.*, 68:576–579, (1992).
17. M. Brauner, J.S. Briggs, and H. Klar, Triply-differential cross sections for ionisation of hydrogen atoms by electrons and positrons. *J. Phys. B*, 22:2265–87, (1989).
18. G.H. Wannier, The Threshold Law for Single Ionization of Atoms or Ions by Electrons. *Phys. Rev.*, 90:817–825, (1953).
19. J. Berakdar and J. S. Briggs, Interference effects in (e,2e)-differential cross sections in doubly symmetric geometry. *J. Phys. B*, 27:4271, (1994).
20. F. Gélébart and R. J. Tweed, Absolute cross sections for 100eV electron impact ionisation of helium in a symmetric energy-sharing geometry. *J. Phys. B*, 23:L641, (1990).

21. A. Pochat, X. Zhang, C. T. Whelan, H. R. J. Walters, R. J. Tweed, F. Gélébart, M. Cherid, and R. J. Allan, Absolute low-energy experimental cross sections for (e,2e) processes on helium. *Phys. Rev. A*, 47:3483, (1993).
22. J. Rasch, M. Zitnik, L. Avaldi, Colm T. Whelan, G. Stefani, R. Camilloni, R. J. Allan, and H. R. J. Walters, A theoretical/experimental investigation of the triple differential cross-sections for electron impact ionisation of Kr(4p) and Xe(5p) at 1keV impact energy. *submitted to Phys. Rev. A*, (1996).
23. J. Rasch. *(e,2e) processes with neutral atom targets*. PhD thesis, University of Cambridge, (1996).

ON THE ADDITION THEOREM FOR JACOBI POLYNOMIALS

J. Rasch

Department of Applied Mathematics and Theoretical Physics
University of Cambridge, Silver Street, Cambridge, CB3 9EW, England

ADDITION THEOREM FOR JACOBI POLYNOMIALS

Jacobi polynomials $P_n^{(\alpha,\beta)}(x)$ form a larger class of orthogonal polynomials than Gegenbauer polynomials $C_n^\lambda(x)$, Chebychev polynomials $T_n(x)$ and Legendre polynomials. They can all be expressed in terms of Jacobi polynomials (see for example equations (8.962.2) and (8.962.3) in Gradshteyn and Ryzhik[1])

$$C_n^{\alpha+\frac{1}{2}}(x) = \frac{\Gamma(n+2\alpha+1)\Gamma(\alpha+1)}{\Gamma(n+\alpha+1)\Gamma(2\alpha+1)} P_n^{(\alpha,\alpha)}(x) \tag{1}$$

$$T_n(x) = \frac{2^{2n}(n!)^2}{(2n)!} P_n^{(-\frac{1}{2},-\frac{1}{2})}(x) \tag{2}$$

$$P_n(x) = P_n^{(0,0)}(x) \tag{3}$$

Addition theorems for Legendre and Gegenbauer polynomials have been known for a long time. This does not seem to be true for Jacobi polynomials. In the beginning of the 70's Koornwinder found an appropriate generalisation for Jacobi polynomials. In order to prove his theorem he had to use group theoretical methods and could prove it only for integer and later on half integer values of α and β [2,3]. Recently Antone[4] proposed an addition theorem which holds for a much wider of values for α and β, namely for any value greater than -1. He achieved this by not resorting to group theoretical methods but finding an analytical proof. The theorem he finds is a bit complicated and also does not exhibit the symmetry between the two angles θ_1 and θ_2.

Here we will give a general addition theorem which holds for any value of α and β and also exhibits clearly the symmetry with respect to the angles. We do this by simply expanding the Jacobi polynomials in terms of Gegenbauer polynomials and applying the well known addition theorem for Gegenbauer polynomials.

Using the techniques described in the book by Rainville[5] it is possible to express every Jacobi polynomial in terms of Gegenbauer polynomials using the relation

$$P_k^{(\alpha,\alpha)}(x) = \frac{(1+\alpha)_k}{(1+2\alpha)_k} C_k^{\alpha+\frac{1}{2}}(x) \tag{4}$$

Coincidence Studies of Electron and Photon Impact Ionization
Edited by Whelan and Walters, Plenum Press, New York, 1997

and the expansion

$$P_n^{(\alpha,\beta)}(x) = \frac{(1+\alpha)_n}{(1+\alpha+\beta)_n} \sum_{k=0}^{n} A(n,k) P_k^{(\alpha,\alpha)}(x) \tag{5}$$

where

$$A(n,k) := \frac{(-)^{n-k}(\beta-\alpha)_{n-k}(1+\alpha+\beta)_{n+k}(1+2\alpha)_k(1+2\alpha+2k)}{(n-k)!(1+2\alpha)_{n+k+1}(1+\alpha)_k} \tag{6}$$

It is also possible to express Gegenbauer polynomials in terms of Jacobi polynomials (see Rasch[6])

$$P_n^{(\alpha,\alpha)}(x) = \frac{(1+\alpha)_n}{(1+2\alpha)_n} \sum_{k=0}^{n} B(n,k) P_k^{(\alpha,\beta)}(x) \tag{7}$$

where

$$B(n,k) := \frac{(-)^{n-k}(\alpha-\beta)_{n-k}(1+2\alpha)_{n+k}(1+\alpha+\beta)_k(1+\alpha+\beta+2k)}{(n-k)!(1+\alpha+\beta)_{n+k+1}(1+\alpha)_k} \tag{8}$$

The well known addition theorem for Gegenbauer polynomials is of the form (see the addition theorem page 223 in Magnus et. al.[7])

$$C_k^\lambda(cos(\gamma)) = \sum_{m=0}^{k} \frac{4^m(2\lambda+2m-1)(k-m)!((\lambda)_m)^2}{(2\lambda-1)_{k+m+1}}$$
$$\sin^m(\theta_1) C_{k-m}^{\lambda+m}(\cos(\theta_1)) \sin^m(\theta_2) C_{k-m}^{\lambda+m}(\cos(\theta_2)) C_m^{\lambda-\frac{1}{2}}(\cos(\phi)) \tag{9}$$

We would like to remark here that although the definition of the Gegenbauer polynomials in Erdelyi et. al.[8,9] agree with the definition in Magnus et. al.[7] none of the two stated addition theorems in Erdelyi et. al. (chapter 3.15.1 equation (19) and chapter 10.9 equation (34)) are actually correct they even contradict each other.

Inserting now the addition theorem (9) in (5) we get

$$P_n^{(\alpha,\beta)}(\cos(\gamma)) = \frac{(1+\alpha)_n}{(1+\alpha+\beta)_n} \sum_{k=0}^{n} A(n,k) \frac{(1+\alpha)_k}{(1+2\alpha)_k} \sum_{m=0}^{k} E(k,m)$$
$$\sin^m(\vartheta_1) C_{k-m}^{\alpha+\frac{1}{2}+m}(\cos(\vartheta_1)) \sin^m(\vartheta_2) C_{k-m}^{\alpha+\frac{1}{2}+m}(\cos\vartheta_2) C_m^{\alpha}(\cos\varphi) \tag{10}$$

where we have adopted the definition

$$E(k,m) := 4^m(2\alpha+2m)(k-m)! \frac{\left(\frac{1}{2}+\alpha\right)_m^2}{(2\alpha)_{k+m+1}} \tag{11}$$

Making now use of the back transformation (7) we finally get for the addition theorem

$$\begin{aligned} &P_n^{(\alpha,\beta)}(\cos(\gamma)) \\ &= \frac{(1+\alpha)_n}{(1+\alpha+\beta)_n} \sum_{k=0}^{n} A(n,k) \frac{(1+\alpha)_k}{(1+2\alpha)_k} \sum_{m=0}^{k} E(k,m) \\ &\times \sin^m(\vartheta_1) \frac{(1+2\alpha+2m)_k(1+\alpha)_k}{(1+\alpha+m)_k(1+2\alpha+k)_m(1+2\alpha)_k} \\ &\times \sum_{l=m}^{k} B(k,l)(l+\alpha+\beta+1)_m P_{l-m}^{(\alpha+m,\beta+m)}(\cos(\vartheta_1)) \end{aligned}$$

$$\times \sin^m(\vartheta_2)\frac{(1+2\alpha+2m)_k(1+\alpha)_k}{(1+\alpha+m)_k(1+2\alpha+k)_m(1+2\alpha)_k}$$
$$\times \sum_{l'=m}^{k} B(k,l')(l'+\alpha+\beta+1)_m P_{l'-m}^{(\alpha+m,\beta+m)}(\cos(\vartheta_2))$$
$$\times \frac{\left(\frac{1}{2}+\alpha\right)_m\left(1+\alpha-\frac{1}{2}\right)_m}{(2\alpha)_m(1+2\alpha-1)_m}\sum_{p=0}^{m} B(m,p)P_p^{(\alpha-\frac{1}{2},\beta-\frac{1}{2})}(\cos(\varphi)) \quad (12)$$

SUMMATION RULE

Comparing Antone's[4] addition theorem with equation (10) we can furthermore obtain a summation rule. We get it by comparing the coefficients of the 0th order Jacobi polynomial in $\cos(\phi)$. Antone obtains (see equation (33) in Antone[4])

$$P_n^{(\alpha,\beta)}(\cos(\gamma)) = \frac{(-)^n\Gamma(2n+\alpha+\beta+1)}{2^n n!\Gamma(n+\alpha+\beta+1)}(\cos(\vartheta_1)\cos(\vartheta_2))^n \quad (13)$$

whilst we get from equation (10)

$$\begin{aligned}
&P_n^{(\alpha,\beta)}(\cos\gamma) \\
&= \frac{(1+\alpha)_n}{(1+\alpha+\beta)_n}\sum_{k=0}^{n} A(n,k)\frac{(1+\alpha)_k}{(1+2\alpha)_k}\sum_{m=0}^{k} E(k,0)C_k^{\alpha+\frac{1}{2}}(\cos(\vartheta_1))C_k^{\alpha+\frac{1}{2}}(\cos(\vartheta_2)) \\
&= \frac{(1+\alpha)_n}{(1+\alpha+\beta)_n}\sum_{k=0}^{n}\frac{(-)^{n-k}(\beta-\alpha)_{n-k}(1+\alpha+\beta)_{n+k}(1+2\alpha+2k)2\alpha k!}{(n-k)!(1+2\alpha)_{n+k+1}(2\alpha)_{k+1}} \\
&\quad C_k^{\alpha+\frac{1}{2}}(\cos(\vartheta_1))C_k^{\alpha+\frac{1}{2}}(\cos(\vartheta_2)) \quad (14)
\end{aligned}$$

so that we finally get

$$\begin{aligned}
(\cos(\vartheta_1)\cos(\vartheta_2))^n = &\frac{2^n n!(1+\alpha)_n\Gamma(1+\alpha+\beta)}{\Gamma(2n+\alpha+\beta+1)} \\
&\times\sum_{k=0}^{n}\frac{(-)^k(\beta-\alpha)_{n-k}(1+\alpha+\beta)_{n+k}(1+2\alpha+2k)2\alpha k!}{(n-k)!(1+2\alpha)_{n+k+1}(2\alpha)_{k+1}} \\
&\times C_k^{\alpha+\frac{1}{2}}(\cos(\vartheta_1))C_k^{\alpha+\frac{1}{2}}(\cos(\vartheta_2)) \quad (15)
\end{aligned}$$

REFERENCES

1. I. S. Gradshteyn and I. N. Ryzhik. "Tables of Integrals, Series and Products". London Academic, (1980).
2. R. H. Koornwinder, The addition formula for Jacobi polynomials, I summary of results. *Indag. Math.*, 34:179, (1972).
3. R. H. Koornwinder, The addition formula for Jacobi polynomials and spherical harmonics. *SIAM J. Appl. Math.*, 25:236, (1973).
4. T. A. Antone, Expansion of Jacobi polynomials in terms of orthogonal functions. *J. Phys. A*, 26:927, (1993).
5. E. D. Rainville. "Special Functions", The Macmillan Company, New York, (1960).
6. J. Rasch. "(e,2e) processes with neutral atom targets", Ph. D. thesis, University of Cambridge, (1996).
7. W. Magnus, F. Oberhettinger, and R. P. Soni. "Formulas and Theorems for the Special Functions of Mathematical Physics". Springer Verlag, Berlin, Heidelberg, New York, (1966).
8. A. Erdelyi, W. Magnus, F. Oberhettinger, and F. G. Tricomi. " Higher Transcendental Functions, Volume I". (1953).
9. A. Erdelyi, W. Magnus, F. Oberhettinger, and F. G. Tricomi. "Higher Transcendental Functions, Volume II". (1953).

PROGRESS REPORT ON RELATIVISTIC (e,2e) PROCESSES

R M Dreizler[1], H R J Walters[2], Colm T Whelan[3], H Ast[1] and S Keller[1]

[1]Institut für Theoretische Physik der Universität, Robert Mayer Straße 8–10, D–60054 Frankfurt am Main, Germany

[2]Department of Applied Mathematics and Theoretical Physics, The Queen's University of Belfast, Belfast BT7 1NN, Northern Ireland

[3]Department of Applied Mathematics and Theoretical Physics, University of Cambridge, Silver Street, Cambridge, CB3 9EW, UK

In this rather compact summary we present an overview of the theoretical investigation of $(e, 2e)$ processes that have been carried out by the

Belfast-Cambridge-Frankfurt Collaboration

during the last five years. This work has profited from close collaboration with the experimental group around W. Nakel in Tübingen.

1. The experiments

Starting in 1982 a wealth of experimental data has been gathered for the inner shell ionisation of heavy atoms by impact of relativistic electrons. Triple differential cross sections for K-shell ionistion have been obtained for the following systems

Coincidence Studies of Electron and Photon Impact Ionization
Edited by Whelan and Walters, Plenum Press, New York, 1997

1. Impact energy: 500 keV
   target system: Ag
   Coplanar asymmetric geometry at fixed angles
   $\theta_1 = -20^\circ, -25^\circ;\ \theta_2 = 40^\circ$
   energy of ejected electron: 210 $keV \leq E_1 \leq 350\ keV$
   (Ref. 1)

2. As in (1) for Ta with the specifications
   $\theta_1 = -20^\circ,\ \theta_2 = 40^\circ\ \ 100\ keV \leq E_2 \leq 360\ keV$
   $\theta_1 = -30^\circ,\ \theta_2 = 30^\circ\ \ 110\ keV \leq E_2 \leq 320\ keV$
   (Ref. 2)

3. Impact energy: 500 keV
   coplanar asymmetric geometry at fixed energy
   target system: Ag
   $E_1 = 375\ keV\ \ E_2 = 100\ keV$
   $\theta_1 = -15^\circ\ \ \ 25^\circ \leq \theta_2 \leq 75^\circ$
   $\theta_1 = -\ 7^\circ\ \ \ -80^\circ \leq \theta_2 \leq -40^\circ$
   $\theta_1 = -\ 7^\circ\ \ \ 25^\circ \leq \theta_2 \leq 75^\circ$
   $25^\circ \leq \theta_2 \leq 75^\circ$
   target system: Au
   $E_1 = 310\ keV\ \ \ E_2 = 100\ keV$
   $\theta_1 = -15^\circ\ \ \ 20^\circ \leq \theta_2 \leq 75^\circ$
   (Ref. 3)

4. Impact energy 300, 500 keV
   energy sharing coplanar symmetric geometry ($\theta_1 = -\theta_2 = \theta$)
   target system: Au
   300 keV : $E_1 = E_2 = 110\ keV\ \ \ 10^\circ \leq \theta \leq 60^\circ$
   500 keV : $E_1 = E_2 = 210\ keV\ \ \ 10^\circ \leq \theta \leq 60^\circ$
   target system: Ag
   300 keV : $E_1 = E_2 = 137\ keV\ \ \ 20^\circ \leq \theta \leq 55^\circ$
   300 keV : $E_1 = E_2 = 237\ keV\ \ \ 10^\circ \leq \theta \leq 50^\circ$
   target system: Cu
   300 keV : $E_1 = E_2 = 145.5\ KeV\ \ \ 30^\circ \leq \theta \leq 50^\circ$
   500 keV : $E_1 = E_2 = 245.5\ keV\ \ \ 24^\circ \leq \theta \leq 56^\circ$
   (Ref. 3,4)

5. Impact energy: 300 keV
   target system: Cu
   coplanar asymmetric geometry
   $E_1 = 220\ keV\ \ E_2 = 71\ keV$
   $\theta_1 = -\ 9^\circ\ \ \ -150^\circ \leq \theta_2 \leq -\ 30^\circ,\ \ \ 15^\circ \leq \theta_2 \leq 135^\circ$
   (Ref. 5)

In addition experimental data are available for the asymmetry parameter and the triple differential cross section by impact of polarised electrons (degree of polarisation: 35 - 40 %) for

6. Impact energy: 300 keV
   target system: Ag
   coplanar asymmetric geometry at fixed energy
   $E_1 = 200\ keV,\quad E_2 = 74.5$
   $\theta_1 = -10^\circ \quad -155^\circ \le \theta_2 \le -30^\circ \qquad 25^\circ \le \theta_2 \le 150^\circ$
   (Ref. 6)

as well as triple differential cross sections near or on Bethe ridge conditions ($\underline{k}_{recoil} = \underline{0}$) for

7. Impact energy: 300 keV
   target system: Ag
   coplanar asymmetric geometry at
   $E_1 = \quad 200\ keV \quad E_2 = 74.5 \quad \theta_1 = -26.1^\circ \quad 30^\circ \le \theta \le 75^\circ$
   $E_1 = E_2 = 137.25\ keV \quad \theta_1 = -37.9^\circ \quad 15^\circ \le \theta \le 60^\circ$
   (Ref. 7)

State selected $L-$shell data for $Au-$ and $U-$targets are available, but have not been published[8].

The measurements for both symmetric and asymmetric geometry at fixed energy cover essentially the region of the binary maximum. In addition, information on the less pronounced recoil peak in asymmetric geometry is available for $Ag-$ and $Cu-$targets.

2. Theoretical Approaches

Earlier attempts to interpret the experimental data for the binary region in terms of semirelativistic approximations

Born-Ochkur[9]: relativistic plane waves, Sommerfeld-Maue functions, nonrelativistic hydrogenic bound state wave function

First Born and Coulomb-Born[10]: relativistic plane waves, Darwin functions or nonrelativistic Coulomb waves with relativistic spinors, Darwin bound state wave function.

were not very successful. In all instances, the theoretical results overestimated the absolute experimental data (estimated experimental error: $\pm 15\%$) by factors between four and ten. In the earlier approaches spin-flip contributions were neglected. It was found that these terms contribute substantially in the case of coplanar symmetric geometry and that they lead to an increased discrepancy (an additional factor of about two) with respect to the experimental results in the binary maximum[4]. In a series of subsequent semirelativistic Coulomb-Born calculations[11,12,13] spin-flip contributions are accounted for.

It was thus necessary to initiate a new theoretical approach in which the following features were incorporated

1. Fully relativistic treatment of all the electron wave functions involved in the evaluation of the matrix elements of the (fully relativstic) photon propagator representing the electron-electron interaction. All exchange contributions are treated exactly.

As any relativistic scattering function, with the exception of the case of plane waves, can only be specified in terms of a partial wave representation, this requirement demands the development of a relativistic partial wave code for the evaluation of the lowest order $S-$matrix element for $K-$ (and $L-$) shell $(e, 2e)$ processes.

2. Incoming as well as outgoing electrons have to traverse the cloud of the remaining target electrons (assumed to be passive). Thus a distorted wave treatment is necessary.

The distorted partial waves are calculated by solution of the radial Dirac equation with a relativistic Kohn-Sham potential[14]. The assembly of the resulting DWBA code is outlined in Keller et al[15].

Particular attention was given to the following points:

- Convergence of the partial wave expansions in four angular momentum quantum numbers (three electron scattering wave functions and the relativistic (free) photon propagator)
- evaluation of radial matrix elements involving highly oscillatory functions (as angular momenta up to $j = 25$ and high linear momenta are involved)
- conformity with nonrelativistic DWBA results
- comparision with plane wave Born results.

In the presently published theoretical results all scattering states are evaluated with short range asymptotic boundary conditions (rather then Coulombic ones for the outgoing electrons). A first indication of the minor changes resulting from Coulomb boundary conditions is found in[16] , a more detailed documentation is in preparation.

3. Relativistic DWBA Results

Theoretical results for asymmetric coplanar geometry are found in Ast et al[17] and Keller et al[15]. They refer to the experimental data for Ag and Au (compare experimental results (3)), in addition predictions for the case of U targets are given. Results for energy sharing symmetric geometry for the cases of Au and U targets (compare experimental results (4)) are presented in Whelan et al[18]. In order to achieve convergence of the partial wave expansions for the case of Ag and Cu targets, the code had to be extended (maximum angular momentum of the partial wave expansion of the Coulomb propagator $l_0 = 70$). The results will be summarised in a forthcoming publication[19].

In comparison with the experimental data, that are available, very satisfactory agreement between the absolute experimental differential cross sections and the relativistic DWBA results has been found. This statement also applies to the (preliminary) investigation of the $L-$shell ionisation[20] (relative triple differential cross section and asymmetry parameter).

Additional investigations of the relativistic $(e, 2e)$ problem can be classified according to the points
(i) search for optimal geometries and finer effects, (ii) theoretical details.
In the second instance we have carried out a number of "model" investigations in order to

- discuss the effect of magnetic and retardation terms of the relativistic photon propagator on the position of the binary peak[21],
- unravel the reasons for the failure of previous theoretical attempts in detail[22],
- provide a fuller understanding of the implications of the relativistic plane wave Born approximation[23],
- take a closer look at the problem of the final state interactions[24] (in coplanar symmetric geometry).

Concerning the first line of research we have

- addressed the question of polarisation experiments by a detailed study of spin sensitive effects in inner shell $(e, 2e)$ processes and a direct comparison with the Tübingen $K-$shell data[25],
- presented an investigation of spin effects in the form of Pauli blocking in $(e, 2e)$ experiments with unpolarised electrons on unpolarised targets[26],
- investigated the coplanar constant θ_{12} geometry and suggested that it is a particularly useful tool for uncovering finer details of relativistic $(e, 2e)$ processes[27,29,30],
- studied the structures observed in the recoil region for Cu targets (compare experiment (5)) more closely[28].

Summary Remarks

Our investigations indicate that a fully relativistic treatment of the innershell ionisation of high-Z atoms by impact of fast electrons is mandatory. The collision can be described in terms of a first order photon exchange process taking place in the electrostatic field generated by the nucleus and the spectator electrons. Application of semirelativistic approximations lead to substantial errors in the results for triply differential cross sections. The use of plane wave models, which have been employed (as well as the semirelativstic formulations) in order to avoid the cumbersome partial wave analysis, is at best sufficient for the prediction of partially integrated cross sections.

We are able to observe a number of manifestations of the relativistic nature of the process, as for instance the shift of the binary maximum towards larger angles, which is not only due to relativistic kinematics but also to the relativistic form of the electron-electron-interaction, or the appearance of substantial continuum spin-orbit coupling effects in the recoil regime.

Our calculations suggest a number of interesting features of the triply differential cross section which have not yet been investigated experimentally

- the magnitude of the large angle recoil peak in symmetric energy sharing geometry becomes comparable to that of the binary peak for very high Z targets and suitably chosen impact energies;
- the constant θ_{12} geometry allows the extraction of information on specific spin channels in a nonspin-resolved setup;
- fine structure effects are clearly apparent in the binary regime for $2p^{\frac{3}{2}}$ substate ionisation.

Acknowledgements

The work on relativistic (e,2e) processes has been supported in part by ARC, DFG, EC, ESF–REHE, and NATO. We thank the experimental group in Tübingen for a fruitful exchange of ideas and data and the theory group of the university Metz (LPMC) for valuable discussions. The substantial computational effort was possible due to the support of the Gesellschaft für Schwerionenforschung (GSI) Darmstadt and the Hochschulrechenzentrum of the university Frankfurt.

References

1 E Schüle and W Nakel, Triply differential cross section for K shell ionization of silver by relativistic electron impact *J. Phys. B: At. Mol. Phys.* **15** , L639 (1982).

2 H Ruoff and W Nakel, Absolute triply differential cross section for K-shell ionisation by relativistic electron impact for high atomic number *J. Phys. B: At. Mol. Phys.* **20** , 2290:1987

3 J Bonfert, H Graf and W Nakel, Relativistic (e,2e) processes on atomic inner shells *J. Phys. B: At. Mol. Opt. Phys.* **24** , 1423:1991

[4] HRJ Walters, H Ast, CT Whelan, RM Dreizler, H Graf, CD Schröter, J Bonfert and W Nakel, Relativistic (e,2e) collisions on atomic inner shells in symmetric geometry *Z. Phys.D* **23** , 253:1992.

[5] KH Besch, M Sauter and W Nakel, Structure of the triply differential cross section in the recoil region in electron impact ionization. I. Experiment *J. Phys. B: At. Mol. Opt. Phys.* **30** , in press: 1997

[6] H-Th Prinz, KH Besch and W Nakel, Spin orbit interaction of the continuum electrons in relativistic (e,2e) measurements *Phys. Rev. Lett.* **74** , 243:1995.

[7] CD Schröter, H-Th Prinz, N Keuler and W Nakel , Relativistic (e,2e) processes under Bethe-ridge conditions *in (e,2e) and related processes* eds. CT Whelan, HRJ Walters, A Lahmam-Bennani and H Ehrhardt p 403 (Kluwer:Dortrecht), 1993

[8] T Kull, CD Schröter and W Nakel, in preparation

[9] JN Das and AN Konar, Inner shell ionization cross section when Sommerfeld-Maue wavefunction is used *J. Phys. B: At. Mol. Phys.* **7** , 2417:1974

[10] D Jakubassa-Amundsen, Relativistic theory of K shell ionization by fast electrons *Z. Phys.D* **11** , 305:1989.

[11] DH Jakubassa-Amundsen, A systematic study of relativistic (e,2e) collisions in comparison with experiment *J. Phys. B: At. Mol. Opt. Phys.* **25** , 1297:1992

[12] DH Jakubassa-Amundsen, Inner shell ionization of heavy targets by polarized and unpolarized electrons *J. Phys. B: At. Mol. Opt. Phys.* **28** , 259:1995

[13] DH Jakubassa-Amundsen, Validity of the Coulomb Born approximation for (e,2e) reactions *Phys. Rev.A* **53** , 2359:1996

[14] Dreizler RM and Gross EKU , Density functional theory *(Springer:Berlin), 1990*

[15] S Keller, CT Whelan , H Ast, HRJ Walters and RM Dreizler, Relativistic distorted wave Born calculations for (e,2e) processes on inner shells of heavy atoms *Phys. Rev.A* **50** , 3865:1994.

[16] LU Ancarani, S Keller, H Ast, CT Whelan, HRJ Walters and RM Dreizler , *this volume*

[17] H Ast, S Keller, CT Whelan, HRJ Walters and RM Dreizler, Electron impact ionization of the K shell of silber and gold in coplanar asymmetric geometry *Phys. Rev.A* **50** , R1:1994.

[18] CT Whelan, H Ast, S Keller , HRJ Walters and RM Dreizler, Triple differential cross sections in energy sharing symmetric geometry for gold and uranium at relativistic impact energies *J. Phys. B: At. Mol. Opt. Phys.* **28** , L33:1995.

[19] to be published,

[20] H-Th Prinz and S Keller, Observation of continuum spin-orbit coupling in L-shell ionization of gold by polaruzed relativistic electrons *J. Phys. B: At. Mol. Opt. Phys.* **29** , L651:1996

[21] H Ast, S Keller, RM Dreizler, CT Whelan, LU Ancarani and HRJ Walters, On the position of the binary peak in relativistic (e,2e) collisions *J. Phys. B: At. Mol. Opt. Phys.* **29** , 585:1996

[22] S Keller, RM Dreizler, LU Ancarani, HRJ Walters, H Ast and CT Whelan, Theoretical analysis of the relativistic first order Born approximation for inner shell (e,2e) processes *Z. Phys.D* **37** , 191:1996

[23] S Keller and CT Whelan, On the plane wave Born approximation for relativistic (e,2e) processes *J. Phys. B: At. Mol. Opt. Phys.* **27** , L771:1994

[24] S Keller, B Joulakian, H Ast and J Hanssen, The effect of final state correlations in the study of electron impact K shell ionization in coplanar symmetric geometry *J. Phys. B: At. Mol. Opt. Phys.* **29** , 2855:1996

[25] S Keller, RM Dreizler, H Ast, CT Whelan and HRJ Walters, Theory of (e,2e) processes with spin polarized relativistic electrons *Phys. Rev.A* **53** , 2295:1996

[26] H Ast, CT Whelan, S Keller, J Rasch, HRJ Walters and RM Dreizler , Pauli Blocking in relativistic (e,2e) collisions *in Selected Topics on Electron Physics, Proceedings of the Farago Symposium, eds DM Campbell and H Kleinpoppen* (Plenum:New York)129:1996

[27] CT Whelan, H Ast, HRJ Walters, S Keller and RM Dreizler, Relativistic energy sharing (e,2e) collisions in coplanar constant $\Theta_{1,2}$ geometry *Phys. Rev.A* **53** , 3262:1996

[28] RM Dreizler, H Ast, S Keller, CT Whelan, LU Ancarani and HRJ Walters, Structure of the triply differential cross section in the recoil region in electron impact ionization. II. Theoretical analysis *J. Phys. B: At. Mol. Opt. Phys.* **30** , in press: 1997

[29] CT Whelan, HRJ Walters, S Keller, H Ast, J Rasch and RM Dreizler, Inner shell (e,2e) processes *Cand. J. Phys.* **74** , 804:1996

[30] CT Whelan, HRJ Walters, J Rasch, S Keller,H Ast and RM Dreizler , Coincidence studies of the electron impact ionization of the inner shells of neutral atoms *in Autoionization in Atoms, ed V. Balashov* Moscow University Press 1996

STRUCTURE OF THE TRIPLY DIFFERENTIAL CROSS SECTION IN THE RECOIL REGION IN RELATIVISTIC ELECTRON IMPACT IONIZATION

K-H Besch, T Kull, M Sauter and W Nakel
Physikalisches Institut der Universität Tübingen, Auf der Morgenstelle 14, 72076 Tübingen, Germany

H Ast, S Keller and R M Dreizler
Institut für Theoretische Physik der Universität, Robert Mayer Straße 8–10, D–60054 Frankfurt am Main, Germany

C T Whelan and L U Ancarani
Department of Applied Mathematics and Theoretical Physics, University of Cambridge, Silver Street, Cambridge, CB3 9EW, UK

H R J Walters
Department of Applied Mathematics and Theoretical Physics, The Queen's University of Belfast, Belfast BT7 1NN, Northern Ireland

Electron impact ionization experiments in which the two outgoing electrons are detected in coincidence after angular and energy analysis have already provided a large body of information in the nonrelativistic energy region (see for instance Lahmam-Bennani 1991[1] and the articles in *(e,2e) and related processes*[2]. In recent years progress has been made in extending the (e,2e) technique to relativistic impact energies and atomic inner shells using polarized[3] and unpolarized electron beams[4,5,6]. Many of the longlasting discrepancies between experimental results and theoretical data (obtained in various calculations[7,8]) have been overcome by calculations carried out in the relativistic distorted wave Born approximation (RDWBA) (for a brief overview of the relevant publications see Dreizler et al , this volume[9]).

In this contribution we report on a combined experimental and theoretical study of the K-shell ionization of copper at relativistic impact energies using coplanar asymmetric kinematics, in which fast outgoing electrons are detected under a small scattering angle with regard to the primary beam. In these kinematics the angular distribution of the slow outgoing electrons consists in general of a binary peak and a recoil peak. The binary peak has a large contribution from a direct binary collision between the incoming electron and the atomic electron with the nucleus in the role of a 'spectator'. This peak is shifted away from the direction of momentum transfer towards larger detection angles. This shift is caused by the relativistic contributions to the electron-electron interaction and the influence of the strong atomic field onto the colliding electrons[10]. A recoil peak, however, cannot be explained unless an electron-nucleus interaction is taken into account. Consequently, the recoil peak is very sensitive to the dynamics of the ionization process. This was the motivation for the present work.

We describe in brief the experimental setup (which is the same as that used by Prinz et al[3], except that it is here used with an unpolarized beam).

The primary electron beam of $E_0 = 300\,keV$ was produced by an electrostatic accelerator and focused to a 1-mm-diameter spot on the target foil placed at the center of a vacuum chamber. As the target we used a self-supporting copper foil with a thickness of $50\,\mu g/cm^2$, for which plural scattering has been found to be small in previous cross section measurements. Each of the two electron detector systems consists of a magnetic spectrometer for the energy analysis combined with a plastic scintillation detector. Each magnet is a doubly focusing homogenous sector field shaped by an iron core.

We chose the following parameters for our measurement: The outgoing fast electrons of $E_1 = 220\,keV$ were observed at a fixed scattering angle of $\theta_1 = -9°$ with respect to the primary beam direction. The detector for the coincident slow electrons was adjusted to an energy of $E_2 = 71\,keV$ in order to select (e,2e) events from the K shell of copper ($E_{bind} = 9\,keV$).

The fast signals from the detectors were fed into a time-to-amplitude converter via constant fraction discriminators. The quantity measured directly is the counting rate of the true coincidences which is proportional to the triply differential cross section. It represents the probability that in a singly ionizing collision an incident electron of energy E_0 will produce two outgoing electrons having energies E_1 and E_2 emitted in the directions θ_1 and θ_2, respectively.

The result of the measurement of the relative triply differential ionization cross section as a function of the scattering angle of the outgoing slow electrons is shown in Fig.1 (symbols, the error bars represent one standard deviation). The solid line represents the data obtained in the relativistic distorted-wave Born approximation (RDWBA) In Figure 1 both experimental and theoretical data are normalized to unity at the binary maximum.

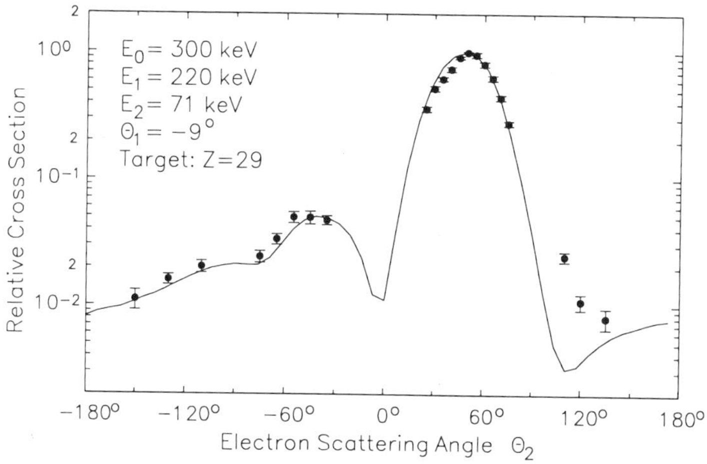

Figure 1. TDCS for the K shell ionization of copper by electron impact plotted against the angle of the slow outgoing electron. Impact energy 300 keV, energy of the slow outgoing electron 71 keV, detection angle of the fast outgoing electron $\theta_1 = -9^0$, • experimental data,— theoretical results obtained in RDWBA, for comparison experimental and theoretical data are normalized to unity at the binary maximum.

The most remarkable feature of the experimental data presented in Fig. 1 is a distinct double peak structure in the recoil region. It is quantitatively reproduced by the theoretical calculation. Previous measurements of K shell ionization cross sections have given no indication of this phenomenon.

The observed phenomenum of a double peak structure in the recoil region is presently not fully understood. We will deal with the first steps towards a detailed analysis. For the sake of clarity the relative maximum detected under θ_2 = -30^0 will be labeled (A), the one detected at θ_2 = -90^0 as (B). It is well known that for inner shell ionization the TDCS are strongly influenced by the interaction of the unbound electrons with the screened target nucleus (see e.g. Whelan et al 1995[11] and references therein), because pictorially speaking, the ionization takes place close to the nucleus. Under the assumption that the target atom can be described by an effective static potential, these effects are accounted for to all orders in the distorted wave approach. In order to investigate the influence of screening effects on the TDCS it is useful to replace the distorted waves one by one by plane waves. This technique has been used with success for the investigation of (e,2e) collisions in symmetric equal energy sharing[11,12] and coplanar constant $\Theta_{1,2}$ scattering geometries[13] as well as for the 2p - ionization of inert gases in asymmetric collisions at nonrelativistic impact energies[14]. The main motivation to apply this technique to the case of strongly asymmetric geometries at relativistic impact energies is the fact that a strong electron - nucleus coupling is a necessary condition for the observation of both electrons on the same side of the beam axis.

In a first calculation, the distorted wave of the incoming electron is replaced by a plane wave (PWIC - plane wave incident channel) so that the distortion effects in the final channel can be studied separately. The results obtained in the PWIC model approximation show a strong peak (A) and only a slight indication of a structure at the position of (B). In a second calculation the distorted waves representing the outgoing electrons are replaced by plane waves (PWFC - plane wave final channel) so that the influence of the atomic potential on the incident electron can be studied. The data obtained in the PWFC model show a marked peak at the position of (B) but no structure at the position of (A). These observations suggest that (A) is due to interaction of the outgoing electrons with the atom. Peak (B) is apparent in both calculations. The distortion in the incident channel enhances (B) notably. Of course the final size and the position of all peaks is influenced by distortion in the incident and in the outgoing channel and their interferences. We note here that a small peak at

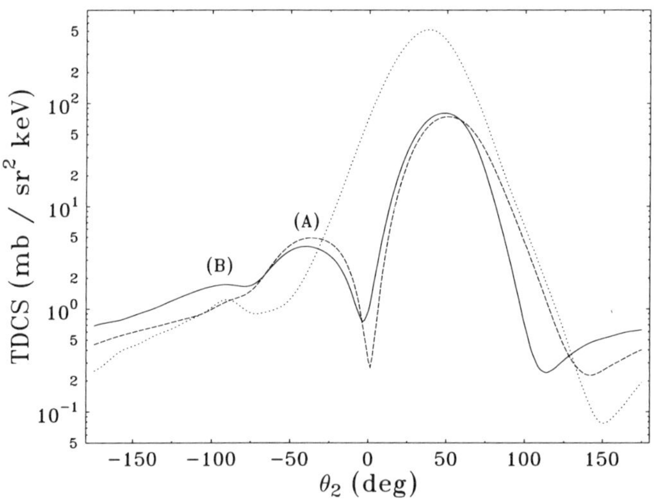

Figure 2. TDCS for the electron impact ionization of the K-shell of copper plotted against the detection angle θ_2 of the slow outgoing electron. E_0 = 300 keV, E_1 = 220 keV, E_2 = 71 keV, $\theta_1^{fixed} = -9^0$. — RDWBA distorted waves used for all unbound electrons, ··· PWFC distorted wave in the incident channel, orthogonalized plane waves in the final channel, - - - PWIC plane wave in the incident channel, distorted waves in the final channel.

position (A) can also be found in the TDCS obtained in the relativistic first order Born approximation[15]. There the incident and the fast outgoing electron are described by plane waves and only the slow outgoing electron is represented by a Coulomb wave. This gives evidence that the recoil peak (A) is dependent on the effective potential of the slow outgoing electron.

It is obvious that a precise understanding of these structures will require a much more detailed analysis. For instance, previous investigations have shown that the absolute magnitude as well as the shape of the secondary peak structure strongly depend on the target charge Z, both at fixed momentum transfer and at fixed ratio of impact to binding energy[16,17]. Here we add only the answer to the most obvious question, namely why the double peak structure has not been seen in previous experiments. Figure 3 shows data for a silver target (investigated by Prinz et al[3]). It is evident that in this case the structure which is visible in the theoretical results was too weak to give a statistically significant signal in the experiment. The measurement of the K shell ionization of silver by Bonfert et al[4] on the other hand was not extented far enough into the range of negative θ_2 (see also Ast et al[18]). Therefore the present experimental results do in no way contradict previous knowledge about the recoil peak region. Rather they should provide an additional stimulus for a more detailed examination of the region of negative θ_2 in relativistic (e,2e) experiments, both theoretical and experimental.

Concerning the absolute size of the cross section, the theoretical prediction obtained in RDWBA yields a TDCS of 79.83 $\left[\frac{mb}{sr^2 keV}\right]$at $\theta_2 = 50^0$, the binary maximum. According to a first measurement the respective experimental value was estimated to be about 35% lower. If one uses scattering eigenfunctions of the remaining ion instead of the neutral atom to describe the outgoing electrons the TDCS will reduce less than 5 % at the binary maximum (TDCS at $\theta_2 = 50\ ^0$ using an ionic potential in the outgoing channel = 76.34 $\left[\frac{mb}{sr^2 keV}\right]$), the curve shapes remain unchanged. This result is in good accord with the findings for heavier targets as discussed by Ancarani et al[19] In order to estimate the lower limit of TDCS that can be obtained with fully relativistic first order theories, it is instructive to compare with data obtained in relativistic Coulomb Born approximation (RCBA) (for details

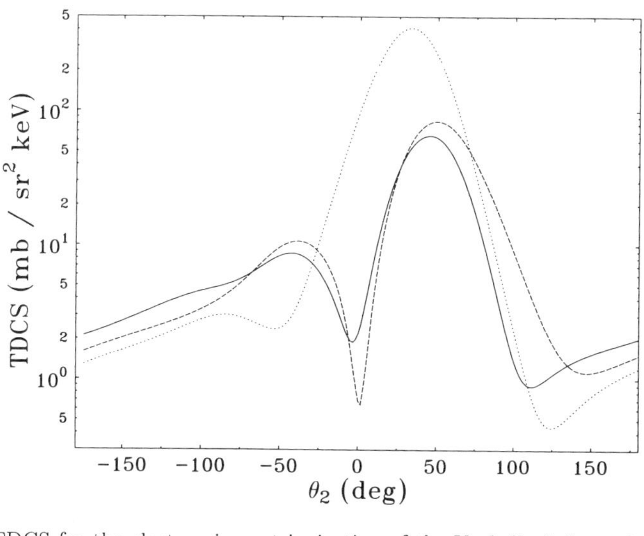

Figure 3. TDCS for the electron impact ionization of the K-shell of silver plotted against the detection angle θ_2 of the slow outgoing electron. E_0 = 300 keV, E_1 = 200 keV, E_2 = 74.5 keV, $\theta_1^{fixed} = -10^0$. —- RDWBA distorted waves used for all unbound electrons, · · · PWFC distorted wave in the incident channel, orthogonalized plane waves in the final channel, - - - PWIC plane wave in the incident channel, distorted waves in the final channel.

see Ancarani et al[19]). Due to the gross overestimation of the influence of the nuclear Coulomb potential by neglection of all electronic screening effects, the TDCS is significantly smaller (TDCS (at $\theta_2 = 50^0$) = 62.66 $\left[\frac{mb}{sr^2 keV}\right]$). However the observed discrepancy in the absolute size of the TDCS between the experimental and theoretical data remains unsolved.

In conclusion, for the first time a double peak structure has been observed experimentally in the recoil region for the K shell ionization of copper at relativistic impact energies. This structure is reproduced qualitatively by calculations obtained in RDWBA. The presented analysis indicates that the different parts of the recoil peak are due to different scattering mechanisms.

Acknowledgements

We are indebted to H. Folger, GSI Darmstadt, for providing us with the copper targets. Financial support from the Deutsche Forschungsgemeinschaft ((Na102/11-2), (As 85/4-1)), the EC (HCM CHRX-CT93-0350) and Nato (CRG 950 407) is gratefully acknowledged. The numerical calculations have been performed on the workstation clusters of the Hochschulrechenzentrum, Universität Frankfurt am Main and the Gesellschaft für Schwerionenforschung (GSI), Darmstadt.

References

1 A Lahmam-Bennani, Recent developments and new trends in (e,2e) and (e,3e) studies *J. Phys. B: At. Mol. Opt. Phys.* **24** , 2401:1991

2 (e,2e) and related processes eds. CT Whelan, HRJ Walters, A Lahmam-Bennani and H Ehrhardt Kluwer: Dordrecht 1993

3 H-Th Prinz, KH Besch and W Nakel, Spin orbit interaction of the continuum electrons in relativistic (e,2e) measurements *Phys. Rev. Lett.* **74** , 243:1995

4 J Bonfert, H Graf and W Nakel, Relativistic (e,2e) processes on atomic inner shells *J. Phys. B: At. Mol. Opt. Phys.* **24** , 1423:1991

5 HRJ Walters, H Ast, CT Whelan, RM Dreizler, H Graf, CD Schröter, J Bonfert and W Nakel, Relativistiv (e,2e) collisions on atomic inner shells in symmetric geometry *Z. Phys.D* **23** , 253:1992

6 CD Schröter, H-Th Prinz, N Keuler and W Nakel , Relativistic (e,2e) processes under Bethe-ridge conditions *in (e,2e) and related processes* eds. CT Whelan, HRJ Walters, A Lahmam-Bennani and H Ehrhardtp 403 (Kluwer:Dortrecht), 1993

7 F Bell, Double and triple differential cross sections for K shell ionization by relativistic electron impact *J. Phys. B: At. Mol. Opt. Phys.* **22** , 287:1989

8 D Jakubassa-Amundsen, Relativistic theory of K shell ionization by fast electrons *Z. Phys.D* **11** , 305:1989

[9] RM Dreizler, HRJ Walters, CT Whelan, H Ast and S Keller, Progress report on relativistic (e,2e) processes *this volume* , :1996
[10] H Ast, S Keller, RM Dreizler, CT Whelan, LU Ancarani and HRJ Walters, On the position of the binary peak in relativistic (e,2e) collisions *J. Phys. B: At. Mol. Opt. Phys.* **29** , 585:1996
[11] CT Whelan, H Ast, S Keller , HRJ Walters and RM Dreizler, Triple differential cross sections in energy sharing symmetric geometry for gold and uranium at relativistic impact energies *J. Phys. B: At. Mol. Opt. Phys.* **28** , L33:1995
[12] X Zhang, CT Whelan and HRJ Walters, (e,2e) cross sections for the ionization of He in coplanar symmetric geometry *J. Phys. B: At. Mol. Opt. Phys.* **23** , L509:1990
[13] CT Whelan, H Ast, HRJ Walters, S Keller and RM Dreizler, Relativistic energy sharing (e,2e) processes in a coplanar constant $\Theta_{1,2}$ geometry *Phys. Rev.A* **53** , 3262:1996
[14] X Zhang, CT Whelan, HRJ Walters, RJ Allan, P Bickert, W Hink and W Schönberger, (e,2e) cross sections for inner shell ionization of argon and neon *J. Phys. B: At. Mol. Opt. Phys.* **25** , 4325:1992
[15] S Keller, RM Dreizler, LU Ancarani, HRJ Walters, H Ast and CT Whelan, Theoretical analysis of the relativistic first order Born approximation for inner shell (e,2e) processes *Z. Phys.D* **37** , 191:1996
[16] S Keller, CT Whelan , H Ast, HRJ Walters and RM Dreizler, Relativistic distorted wave Born calculations for (e,2e) processes on inner shells of heavy atoms *Phys. Rev.A* **50** , 3865:1994
[17] S Keller, RM Dreizler, H Ast, CT Whelan and HRJ Walters, Theory of (e,2e) processes with spin polarized relativistic electrons *Phys. Rev.A* **53** , 2295:1996
[18] H Ast, S Keller, CT Whelan, HRJ Walters and RM Dreizler, Electron impact ionization of the K shell of silber and gold in coplanar asymmetric geometry *Phys. Rev.A* **50** , R1:1994
[19] LU Ancarani et al, in preparation :1997

COULOMB BOUNDARY CONDITIONS FOR RELATIVISTIC (e,2e) PROCESSES

L.U. Ancarani [1], S. Keller [2], H. Ast [2],
Colm T. Whelan [1], H.R.J. Walters [3], and R.M. Dreizler [2]

[1]Department of Applied Mathematics and Theoretical Physics
University of Cambridge, Cambridge CB3 9EW, England
[2]Institut für Theoretische Physik der Universität
Robert Mayer Straße 8–10, D - 60054 Frankfurt am Main, Germany
[3]Department of Applied Mathematics and Theoretical Physics
The Queen's University of Belfast, Belfast BT7 1NN
Northern Ireland

INTRODUCTION

The relativistic (e,2e) experiments on high-Z atoms performed in the last years by the Tübingen group[1,2] have stimulated theoreticians to provide an acceptable model to explain the ionization processes involved. The relativistic distorted wave Born approximation (rDWBA)[3,4,5] seems the one which yields the best results and has produced data in qualitative, and in most cases quantitative, agreement with the absolute measurements[1,2]. In this contribution we study, theoretically, the relevance of Coulomb boundary conditions for (e,2e) processes on the K-shell of high-Z atoms. We consider several fully relativistic approximations and compare calculated triple differential cross sections (TDCSs) with the measurements[1].

In previous rDWBA results the two outgoing electrons are described as escaping in the field of the neutral atom rather than in that of a positive ion. We have now modified the theoretical model to include Coulomb boundary conditions in both the initial and final channels. This involves matching, at large distances from the nucleus, the Dirac wavefunctions describing the outgoing electrons to relativistic Coulomb wavefunctions rather than to Bessel functions (see equation (14) in reference 4). For neutral targets, the residual ion is singly charged: the short range effective potential of the atom should be replaced by a long range potential which reaches asymptotically the value $-1/r$ (the atomic/ionic potentials are obtained in a self-consistent relativistic Kohn-Sham local density approximation[6]). In other words, the outgoing electrons see, as they reach the asymptotic region, a single positive charge on the target. In order to provide a realistic post-ionization picture, Coulomb boundary conditions should therefore be imposed in the final channel. However, at the relativistic energies of the experiments[1,2], one does not expect the asymptotic charge $z = +1$ to play any substantial rôle since $z/k << 1$ where k is the typical electron momentum.

The extension of the rDWBA code to include Coulomb boundary conditions is not just a purely academic exercise to confirm what is expected. It provides us also with the first fully relativistic Coulomb Born approximation for (e,2e) processes, where the incident and outgoing electrons are described by pure relativistic Coulomb waves, elastically scattered from the Coulomb potential Z_{at}/r where Z_{at} is the atomic charge of the neutral, screened or unscreened, atom. The approximation is the only proper one to assess the validity of the calculations based on semi-relativistic Coulomb wavefunctions[7,8]. Moreover, if (e,2e) processes on positive ions were to be studied, Coulomb boundary conditions would have to be included in both initial and final channels (asymptotic charge $+z_{\rm ion}$ and $+z_{\rm ion} + 1$, respectively) for the theoretical model to describe the physical post-ionization picture.

RESULTS AND DISCUSSION

The extended code has been used to calculate TDCSs for (e,2e) processes on the K-shell of silver and gold, at 300 and 500 keV impact energy, in both coplanar symmetric and asymmetric geometry. A selection of results is illustrated below, while a more comprehensive report will be given in a forthcoming contribution[9].

To name the different types of relativistic approximations considered here we label each of the three unbound electrons, incoming and two outgoing, with a letter and a subscript (physically, the two outgoing electrons should be treated on an equal footing). The letter, D or C, shall indicate whether the electron is described by a distorted wave or a pure Coulomb wave, while the subscript labels the asymptotic charge seen by the outgoing electrons. The previous rDWBA calculations will therefore be denoted by $D_0D_0D_0$, the same approximation with Coulomb boundary conditions by $D_0D_1D_1$, while the Coulomb Born approximation by $C_{Z_{at}}C_{Z_{at}}C_{Z_{at}}$. Of these, only $D_0D_1D_1$ describes the physical long range behaviour, where the incident electron is elastically scattered from the neutral atom, while the two outgoing electrons are scattered from the singly charged residual ion.

TDCSs calculated in these approximations are compared with each other, with the results obtained in a semi-relativistic approach[8], and with the experimental data. In Figure 1, we show the TDCS for the ionization of the K-shell of gold in symmetric geometry (indices $0, 1$ and 2 label the energies and the scattering angles of, respectively, the incoming and the two outgoing electrons). One observes that the results in the $D_0D_0D_0$ and $D_0D_1D_1$ approximations differ only slightly (about 2% at 30^0); the inclusion of physical boundary conditions in the rDWBA model improves

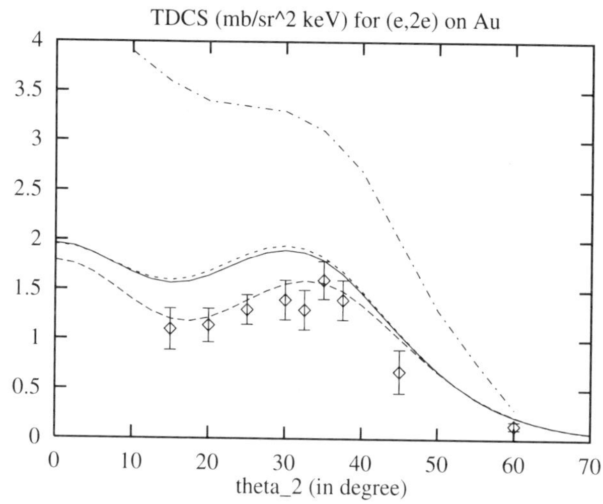

Figure 1. TDCS (in mb/sr^2keV) for ionization of the K-shell of gold ($Z_{at} = 79$) as a function of $\theta_2 = \theta_1$ for $E_0 = 300$ keV, $E_1 = E_2 = 109.65$ keV. Full curve, $D_0D_1D_1$; dotted curve, $D_0D_0D_0$; dashed curve, $C_{79}C_{79}C_{79}$; dot-dashed curve, semi-relativistic Coulomb-Born approximation[8]. The experimental points are given with their error bars[1].

only minutely the agreement with experimental data. This can be understood from the fact that the ionization of a K-shell electron takes place at small radii; hence, a change in the large distance behaviour of the effective potential is not expected to modify substantially the TDCS. The fully relativistic Coulomb Born result is similar in shape, and is very good in magnitude; the strong Coulomb field brings the TDCS down by circa 20% in proximity of the binary maximum. The quality of the result is quite surprising since the $C_{79}C_{79}C_{79}$ approximation is far from satisfying the physical boundary conditions. When comparing the semi- and fully relativistic Coulomb Born results, one sees how the use of approximate wavefunctions (the multiplication of non-relativistic Coulomb wavefunctions by free Dirac spinors[7,8]) yields a TDCS with quite different shape and magnitude.

In Figure 2, we show the TDCSs versus the angle of the slow outgoing electron, for 500 keV incident energy and $E_2 = 100$ keV, in an asymmetric geometry where the fast outgoing electron is detected at a fixed angle of -15^0 . The analysis of the results is similar to that of Figure 1 with the exception that, in this case, the $C_{79}C_{79}C_{79}$ cross section is surprisingly seen to underestimate the experimental data. At the maximum position the difference between the $D_0D_0D_0$ and $D_0D_1D_1$ results is less than 1%. This is smaller than in the case dealt with in Figure 1; in spite of the different geometry, this decrease was to be expected as the energy of the outgoing electrons is larger.

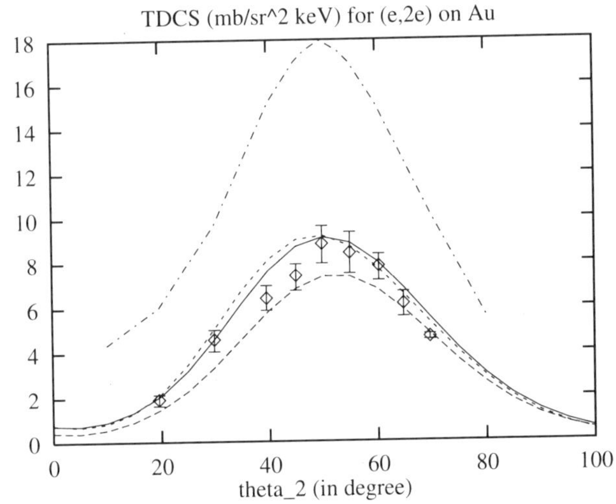

Figure 2. TDCS for ionization of gold as a function of θ_2 for $E_0 = 500$ keV, $E_2 = 100$ keV, and θ_1 fixed at -15^0 . Same key as in Figure 1.

Finally, in Figure 3, we look at the same kinematics and geometry as in Figure 2, but for silver. Here, we observe that the improvement of $D_0D_1D_1$ as compared to $D_0D_0D_0$ is larger than in Figure 2, about 5% at the maximum. This is not surprising since the target is lighter, although a proper comparison should be made with sets of kinematics similar with respect to the ionization potentials. Once again, the fully relativistic Coulomb Born result is, strikingly, in very good agreement with the experimental data.

The TDCSs for the K-shell ionization of high-Z atoms, by high incident energies, is not very sensitive to the inclusion of the physical asymptotic behaviour in the effective potential. In the kinematics and geometry sets studied in the present contribution, the fully relativistic Coulomb Born approximation is seen to yield very good results, in spite of its unphysical nature, while the semi-relativistic approach[8]

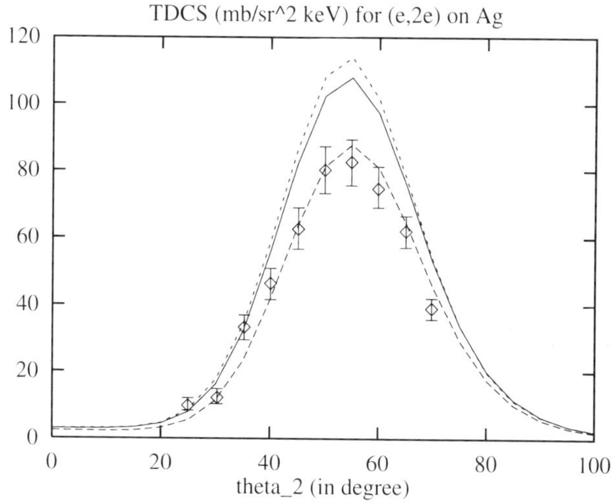

Figure 3. Same as for Figure 2, but for silver ($Z_{at} = 47$); $C_{47}C_{47}C_{47}$ replaces $C_{79}C_{79}C_{79}$.

largely overestimates the cross sections. A systematic study of the approximations considered here, and others, will be reported elsewhere in greater detail[9].

Acknowledgements We would like to thank Professor Nakel and his colleagues for many fruitful discussions. We gratefully acknowledge financial support for this work by the EC, DFG and Nato. The numerical calculations have been performed on the workstation clusters of the Gesellschaft für Schwerionenforschung (GSI) Darmstadt.

REFERENCES

1. J Bonfert, H Graf and W Nakel, Relativistic (e,2e) processes on atomic inner shells *J. Phys. B: At. Mol. Opt. Phys.* **24**, 1423:1991
2. H-Th Prinz, KH Besch and W Nakel, Spin orbit interaction of the continuum electrons in relativistic (e,2e) measurements *Phys. Rev. Lett.* **74**, 243 :1995
3. H Ast, S Keller, CT Whelan, HRJ Walters and RM Dreizler, Electron impact ionization of the K shell of silver and gold in coplanar asymmetric geometry *Phys. Rev.A* **50**, R1:1994
4. S Keller, CT Whelan , H Ast, HRJ Walters and RM Dreizler, Relativistic distorted wave Born calculations for (e,2e) processes on inner shells of heavy atoms *Phys. Rev.A* **50**, 3865:1994
5. CT Whelan, H Ast, S Keller, HRJ Walters and RM Dreizler, Triple differential cross sections in energy sharing symmetric geometry for gold and uranium at relativistic impact energies *J. Phys. B: At. Mol. Opt. Phys.* **28**, L33:1995
6. RM Dreizler and EKU Gross , Density functional theory *(Springer:Berlin) 1990*
7. DH Jakubassa-Amundsen, Relativistic theory of K shell ionization by fast electrons *Z. Phys.D* **11**, 305:1989
8. DH Jakubassa-Amundsen, A systematic study of relativistic (e,2e) collisions in comparison with experiments *J. Phys. B: At. Mol. Opt. Phys.* **25**, 1297:1992
9. LU Ancarani et al., in preparation :1997

SPIN EFFECTS IN THE (e,2e) CROSS SECTIONS OF XENON

A. Dorn, A. Elliot, X. Guo, J. Hurn, J. Lower, S. Mazevet,
I.E. McCarthy†, Y. Shen and E. Weigold

Atomic and Molecular Physics Laboratories,
Institute of Advance Studies,
RSPhySSE, ANU, ACT 0200, Australia.

†E.S.M. Center,
The Flinders University of South Australia,
Adelaide SA5001, Australia.

INTRODUCTION

The study of electron atom collisions involving polarised partners have been of major interest in the past ten years. These experiments when analysed using the density matrix formalism reveal the most detailed information on the role of the spin interactions during the process (Bartschat, 89). This experimental technique has been used for the study of elastic (Kessler, 85) and inelastic collisions (Hanne, 83) and only recently in (e,2e) experiments involving a polarised electron beam and/or a polarised atomic beam. The first one involved a polarised electron beam and polarised lithium atoms at intermediate energy ($E_0 = 54.4$eV) and concentrated on the study of the exchange process (Baum et al., 92). This was followed by a preliminary report on (e,2e) measurements on xenon using polarised electrons (Granitza et al., 93). Quite recently, a relativistic polarised electron beam ($E_0 = 300$ keV) was used in the K-shell ionization of silver and gold (Prinz et al., 95) and as such focused on the influence of the relativistic interactions.

Recently, we reported detailed (e,2e) measurements involving a polarised electron beam on xenon atoms at intermediate energy ($E_0 = 147$eV) and for coplanar asymmetric kinematics (Guo et al., 96). A spin up-down asymmetry was clearly observed when the fine structure of the ion was resolved experimentally. For xenon the separation in energy of the $^2P_{1/2}$ and $^2P_{3/2}$ fine structure states is 1.3eV. This effect, which was postulated by Hanne (Hanne, 91; Jones et al., 94) is understood as the influence of the fine structure of the ion via exchange and orientation of the target during the process (Madison et al., 96).

To analyse the parameters of such an experiment we applied the density matrix formalism to the case of ionization of a closed shell system by a polarised electron beam. This formulation allows a definition of the asymmetry from general principles which accounts for all the exchange interactions as well as for the relativistic interactions (Guo et al., 95; Dorn et al., 96). In the previous chapter the influence of the exchange between the two outgoing electrons and between each outgoing electron with the remaining electrons of the ion in the behaviour of the spin-up down asymmetry was clearly demonstrated. However, as a heavy target is required experimentally to observed such an effect, in the following chapter we will concentrate on the influence of the relativistic interactions in the behaviour of the asymmetry function and triple differential cross section (TDCS) at intermediate energy.

THEORY

First one needs to define a frame in which the different parameters of the (e,2e) experiment are to be expressed. In the present study the collision frame, which is commonly used in collision theory, will be employed. This frame presents the advantage of a unique definition for both the experimental and calculated asymmetry.

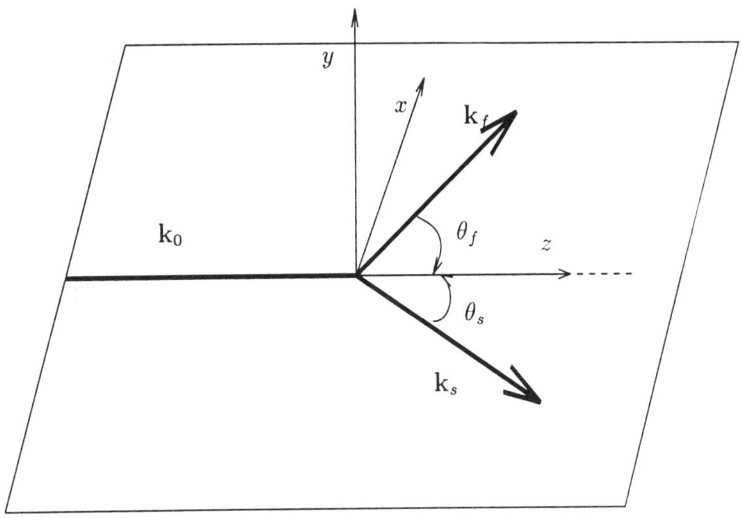

Figure 1. The collision frame

In the collision frame (figure 1) the z^c axis is along the direction of the incident electron beam while the y^c axis is perpendicular to the scattering plane containing the momenta $\mathbf{k}_f$ and $\mathbf{k}_s$ of the fast and slow electron (also sometimes called respectively the "scattered" and "ejected" electron) as we are considering coplanar kinematics.

Parameters of the experiment

When the density matrix formalism is applied to the case of ionization of closed shell atoms by polarised electrons the polarised TDCS is expressed as (Guo et al., 95; Dorn et al., 96)

$$\frac{d^3\sigma(\theta)}{d\Omega_f d\Omega_s dE_f} = \left(\frac{d^3\sigma(\theta)}{d\Omega_f d\Omega_s dE_f}\right)_{unpol.} [1 + P_y S_a]. \tag{1}$$

The relation (1) indicates that the TDCS corresponding to an ionization process where the incident electron beam has a polarisation $\mathbf{P} \equiv (P_x, P_y, P_z)$ depends directly on the component of the polarisation perpendicular to the scattering plane P_y. The polarised TDCS does not show any direct dependence on the components of the polarisation in plane P_x and P_z due to the reflection invariance of the scattering amplitudes in the scattering plane (Guo et al., 95).

Further, from the relation (1), two independent parameters of this particular ionization process are identified: the unpolarised TDCS $\left(\frac{d^3\sigma(\theta)}{d\Omega_f d\Omega_s dE_f}\right)_{unpol.}$ and the asymmetry function $\mathcal{S}_a$. When an unpolarised electron beam or an electron beam with no polarisation perpendicular to the scattering plane is used in the entrance channel, the TDCS (1) is equal to the unpolarised TDCS. An explicit expression for the unpolarised TDCS is (Guo et al., 95; Dorn et al., 96)

$$\left(\frac{d^3\sigma(\theta)}{d\Omega_f d\Omega_s dE_f}\right)_{unpol} = (2\pi)^4 \frac{k_f k_s}{k_0} \tag{2}$$

$$\sum_{M_i} \left\{ f\left(\frac{1}{2}, \frac{1}{2}, M_i, \frac{1}{2}\right) f^*\left(\frac{1}{2}, \frac{1}{2}, M_i, \frac{1}{2}\right) + f\left(\frac{-1}{2}, \frac{-1}{2}, M_i, \frac{1}{2}\right) f^*\left(\frac{-1}{2}, \frac{-1}{2}, M_i, \frac{1}{2}\right) \right.$$

$$\left. + f\left(\frac{-1}{2}, \frac{1}{2}, M_i, \frac{1}{2}\right) f^*\left(\frac{-1}{2}, \frac{1}{2}, M_i, \frac{1}{2}\right) + f\left(\frac{1}{2}, \frac{-1}{2}, M_i, \frac{1}{2}\right) f^*\left(\frac{1}{2}, \frac{-1}{2}, M_i, \frac{1}{2}\right) \right\}.$$

where the scattering amplitude $f(\nu_f, \nu_s, M_i, \nu_0)$ and its complex conjugate $f^*(\nu_f, \nu_s, M_i, \nu_0)$ are defined as

$$(2\pi)^2 \sqrt{\frac{k_f k_s}{k_0}} f(\nu_f, \nu_s, M_i, \nu_0) = \langle \nu_f, \nu_s, J_i, M_i | T | J_0, M_0, \nu_0 \rangle, \tag{3}$$

$$(2\pi)^2 \sqrt{\frac{k_f k_s}{k_0}} f^*(\nu_f, \nu_s, M_i, \nu_0) = \langle \nu_0, J_0, M_0 | T^\dagger | \nu_f, \nu_s, J_i, M_i \rangle. \tag{4}$$

$\langle \nu_f, \nu_s, J_i, M_i | T | J_0, M_0, \nu_0 \rangle$ is the unsymmetrysed T-matrix element. The unpolarised TDCS is, as expected, the sum of different TDCS corresponding to different combinations of spin projections ν_f, ν_s and projection of the total angular momentum of the ion M_i in the final state. For a particular projection of the total angular momentum and spin up for the incident electron we identify the sum of four partial TDCS corresponding to a particular combination of spin projections of the two outgoing electrons in the final state

$$\left.\frac{d^3\sigma(\theta)}{d\Omega_f d\Omega_s dE_f}\right|_{(\nu_f=\uparrow, \nu_s=\uparrow)}^{(J_i, M_i)} = (2\pi)^4 \frac{k_f k_s}{k_0} f\left(\frac{1}{2}, \frac{1}{2}, M_i, \frac{1}{2}\right) f^*\left(\frac{1}{2}, \frac{1}{2}, M_i, \frac{1}{2}\right), \tag{5}$$

$$\left.\frac{d^3\sigma(\theta)}{d\Omega_f d\Omega_s dE_f}\right|_{(\nu_f=\downarrow, \nu_s=\uparrow)}^{(J_i, M_i)} = (2\pi)^4 \frac{k_f k_s}{k_0} f\left(\frac{-1}{2}, \frac{1}{2}, M_i, \frac{1}{2}\right) f^*\left(\frac{-1}{2}, \frac{1}{2}, M_i, \frac{1}{2}\right), \tag{6}$$

$$\left.\frac{d^3\sigma(\theta)}{d\Omega_f d\Omega_s dE_f}\right|_{(\nu_f=\uparrow, \nu_s=\downarrow)}^{(J_i, M_i)} = (2\pi)^4 \frac{k_f k_s}{k_0} f\left(\frac{1}{2}, \frac{-1}{2}, M_i, \frac{1}{2}\right) f^*\left(\frac{1}{2}, \frac{-1}{2}, M_i, \frac{1}{2}\right), \tag{7}$$

$$\left.\frac{d^3\sigma(\theta)}{d\Omega_f d\Omega_s dE_f}\right|_{(\nu_f=\downarrow, \nu_s=\downarrow)}^{(J_i, M_i)} = (2\pi)^4 \frac{k_f k_s}{k_0} f\left(\frac{-1}{2}, \frac{-1}{2}, M_i, \frac{1}{2}\right) f^*\left(\frac{-1}{2}, \frac{-1}{2}, M_i, \frac{1}{2}\right). \tag{8}$$

A simple interpretation of these reactions can be obtained by considering the spin projection of the incident electron only and by assuming that the spin projection of the slow electron can always be obtained, since we are considering the ionization of a spinless target. Under the LS coupling scheme, the target atoms contain the same number of electrons with spin projection up and down. The TDCS (5), (6), (7) and (8) correspond respectively to the following reactions, with direct, direct spin-flip, exchange and exchange spin-flip amplitudes as indicated.

$$e_i(\uparrow) + A \longrightarrow e_f(\uparrow) + e_s(\uparrow) + A^+(J_i, M_i) \qquad f_{(J_i,M_i;\uparrow)} - g_{(J_i,M_i;\uparrow)}, \tag{9}$$
$$e_i(\uparrow) + A \longrightarrow e_f(\downarrow) + e_s(\uparrow) + A^+(J_i, M_i) \qquad F_{(J_i,M_i;\uparrow)} - g_{(J_i,M_i;\downarrow)}, \tag{10}$$
$$e_i(\uparrow) + A \longrightarrow e_f(\uparrow) + e_s(\downarrow) + A^+(J_i, M_i) \qquad f_{(J_i,M_i;\downarrow)} - G_{(J_i,M_i;\uparrow)}. \tag{11}$$
$$e_i(\uparrow) + A \longrightarrow e_f(\downarrow) + e_s(\downarrow) + A^+(J_i, M_i) \qquad F_{(J_i,M_i;\downarrow)} - G_{(J_i,M_i;\downarrow)}. \tag{12}$$

The first reaction (9) occurs through a direct process f_{J_i,M_i} only or through direct f_{J_i,M_i} and exchange g_{J_i,M_i} processes if exchange between the two outgoing electrons is considered by using a symmetrised T-matrix element in the relations (3–4). These amplitudes correspond to an ion left in the final state $|J_i, M_i\rangle$. The second reaction (10) occurs through a direct spin flip F_{J_i,M_i} for the incident electron, or through exchange g_{J_i,M_i} and direct spin flip F_{J_i,M_i} when exchange is considered. The direct spin-flip amplitude F_{J_i,M_i} represents a process where the incident electron is detected as the fast electron but its spin projection changes during the process without conservation of the total spin for the collision state. This relativistic effect (Mott scattering) has been demonstrated for elastic and inelastic processes at low energy and has to be taken into account for the scattering of electrons by a heavy target. The third reaction (11) occurs through a direct process f_{J_i,M_i} for the incident electron or a direct process f_{J_i,M_i} and an exchange spin-flip process G_{J_i,M_i} if we consider exchange. The exchange spin-flip amplitude G_{J_i,M_i} corresponds to a process where exchange occurs between the two outgoing electrons, the incident electron is detected as the slow one while its spin projection has changed during the scattering due to relativistic effects. The last reaction (12) occurs through a direct spin-flip process F_{J_i,M_i} and an exchange spin-flip process G_{J_i,M_i}.

It is also possible to give the same argument for the ejected electron if the final state of the ion is considered in conjunction with the spin of the outgoing electrons, assuming that LS coupling is valid for the ion state. For the case where the spin projection of the ejected electron is not equal to the negative spin projection of the ion, this electron has experienced a spin-flip process due to relativistic interactions. We will therefore indicate the spin projection of the ejected electron in conjunction with the total angular momentum projection of the ion as indices of the different scattering amplitudes. For a particular state $|J_i, M_i\rangle$ of the ion, we thus consider eight different scattering amplitudes when the exchange process between the two outgoing electrons is considered.

The asymmetry function $\mathcal{S}_a$ is expressed as a function of independent scattering amplitudes as (Guo et al., 95; Dorn et al., 96)

$$\mathcal{S}_a = K \sum_{M_i} (-1)^{M_i - J_i} \mathrm{Im} \left\{ f\left(\frac{1}{2}, \frac{1}{2}, M_i, \frac{1}{2}\right) f^*\left(\frac{-1}{2}, \frac{-1}{2}, -M_i, \frac{1}{2}\right) \right. \tag{13}$$
$$\left. - f\left(\frac{1}{2}, \frac{-1}{2}, M_i, \frac{1}{2}\right) f^*\left(\frac{-1}{2}, \frac{1}{2}, -M_i, \frac{1}{2}\right) \right\},$$

where the normalisation factor K is

$$K = 2(2\pi)^4 \frac{k_f k_s}{k_0} \left(\frac{d^3\sigma(\theta)}{d\Omega_f d\Omega_s dE_f} \right)^{-1}_{unpol.} . \tag{14}$$

The expression (13) shows explicitly the dependence of the polarised TDCS on the component of the polarisation of the initial-electron beam perpendicular to the scattering plane. Using the relations (9–12) the asymmetry function (13) can be expressed as a function of the direct, exchange, direct spin-flip and exchange spin-flip amplitudes.

$$\begin{aligned} \mathcal{S}_a = K \sum_{M_i} (-1)^{J_i - M_i} & \\ \mathrm{Im} \{ & f_{(M_i,\uparrow)} F^*_{(-M_i,\downarrow)} - f_{(M_i,\downarrow)} F^*_{(-M_i,\uparrow)} + f_{(M_i,\downarrow)} g^*_{(-M_i,\downarrow)} + G_{(M_i,\uparrow)} F^*_{(-M_i,\uparrow)} \\ & - G_{(M_i,\uparrow)} g^*_{(-M_i,\downarrow)} - g_{(M_i,\uparrow)} F^*_{(-M_i,\downarrow)} + g_{(M_i,\uparrow)} G^*_{(-M_i,\downarrow)} - f_{(M_i,\uparrow)} G^*_{(-M_i,\downarrow)} \} . \end{aligned} \tag{15}$$

Two spin-dependent interactions contribute to the asymmetry function $\mathcal{S}_a$: the relativistic interaction or spin-orbit interaction for the continuum electrons and the exchange between these two electrons. Further, in the expression (15) it appears that the asymmetry function $\mathcal{S}_a$ is in general a function of the interference between the direct, exchange, direct spin-flip and exchange spin-flip amplitude corresponding to different projections of the ion. Also, in the situation where the exchange process becomes negligible ($G = g = 0$), the expression (15) shows that the asymmetry function $\mathcal{S}_a$ tends to the value

$$\mathcal{S}_a = K \sum_{M_i} (-1)^{J_i - M_i} \mathrm{Im} \{ f_{(M_i,\uparrow)} F^*_{(-M_i,\downarrow)} - f_{(M_i,\downarrow)} F^*_{(-M_i,\uparrow)} \} . \tag{16}$$

In addition, due to the summation over M_i in the relation (15), the asymmetry function $\mathcal{S}_a$ will be equal to zero when the scattering amplitudes satisfy the relations

$$F_{(M_i,\uparrow/\downarrow)} = F_{(-M_i,\uparrow/\downarrow)}, \qquad G_{(M_i,\uparrow/\downarrow)} = G_{(-M_i,\uparrow/\downarrow)}, \tag{17}$$

$$f_{(M_i,\uparrow/\downarrow)} = f_{(-M_i,\uparrow/\downarrow)}, \qquad g_{(M_i,\uparrow/\downarrow)} = g_{(-M_i,\uparrow/\downarrow)}. \tag{18}$$

This indicates that the asymmetry function $\mathcal{S}_a$ is strongly correlated to the state of the ion after the collision. A non-zero orientation of the ion after the collision is a necessary condition to observe a non-zero asymmetry. The situation can be associated with an (e,2e) experiment that uses polarised electron and atomic beams. In that case the atoms are initially oriented through, say, laser pumping, while in the present case the ions are naturally oriented during the process through the Coulomb interaction. The expressions (2) and (13) of respectively the unpolarised TDCS and asymmetry function are obtained without approximation. The approximation in the evaluation of these parameters will appear through the scattering method used to evaluated the T-matrix elements (3, 4).

Scattering approximation

To evaluate the T-matrix elements (3) we used a semi relativistic DWBA where the distorted waves are calculated using the Schrödinger form of the Dirac equation (McEachran et al., 86; Hasenburg et al., 86). The set of two uncoupled second order differential equations solved for the radial part of the distorted waves is

$$\left[\frac{d^2}{dr^2} - \frac{l(l+1)}{r^2} - 2\alpha V(r) - 2V_R(r) + k^2\right] u_\kappa(r) = 0, \tag{19}$$

where

$$V_R(r) = -\frac{\alpha^2}{2}\left[V^2(r) + \frac{\kappa}{r(\gamma+1-\alpha^2 V)}\frac{dV}{dr} - \frac{1}{2(\gamma+1-\alpha^2 V)}\frac{d^2V}{dr^2} - \frac{3\alpha^2}{4(\gamma+1-\alpha^2 V)^2}\left(\frac{dV}{dr}\right)^2\right]. \tag{20}$$

and

$$\kappa = l, \qquad J = l + 1/2 \tag{21}$$

$$\kappa = -(l+1), \qquad J = l - 1/2. \tag{22}$$

α is the fine structure constant which is the inverse of the speed of light in atomic units and $\gamma = (1+\alpha^2 k^2)^{\frac{1}{2}}$ (Hasenburg et al., 86). $V_R(r)$ is the so-called relativistic potential and includes the relativistic corrections to the Schrödinger equation. $V(r)$ is chosen as the static exchange potential of Furness and McCarthy (Furness and McCarthy, 73) for the calculation of the distorted waves in the entrance channel. The Coulomb potential is added to the static-exchange potential for the calculation of both distorted waves in the exit channel. The semi-relativistic DWBA method approximates the exact T-matrix elements (3) as

$$f(\nu_f, \nu_s, M_i, \nu_0) \equiv (-1)^{J_i - M_i}\frac{1}{2\sqrt{2J_i+1}} \left\{\langle\chi_{\nu_f}^{(-)}(k_f)\chi_{\nu_s}^{(-)}(k_s)|V|\varphi_{L_i,J_i,-M_i}\chi_{\nu_0}^{(+)}(k_0)\rangle - \langle\chi_{\nu_s}^{(-)}(k_s)\chi_{\nu_f}^{(-)}(k_f)|V|\varphi_{L_i,J_i,-M_i}\chi_{\nu_0}^{(+)}(k_0)\rangle\right\}. \tag{23}$$

where $\chi_\nu(k)$ are the distorted waves with momentum k and spin projection ν obtained from equation (19). $\varphi_{L_i,J_i,-M_i}$ is a one electron orbital of the atomic electron involved in the collision with the reversed quantum numbers of the ion as we are considering a closed shell atom as a target. L_i, J_i and M_i are respectively the orbital angular momentum of the ion, its total angular momentum and the corresponding projection. To account for relativistic effects within the atom two different Dirac-Fock one electron orbitals (Grant et al. 80) are used for the two different final states of the ion resolved experimentally.

As we are considering relativistic effects, LS coupling no longer holds for the scattering state, and the direct and exchange potential matrix elements in equation (23) are therefore obtained in a j-j coupling scheme (McCarthy and Weigold, 95; Dorn et al., 96). The semi-relativistic DWBA method gives an approximate evaluation of the potential matrix elements that do not conserve the total spin of the system, even if the interaction potential is not spin dependent (in our case the interaction potential reduces to the e-e interaction). The relativistic interactions are directly introduced in the calculation of the distorted waves.

COMPARISON WITH EXPERIMENT

Before investigating the influence of relativistic interactions in the behaviour of the asymmetry function $\mathcal{S}_a$ it is important to quantify the ability of the semi-relativistic DWBA to describe the unpolarised TDCS for each final states of the ion. Figure 2 shows a comparison of the experimental unpolarised TDCS for both states of the ion with a semi-relativistic and non-relativistic calculation for a coplanar asymmetric kinematics (incident energy $E_0 = 147$eV, fast electron energy $E_f = 100$eV, fast electron angle $\theta_f = 28$ degrees). In the non-relativistic calculation a one-electron Harthree-Fock orbital (Clementi and Roetti, 74) is used for the bound electron while the static-exchange potential is used to calculate the distorted waves. The experiment is normalised to the semi-relativistic calculation for the $J_i = 1/2$ state of the ion at a slow electron angle $\theta_s = 40$ degrees. The non-relativistic calculations are divided by 1.2 to force the normalisation at that point. From this graph it is apparent that the shape of the TDCS are similar to the semi-relativistic and non-relativistic calculations for both final states of the ion. For both calculations we also notice a poor agreement with the experimental values for large values of the slow electron angle θ_s, while the shape of the TDCS is well described for small values of θ_s. However, a significant difference between the two calculations is observable in the prediction of the relative amplitudes of the TDCS of each final state of the ion. This later feature can be analysed in terms of branching ratio defined as

$$R = \frac{\sigma_{J_i=3/2}}{\sigma_{J_i=1/2}}. \tag{24}$$

The branching ratio for these transitions was studied under electron momentum spectroscopy conditions which measured the TDCS in non-coplanar symmetric kinematics near the Bethe-ridge (Cook et al., 86). The branching ratio in these measurements was dominated by the statistical weighting of the final states and the difference in the $^2P_{1/2}$ and $^2P_{3/2}$ wavefunctions due to relativistic structure effects. In the present case, due to the kinematics used, this quantity generally not only contains information on the structure of the target but also on the scattering approximation used (Dorn et al., 1996). This quantity is a more severe test for the accuracy of the evaluation of the unpolarised TDCS as it does not suffer from any normalisation procedure. Figure 2b shows a comparison of the experimental branching ratio with the semi-relativistic and non-relativistic calculations. It appears clear that only the semi-relativistic model is able to reproduce the experimental values of this quantity. The Dirac-Fock wave function used in the relativistic case is the cause of this improvement. The relativistic interactions for the continuum electrons have a small contribution to the branching ratio in the present kinematics. We also notice that at this particular scattering angle. $\theta_f = 28$ degrees, the measurements are performed near the Bethe-ridge E.M.S. condition. This suggests that the relativistic interactions within the atom need to be taken into account for a description of the unpolarised TDCS while the relativistic interaction for the continuum electrons can be neglected to a first approximation.

We will turn now to the main motivation of the present work, the asymmetry function. In figures 3a and b we compare the semi-relativistic and non-relativistic

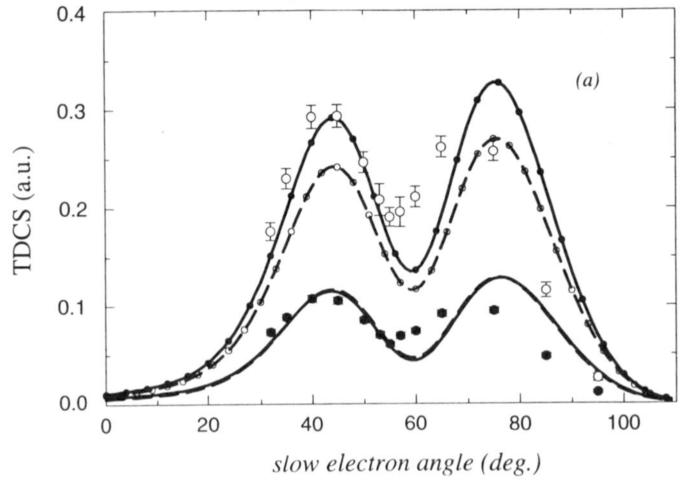

Figure 2a. Results for the unpolarised TDCS for the $J_i = 1/2$ (open circle) and $J_i = 3/2$ (full circle) final states of the ion, compared with the semi-relativistic calculations (solid lines) and non-relativistic calculations (dashed lines). The kinematics and normalisations are discussed in the text.

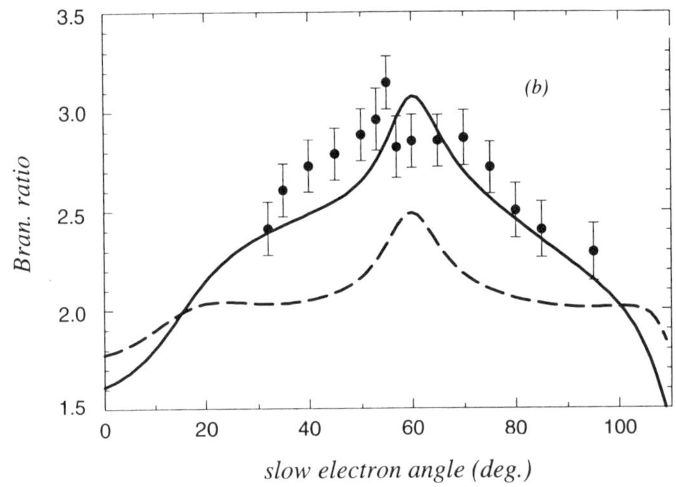

Figure 2b. Results for the branching ratio for the data in fig 2a compared with a semi-relativistic calculation (solid line) and a non-relativistic calculation (dashed line).

calculations of the asymmetry function with the experimental measurements for respectively the $J_i = 1/2$ and $J_i = 3/2$ states of the ion. Despite a marked improvement in the unpolarised TDCS and the branching ratio by the semi-relativistic calculation, the semi-relativistic and non-relativistic calculations predict a similar behaviour for the asymmetry function for both states of the ion. This is especially the case for the $^2P_{3/2}$ state of the ion where the two calculations overlap for all values of the slow electron angle θ_s. The situation is similar for the $^2P_{1/2}$ state of the ion. We notice however that

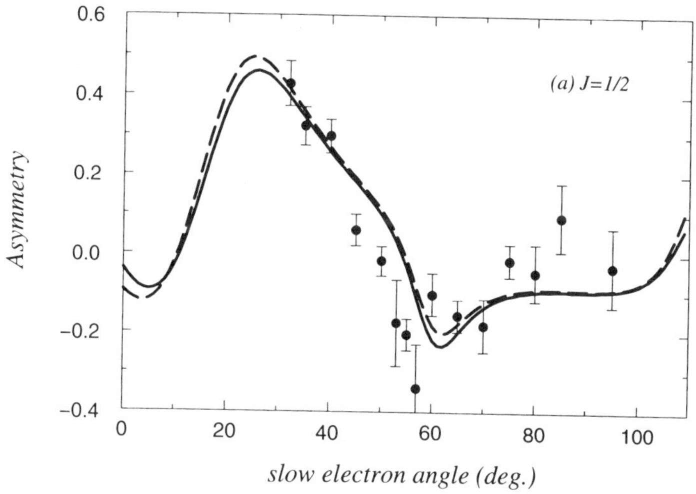

Figure 3a. Results for the asymmetry function for the $J_i = \frac{1}{2}$ state of the ion for the kinematics described in the text compared with that given by the semi-relativistic calculation (solid line) and the non-relativistic calculation (dashed line).

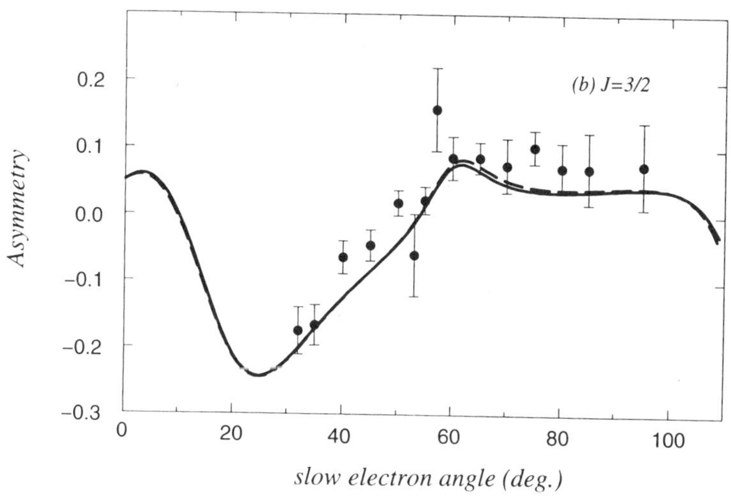

Figure 3b. Results for the asymmetry function for the $J_i = \frac{3}{2}$ state of the ion for the kinematics described in the text compared with that given by the semi-relativistic calculation (solid line) and non-relativistic calculation (dashed line).

small differences between the two calculations are observable when the TDCS is small. This can be understood from the expression (15) of the asymmetry function where it is shown that the relativistic interactions contribute to the asymmetry function as an interference between the direct and spin-flip amplitudes. This contribution is expected to be significant when these two amplitudes have magnitudes of the same order. As the direct amplitude has the largest contribution for the unpolarised TDCS, we can thus expect the most significant modifications of the asymmetry function due to the relativistic effects when the TDCS is small. In the present situation the asymmetry function is measured in the binary peak and the relativistic interactions do not modify dramatically the asymmetry function.

CONCLUSION

We have, by using incident polarised electrons, investigated the influence of relativistic interactions within the atom and for the continuum electrons on the behaviour of the asymmetry function at intermediate energy in a coplanar asymmetric kinematics. This study shows that a Dirac-Fock wave function is needed for a description of the unpolarised TDCS. In the present kinematics the exchange process is the dominant contribution for the asymmetry function while the relativistic interactions have little influence.

REFERENCES

K. Bartschat, *Phys. Rep.* **180**, (1989).

G. Baum, W. Blask, P. Freienstein, L. Frost, S. Hesse, W. Raith, P. Rappolt and M. Streun, *Phys. Rev. Lett.* **69**, 3037 (1992).

E. Clementi and C. Roetti, *At. Dat. Nucl. Dat. Tab.* **14**, 177 (1974).

J.P.D. Cook, I.E. McCarthy, J. Mitroy and E.Weigold, *Phys. Rev. A* **33**, 211 (1986).

J.P. Grant, B.J. McKenzie, P.H. Norrington, D.F. Mayers and N.C. Pyper, *Com. Phys. Com.* **21**, 207 (1980).

X. Guo, J. Hurn, J. Lower, S. Mazevet, Y. Shen, I.E. McCarthy and E. Weigold, *Abstract of Contributed Papers, The Physics of Electronic and Atomic Collisions, XIX International conference*, eds. L.J. Dubé, J.B.A. Mitchell, J.W. McConkey, C.E. Brion, 795 (AIP conf. proc., New York, 1995).

X. Guo, J. Hurn, J. Lower, S. Mazevet, I.E. McCarthy, Y. Shen, E. Weigold, *Aust. J. Phys.* **49**, 383 (1996).

X. Guo, J. Hurn, J. Lower, S. Mazevet, Y. Shen, E. Weigold, B. Granitza and I.E. McCarthy, *Phys. Rev. Lett.* **76**, 1228 (1996).

A. Dorn, A. Elliot, X. Guo, J. Hurn, J. Lower, S. Mazevet, I.E. McCarthy, Y. Shen and E. Weigold, to be published.

B. Granitza, X. Guo, J. Hurn, Y. Shen and E. Weigold *Abstract of Contributed Papers XVIII International Conference on the Physics of Electronic and Atomic Collisions*, Aarthus, 201 (1993).

J.B. Furness and I.E. McCarthy, *J. Phys. B* **6**, 2280 (1973).

G.F. Hanne, *Phys. Rep.* **95**, 95 (1983).

G.F. Hanne, *Abstract of Contributed Papers, XVII International Conference on the Physics of Electronic and Atomic Collisions*, W.R. MacGillivray, I.E. McCarthy, M.C. Standage, 199 (1991).

K. Hasenburg, D.H. Madison, K. Bartschat and K. Blum, *J. Phys. B* **19**, 1803 (1986).

S. Jones, D.H. Madison, G.F. Hanne, *Phys. Rev. Lett.* **72**, 2554 (1994).

J. Kessler, *Polarised Electrons, second editions*, Springer–Verlag, Berlin 1985.

D.H. Madison, V.D. Kravtsov, S. Jones and R.P. McEachran, *Phys. Rev. A* **53**, 2399 (1996).

I.E. McCarthy and E. Weigold, *Electron–atoms collisions*, Cambridge University Press, 1995.

R.P. McEachran and A.D. Stauffer, *J. Phys. B* **19**, 3523 (1986).

H.Th. Prinz, K.H. Besch and W. Nakel, *Phys. Rev. Lett.* **74**, 243 (1995).

MEASUREMENTS OF SPIN POLARIZED (e,2e) PROCESSES AT 54.4 eV

M. Streun, G. Baum, W. Blask, S. Förster, W. Raith, I. Schmähl

Fakultät für Physik, Universität Bielefeld, D-33501 Bielefeld, Germany

INTRODUCTION

The emphasis of our (e,2e) studies is directed towards understanding the spin behavior of the three–body coloumb problem when two continuum electrons emerge from a positively charged ion. At low energies distinct correlations are present between spin and momentum of these electrons and interesting effects are expected for the observable spin asymmetries. In coplanar geometry we extended our first measurements with equal energy of the outgoing electrons to asymmetric energy–sharing conditions using lithium atoms as target.

Our experimental setup contains a polarized electron beam which we cross with a spin-polarized Li atomic beam. Two electrostatic spectrometers are used to detect the (e,2e) events in coincidence technique. The angle of one of the electrons (A) is kept fixed and the angle of the other electron (B) is varied.

The spin asymmetry is defined by the ratio of triple–differential cross sections (σ)

$$A = \frac{\sigma^{\uparrow\downarrow} - \sigma^{\uparrow\uparrow}}{\sigma^{\uparrow\downarrow} + \sigma^{\uparrow\uparrow}}.$$

The arrows indicate the relative spin directions in the two beams, being antiparallel or parallel to each other. For light Li atoms the asymmetry is solely produced by exchange effects, therefore the orientation of the polarization direction with respect to the scattering plane is irrelevant. Parallel spins correspond to triplet states (σ_t) and antiparallel spins correspond 50% to triplet and 50% to singlet (σ_s) states, thus $A = (\sigma_s - \sigma_t)/(\sigma_s + 3\sigma_t)$. Experimentally the asymmetry is determined by measuring the quantities contained in the relation

$$A = \frac{1}{P_a\, P_e}\,\frac{N^{\uparrow\downarrow} - N^{\uparrow\uparrow}}{N^{\uparrow\downarrow} + N^{\uparrow\uparrow}}.$$

Here, P_a and P_e are the polarizations of atomic and electron beams; $N^{\uparrow\downarrow}$ and $N^{\uparrow\uparrow}$ are the coincidence counting rates for the indicated spin orientations which are measured in succession.

Coincidence Studies of Electron and Photon Impact Ionization
Edited by Whelan and Walters, Plenum Press, New York, 1997

Other (e,2e) experiments which study spin effects also use polarized electrons but unpolarized target atoms: The spin asymmetries which are observed in the experiment of the Tübingen group are the result of the spin–orbit interaction of continuum electrons [1]. Spin effects can also be observed for light or heavy atoms if the final fine–structure levels of the residual ion are resolved. Recent results have been reported by X. Guo et al. [2] on xenon atoms, by K.-H. Besch et al. [3] on uranium inner–shell ionization, and work with xenon is also in progress at Münster [4]. Depending on the size of the atom and on the spin orientation to the scattering geometry, in general an interplay of exchange and spin–orbit mechanism will be present for excitation amplitudes to the $m_l = -1$ and $m_l = +1$ orbital angular momentum states [5].

APPARATUS

In our first studies [6] we investigated the scattering at 54.4 eV with equal energy sharing ($E_A = E_B = 24.5$ eV) under $\Theta_A = 45°$ and variation of Θ_B from 35° to 80°. Thereafter, to improve on the precision of the experiment, we tried to upgrade the apparatus, mainly concentrating on increasing the polarizations of the two beams.

The polarized electrons are now produced with a strained GaAs crystal [7]. The strain causes a splitting of the degenerated valence bands and thus allows single band excitation with a high degree of polarization of the emitted electrons [8].

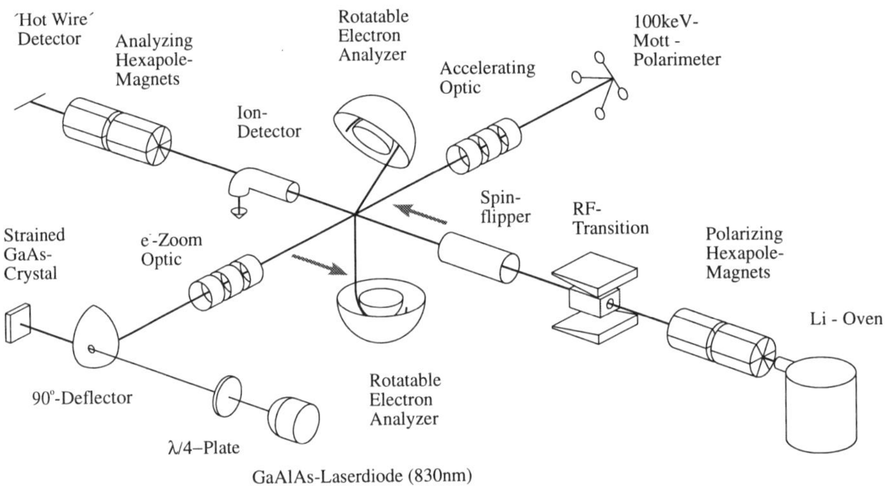

Figure 1: Schematic overview of the apparatus. Beam-flag, guiding–field coils, and spectrometer optics are not shown.

The crystal is illuminated with light from a GaAlAs laserdiode at a wavelength of 830 nm. The linear polarization of the light is transformed to circular polarization by means of a quarter–wave plate. Rotation of the quarter–wave plate by 90° allows to reverse the helicity of the light and with it the direction of the electron spin.

The surface of a newly installed crystal is treated with oxygen and cesium vapor alternately, until the current of the electron beam reaches a maximum. Subsequently, the Cs evaporator is set at a constant current by which the electron intensity is kept stable.

The electrons leave the crystal with spin orientation perpendicular to the crystal surface. After passing a 90° electrostatic deflector and a seven–element zoom optics, they reach the scattering region with a transverse spin orientation, parallel to the atomic beam. Electron optical elements immediately behind the scattering region ensure a proper guidance of the electron beam out of the scattering chamber, reducing the number of background counts. Finally the beam reaches the Mott polarimeter where the degree and direction of the polarization is measured. With the strained crystals we obtain a polarization of $P_e = 0.6$.

The atomic beam is produced by evaporating lithium from an oven and by collimating it by means of three heatable diaphragms. The temperature of the oven orifice of about 1100 K is sufficently high to dissociate Li_2 molecules.

For polarizing, the atoms first traverse a pair of sextupole magnets. In the strong field of these magnets the spins of nucleus and electron are decoupled. Because of the inhomogenity of the magnetic field, atoms with $m_J = +1/2$ will be focused and with $m_J = -1/2$ will be defocused. Thus all magnetic hyper-fine substates belonging to $m_J = -1/2$ will be eliminated from the beam. The weak–field spin polarization is given by the average of the polarization of the substates remaining in the beam. It is advantageous to use the isotope ^{6}Li instead of ^{7}Li. The nuclear spin of ^{6}Li is 1 instead of 3/2 and the remaining hfs substates give a polarization of 33%, instead of 25% for ^{7}Li.

Behind the sextupole magnets a newly installed rf-transition unit is located. Within a static magnetic field of about 10 mT, which is slightly decreasing along the beam axis, a magnetic rf field is produced in a rf resonator. The frequency (295 megacycles) and the static field strength are tuned such that transitions from the hfs substate $|F=3/2, m_F=-1/2\rangle$ to $|F=1/2, m_F=-1/2\rangle$ are induced (see Fig. 2). Here a state with polarization $-1/3$ is replaced by a state with polarization $+1/3$. This increases the beam polarization from 33% to 56% without loss in intensity. The obtainable figure of merit is nearly the same as for the case when all atoms but the ones in the single hfs substate with polarization +1 are removed from the beam.

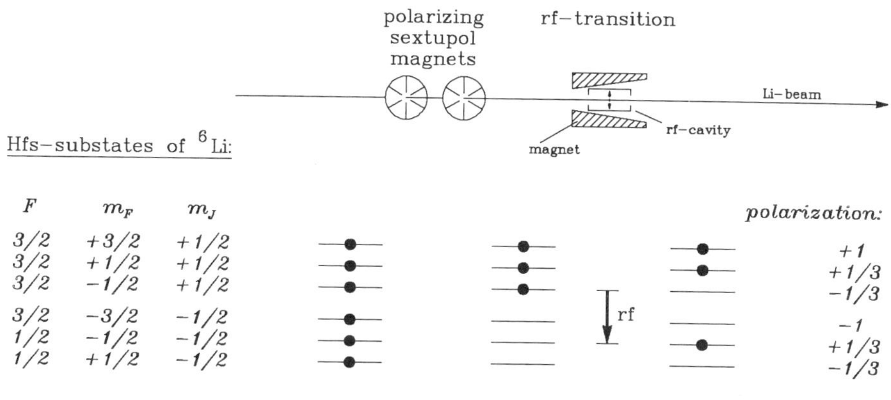

Figure 2: Mode of operation of the polarization unit for the atomic beam. The population of the hfs–substates is illustrated.

The polarizing unit is followed by the spin–flipper which allows to reverse the polarization by choosing a diabatic or adiabatic passage [9]. A beam flag at the entrance to the scattering chamber allows to stop the atomic beam for background measurements.

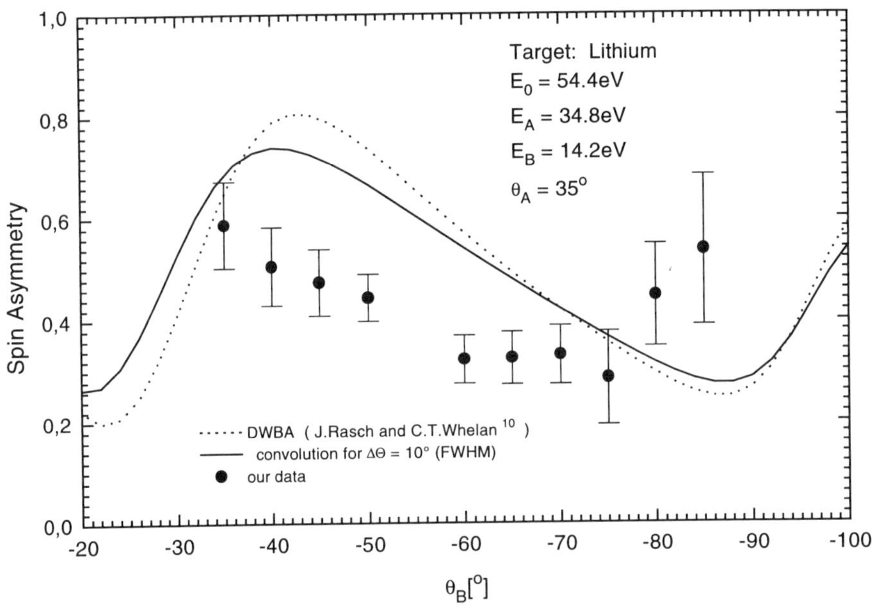

Figure 3: Measurements for asymmetric energy sharing in comparison with theoretical data.

Along the atomic beam–line there are several coils producing magnetic guiding fields parallel to the beam axis to avoid depolarization of the atomic spins by uncontrolled magnetic fields. In the scattering region the guiding field must be weak in order that its influence on the path of the scattered electrons can be neglected.

The ions produced in the scattering region are collected for monitoring measurements. They are separated from the atomic beam by means of an electrostatic deflector and detected in a channel electron multiplier. The ion count rate is very helpful to check for the proper setting of the system. The spin asymmetry of the ion signal allows to determine the asymmetry of the total cross section, which should be constant for fixed incident electron energy.

Before the atomic beam reaches the hot–wire detector, where its intensity is recorded, it has to pass a second pair of sextupole magnets. The state selection of these magnets leads to different detector signals whether the spin is flipped or not. This allows to determine the atomic beam polarization, as described in its principle in [9]. With employing the rf transition we obtain $P_a = 0.5$.

RESULTS AND DISCUSSION

For investigations with asymmetric energy sharing at $E_A = 34.8$ eV, $E_B = 14.2$ eV and $\Theta_A = 35°$ new results are shown in Fig. 3. The angular range covered extends from $\Theta_B = 35°$, given through the limits of the apparatus, to $\Theta_B = 85°$, limited by the smallness of the cross section. Within this angular range a nearly linear decrease of the asymmetry is seen from 0.6 to 0.3, corresponding to an increase in the triplet to singlet cross section ratio σ_t/σ_s from 0.14 to 0.37.

In order to compare our data with calculations, theoretical results have to be convoluted with the experimental angular resolution. In particular this is necessary as our angular resolution is rather modest and corresponds to a width of $\Delta\Theta_A = \Delta\Theta_B =$ $10°$ (FWHM) in the detector acceptances, which are determined by electron optical simmulations. To get the convoluted asymmetry function $\bar{A}$ which can be compared with the experimental data, the theoretical asymmetry data A must be averaged over the angular range by weighting with the theoretical cross section σ and with the angular acceptance functions ξ_A, ξ_B of the detectors:

$$\bar{A}(\Theta_A,\Theta_B) = \frac{\iint A(\vartheta_A,\vartheta_B)\ \sigma(\vartheta_A,\vartheta_B)\ \xi_A(\vartheta_A,\Theta_A)\ \xi_B(\vartheta_B,\Theta_B)\ d\vartheta_A\, d\vartheta_B}{\iint \sigma(\vartheta_A,\vartheta_B)\ \xi_A(\vartheta_A,\Theta_A)\ \xi_B(\vartheta_B,\Theta_B)\ d\vartheta_A\, d\vartheta_B}$$

$$\approx \frac{\sum_{\vartheta_A}\sum_{\vartheta_B} A(\vartheta_A,\vartheta_B)\ \sigma(\vartheta_A,\vartheta_B)\ \xi_A(\vartheta_A,\Theta_A)\ \xi_B(\vartheta_B,\Theta_B)}{\sum_{\vartheta_A}\sum_{\vartheta_B} \sigma(\vartheta_A,\vartheta_B)\ \xi_A(\vartheta_A,\Theta_A)\ \xi_B(\vartheta_B,\Theta_B)}$$

Here Θ_A, Θ_B indicate the angular positions of the two detectors, and ϑ_A, ϑ_B are the scattering angles. We have calculated the sums with an angular step size of $1°$. If the theoretical values are available only with a larger step size, the values in between are obtained by linear interpolation. The angular acceptance functions ξ_A, ξ_B were approximated with a triangle function:

$$\xi_i(\vartheta_i,\Theta_i) = \begin{cases} 1 - \frac{|\vartheta_i - \Theta_i|}{\Delta\Theta_i} & \text{for} \quad |\vartheta_i - \Theta_i| \le \Delta\Theta_i \\ 0 & \text{else.} \end{cases}$$

Based on electron optical studies, this approximation is considered to be quite accurate. It has to be noted that this convolution is only an averaging over the angles in the scattering plane. It does not take into account effects of the the angular acceptance of the detectors in out–of–plane direction. Studies with available out–of–plane data showed, that their influence on the value of $\bar{A}$ is negligible.

As can be seen from Fig. 3, the convoluted theoretical asymmetry data with which we compare deviate less than 0.1 from the unconvoluted one. Theory places the maximum of the triple differential cross section near $60°$. The maximum of the theoretical asymmetry occurs near $45°$, indicating that the singlet part of the cross section has a relative stronger contribution on the forward part of the binary peak than the triplet part. In the angular range from $45°$ to $65°$ our asymmetry data are about 0.2 lower than the theoretical expectations. The maximum seems to lie below $35°$, a region which we can not access due to mechanical constrains.

In Fig. 4 we show again our earlier results which were obtained under equal energy sharing conditions [6]. At $\Theta_B = 45°$ a new point is added which has been obtained with the improved apparatus. In our earlier work we had carried out the comparison with theory by computing an average of the theoretical asymmetry over the angular acceptance range of the detectors, without weighting the asymmetry by the cross section. If the convolution is done more correctly with the procedure which we have described in detail above it leads to the result shown in Fig. 4. The agreement with the data is now not very good in contrast to the comparison before. In the range where the cross section is large three points lie below the theory by one standard deviation.

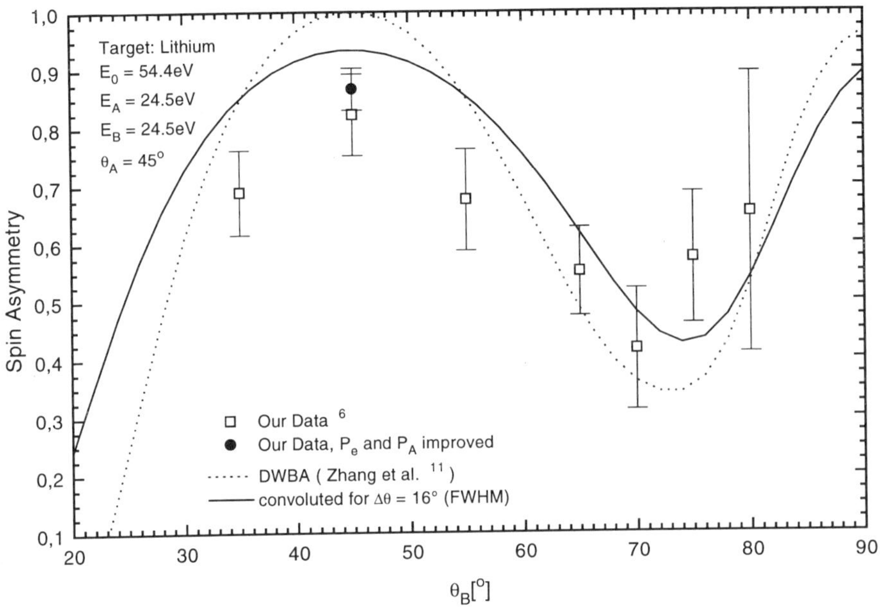

Figure 4: Earlier measurements, but now with new convolution of the theoretical data.

CONCLUSION

Enhanced polarization values in our two beams enables us to measure asymmetries in the triple differential cross section with reasonable statistical accuracy (± 0.05). These data provide new experimental information which is complementing absolute cross section data and which will be useful in understanding the three–body Coulomb problem and the electron impact ionization process at low energies. We have obtained results at 54.4 eV incident energy in symmetric and asymmetric energy sharing conditions. Our studies at this energy will be completed soon and we will then attack measurements at lower incident energies.

ACKNOWLEDGEMENTS

We thank Drs. J. Rasch and C. T. Whelan for providing us with their unpublished theoretical results. Our research is supported by the DFG (SFB 216).

REFERENCES

1. H.-Th. Prinz, K.-H. Besch, and W. Nakel, Phys. Rev. Lett. **74**, 243 (1995)
2. X. Guo, J. M. Horn, J. Lower, S. Mazevet, Y. Shen, E. Weigold, B. Granitza, and I. E. McCarthy, Phys. Rev. Lett. **76**, 1228 (1996)
3. K.-H. Besch, M. Sauter, and W. Nakel contribution to *Euroconference on Coincidence Studies of Electron and Photon Impact Ionization*, Belfast (1996)
4. T. Simon, S. Mette, and G. F. Hanne, XIX. ICPEAC Book of Abstracts (1995) 223

5. S. Jones, D. H. Madison, and G. F. Hanne Phys. Rev Lett **72**, 2443 (1994), and D. H. Madison, V. D. Kravtsov, S. Jones, and R. P. McEachran, Phys. Rev. A **53**, 2399 (1996)

6. G. Baum, W. Blask, P. Freienstein, S. Hesse, W. Raith, P. Rappolt, and M. Streun Phys. Rev. Lett. **69**, 3037 (1992)

7. G. Baum in *Proceedings of the Workshop on Photocathodes for Polarized Electron Sources for Acceleration*, M. Chatwell et al., eds, SLAC Report 432 Rev. (1994) p.13

8. R. Prepost and T. Muruyama, Annu. Rev. Nucl. Part. Sci. **45**, 41 (1995)

9. W. Schröder and G. Baum J. Phys. E **16**, 52 (1983)

10. J. Rasch and C. T. Whelan, priv. comm. 1996

11. X. Zhang, C. T. Whelan, and H .R. J. Walters, J. Phys. B **25**, 17 (1992)

IS THERE ANY NEW PHYSICS IN SPIN DEPENDENT (e,2e) COLLISIONS?

Don H. Madison[1], V.D. Kravtsov[1], S. Jones[1] and S. Mazevet[2]

[1]Department of Physics
University of Missouri-Rolla
Rolla, MO 65401, USA

[2]Research School of Physical Sciences and Engineering
Australian National University
Canberra, ACT 0200, Australia

INTRODUCTION

The technology for producing beams of spin polarized electrons has been developing for the last two decades. Until recently, most of the experiments which were performed using spin polarized electrons concentrated on elastic and inelastic scattering from atoms[1-6]. Quite recently, it was suggested that spin polarized electrons could also be used to examine ionizing collisions[7-9].

Of course, the most important question concerns whether or not any new physics could be learned with spin-polarized electrons. It was felt by many that such experiments would not reveal any new physical phenomena and consequently would not be of value. However, for the case of inelastic scattering, it was well known that spin-polarized electrons provided sensitive probes for exchange scattering if atoms were excited which were heavy enough that the final J-levels could be distinguished experimentally[4,5]. This effect, which has become known as the fine-structure-effect, predicts that large spin-effects are possible even if one neglects spin-dependent forces on the spin polarized projectile under the assumption that LS coupling is valid. Hanne[7] suggested that it should be possible to obtain comparable information for ionization if the final J-state of the ion could be experimentally resolved. On the other hand, ionization represents the equivalent of a coherent sum over an infinite number of discrete excitations and interference effects might tend to either diminish or even eliminate possible spin effects. This chapter will examine the current status of the spin-polarized work.

HANNE EFFECT

Figure 1 illustrates the spin-polarized scattering experiments. Non-relativistic experiments are currently being performed at Münster, Germany and Canberra, Australia and relativistic experiments are being performed at Tübingen, Germany. For the non-relativistic experiments, the direction of polarization for the incident beam is perpendicular to the scattering plane so the initial arrow in Fig. 1 should be considered as perpendicular to the scattering plane. After the atom is ionized, the two final state electrons with energy E_a and E_b are detected in coincidence in the scattering plane and the ion (designated as I in the figure) is left in a state of definite total angular momentum J.

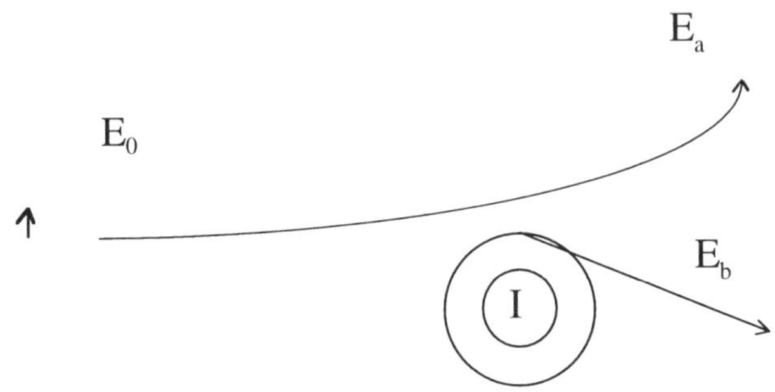

Figure 1. Schematic representation of the spin-polarized experiments. The incident electron with energy E_0 has a spin polarization perpendicular to the scattering plane. As a result of the ionizing collision, electrons with energy E_a and E_b are measured in coincidence. The residual ion I has a definite angular momentum J.

The experimentally measurable quantity is the asymmetry A which is the difference in coincidence counts for spin up and spin down electrons divided by the sum

$$A = \frac{Counts_{\uparrow} - Counts_{\downarrow}}{Counts_{\uparrow} + Counts_{\downarrow}} \tag{1}$$

In terms of Cross sections, A is given by

$$A_J = \frac{\sigma_J(\uparrow) - \sigma_J(\downarrow)}{\sigma_J(\uparrow) + \sigma_J(\downarrow)} \tag{2}$$

where $\sigma_J(\uparrow)$ is the cross section for leaving the ion in angular momentum state J as a result of collisions with electrons with spin-polarization up perpendicular to the scattering plane. This cross section depends on the incident energy as well as the energies and observation angles for the two final state ionized electrons.

The primary question concerns what can cause the asymmetry to be non-zero. The obvious answer is spin dependent forces acting on the spin-polarized electron which, of course, means relativistic effects. As mentioned above, Hanne[7] pointed out that it is possible to have spin dependent effects even if there are *no spin-dependent forces* on the projectile. This effect has been known and studied for some time for the case of inelastic electron-atom scattering and for this case it has become known as the fine-structure effect. The two requirements for the fine structure effect are that the final J-state of the atom be experimentally resolvable and that LS coupling be assumed for the atom.

The essence of the fine structure effect can be seen from Fig. 2. The target gas used in current experiments is xenon in which case the final ionic state contains a p-hole. For this case, there are two resolvable J-states of 1/2 and 3/2. For simplicity, let's consider the J= 1/2 case in this chapter. The top part of Fig. 2 illustrates the J=1/2, M_J=1/2 possibility. Within LS coupling, the only way the ion can have M_J=1/2 is to have M_L=1 and M_S=-1/2 where M_L and M_S are the projections of orbital and spin angular momentum (perpendicular to the scattering plane) respectively. If the ion has a spin projection of M_S=-1/2, the ejected electron must have a spin up (relative to the scattering plane) since total spin is conserved and the initial spin was zero. As a result of the Pauli principle, the collision can proceed either via the direct or exchange mechanisms. The direct collision is represented by the left-hand-side of Fig. 2. For this case, the projectile electron (solid line) reaches detector 'a' and the atomic electron (dashed line) reaches detector 'b'. The scattering amplitude for this process is f_1 where f is the direct amplitude for leaving the ion in a state with angular momentum projection 1. The right-hand-side of Fig. 2 illustrates the exchange process. For this case the projectile and atomic electron switch places and the projectile reaches detector 'b' and the atomic electron reaches detector 'a'. The exchange scattering amplitude is designated as g_1.

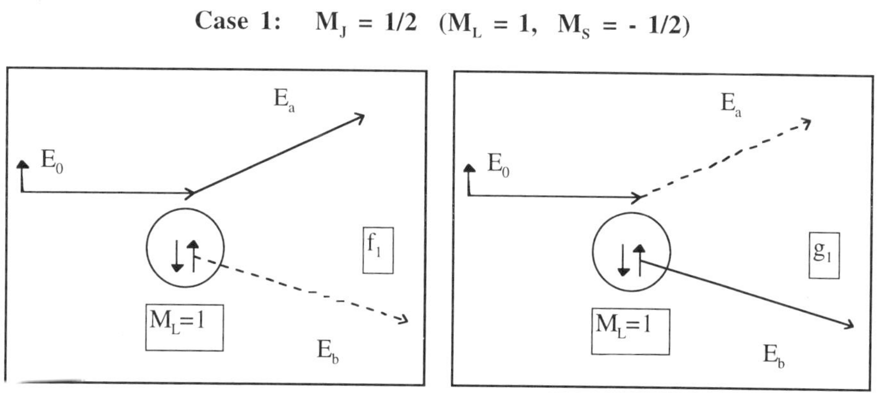

Figure 2. Schematic representation of a spin-up electron leaving the final ion in a state with M_J = 1/2, M_L = 1 and M_S = - 1/2. The solid line represents the path of the projectile electron and the dashed line represents the path of the atomic electron and f_M (g_M) is the direct (exchange) scattering amplitude for leaving the ion in a state with angular momentum projection M perpendicular to the scattering plane.

If the incident electron is spin-up, then both the direct and exchange processes result in spin-up electrons in both detectors which is experimentally indistinguishable. As a result, the amplitudes add coherently and the triple differential cross section (TDCS) is given by

$$\sigma_{\frac{1}{2},M_J=\frac{1}{2}}(\uparrow) = \tfrac{2}{3}\left|f_1 - g_1\right|^2 \qquad (3)$$

where the 2/3 in front stems from the angular momentum coupling coefficients. Figure 3 represents the second possibility of leaving the ion in the M_J = - 1/2 state.

Case 2: $M_J = -1/2$ ($M_L = -1$, $M_S = 1/2$)

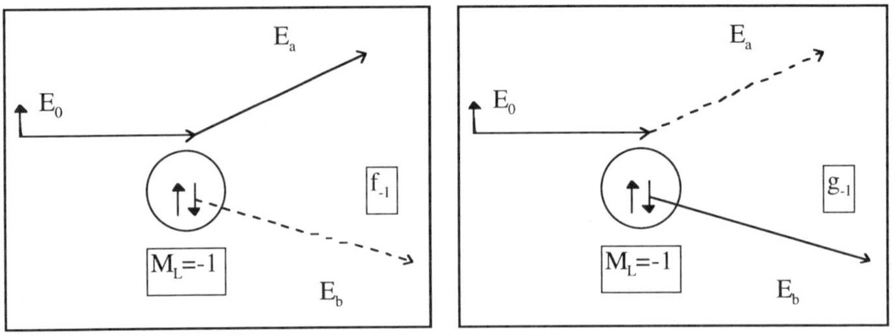

Figure 3. Schematic representation of a spin-up electron leaving the final ion in a state with $M_J = -1/2$, $M_L = -1$ and $M_S = 1/2$. The designations are the same as Fig. 2

For this case, the ejected electron must be a spin-down and the direct case results in a spin-up electron in detector 'a' and a spin-down in detector 'b'. The exchange case, on the other hand, results in a spin-down in detector 'a' and a spin-up in detector 'b'. Since these two cases are experimentally distinguishable, the amplitudes add incoherently and the TDCS is given by

$$\sigma_{\frac{1}{2},M_J=-\frac{1}{2}}(\uparrow) \quad = \quad \tfrac{2}{3}\left(|f_{-1}|^2 + |g_{-1}|^2\right) \tag{4}$$

Since the experiments do not resolve the angular momentum projection of the ion, the two cases represented in Figs. 2 and 3 must be summed over and the spin-up cross section for leaving the ion in J= 1/2 is given by

$$\sigma_{\frac{1}{2}}(\uparrow) \quad = \quad \tfrac{2}{3}\left(|f_{-1}|^2 + |g_{-1}|^2 + |f_{+1} - g_{+1}|^2\right) \tag{5}$$

The cross sections for the spin down case may be obtained using similar arguments

$$\sigma_{\frac{1}{2}}(\downarrow) \quad = \quad \tfrac{2}{3}\left(|f_{+1}|^2 + |g_{+1}|^2 + |f_{-1} - g_{-1}|^2\right) \tag{6}$$

Since amplitudes with different angular momentum projection perpendicular to the scattering plane should be different, Eqs. (5) and (6) reveals that the asymmetry would be expected to be zero only if the exchange amplitude vanishes. As a result, in this simple model, non-zero asymmetry is measure of the importance of the exchange amplitude.

THEORETICAL MODEL

The calculations reported here have been performed within the framework of the first order distorted wave Born approximation (DWBA). The details of this method may be found elsewhere[10-14]. Here we will only provide a brief overview of the main points of the theory. In the DWBA, the direct scattering amplitude may be expressed as

$$f \quad = \quad \left\langle \chi_a^- \, \chi_b^- \, \psi_I \,\middle|\, V - U_i \,\middle|\, \psi_i \, \chi_i^+ \right\rangle \tag{7}$$

where χ_i is the wavefunction for the incoming electron, χ_a and χ_b are the wavefunctions for the two final state electrons, ψ_A is the initial atomic wavefunction, ψ_I is the wavefunction for the final state ion and V is the full interaction between the incident electron and the atom. In the direct amplitude, the incident electron is finally in χ_a and in the exchange amplitude, this electron is finally in χ_b.

For the atomic wavefunctions, we have used the multiconfiguration Dirac-Fock code of Grant[15] which produces both $5p_{\frac{1}{2}}$ and $5p_{\frac{3}{2}}$ orbitals for the ground state of xenon. For the case of interest here where the residual ion has J = 1/2, the six electrons in the outermost shell were treated as a p-orbital and a p-hole and both the hole and orbital were assumed to have a J of 1/2. The corresponding J-dependent p-orbitals from Grant's code were then used for the active electron. These relativistic wavefunctions, of course, have large and small components whereas the rest of the theory is non-relativistic. To use the relativistic wavefunctions, we dropped the small component and renormalized the large component to unity. For xenon, the large component represented 99.99% of total wavefunction prior to renormalization.

The initial state distorting potential for the atom U_i is calculated from the atomic orbitals in the standard static approximation and the initial state wavefunction χ_a is a solution of Schrödinger's equation using this distorting potential

$$\left(T_0 + U_i - \varepsilon_i\right)\chi_i^+ = 0 \tag{8}$$

Here T_0 is the kinetic energy operator and ε_i is the energy of the incident electron. The wavefunctions for the two final state continuum electrons are calculated in a similar manner from the final state potential U_f

$$\left(T_0 + U_f - \varepsilon_{a(b)}\right)\chi_{a(b)}^+ = 0 \tag{9}$$

One of the issues for an ionizing calculation concerns the proper asymptotic charge for each of the two continuum electrons. In principle, the appropriate charge will depend both on the relative energies and angular separations of the two electrons. We have investigated the different possibilities[11] for the final state distorting potential and have found that it is best to use the same potential for both electrons (i.e. symmetric treatment) with an angle dependent effective charge.

$$U_f = Z_{eff}\, U_I + (1 - Z_{eff})\, U_i \tag{10}$$

where

$$Z_{eff} = 1 - \frac{1}{2\sin\left(\frac{\theta_{12}}{2}\right)} \tag{11}$$

Here U_I is the static potential for the final state ion and θ_{12} is the angle between the two continuum electrons. Close to the nucleus, U_i and U_I are essentially identical and far from the nucleus U_i vanishes and U_I becomes the potential for a single positive charge. As a result, the potential (10) has the proper form near the nucleus and asymptotically becomes the potential of a point charge Z_{eff}. The particular form (11) for Z_{eff} insures that the two electrons will have the proper potential energy when the two electrons leave the ion in opposite directions. Finally, we would note that the distorted waves are orthogonalized to the atomic wavefunctions which causes the capture and shake-off amplitudes to vanish.

RESULTS

Before we look at the spin dependent results, it is instructive to examine how well the theory works for spin-averaged differential cross sections (DCS). In Fig. 4, the present theoretical results are compared with the experimental spin averaged DCS measurements of Granitza *et al.*[16] for ionization of xenon by 147 eV incident electrons for the case in which the

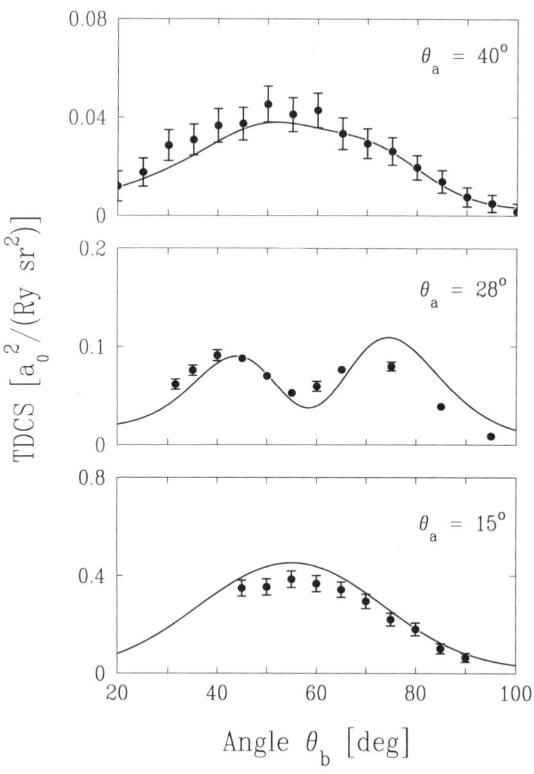

Figure 4. Triple differential cross section for ionization of xenon by 147 eV unpolarized electrons with the residual ion being left in the J=3/2 state. The faster final state electron has an energy of 100 eV and it is observed at an angle θ_a measured counterclockwise relative to the beam direction. The angle θ_b corresponds to the angle of observation for the slower electron which is measured clockwise relative to the beam direction. The experimental data are those of Granitza *et al.*[16] normalized to obtain the best overall agreement with theory. The theoretical curves are DWBA results as described in the text.

final ion is left in the J = 1/2 state. The unnormalized experimental data were normalized to give the best fit to the theoretical calculation. In our convention, θ_a is the angle of observation for the faster final state electron which is measured counterclockwise from the beam direction when viewing the scattering plane from above and θ_b is the observation angle for the slower electron which is measured clockwise. The three parts of Fig. 4 correspond to three different θ_a. It is seen that overall there is very good agreement between experiment and theory with the largest differences occurring for $\theta_a = 28°$.

Figure 5 compares theory and experiment for the spin asymmetry. The symbols and curves are the same as Fig. 4. In fact Fig. 4 and Fig. 5 represent the same data set. The DCS presented in Fig. 4 were obtained by averaging over the spin-dependent results of Fig. 5. For the spin-dependent results, we see a much different picture than was seen for the spin-averaged DCS. Instead of very good agreement between experiment and theory, the agreement is quite bad! Interestingly, the only case where there is even qualitative agreement between experiment and theory is the $\theta_a = 28°$ case which was the worst case for the spin-averaged DCS. From this observation, it is very clear that the spin-dependent measurements are sensitive to physical effects that cannot be seen in the spin-averaged DCS measurements.

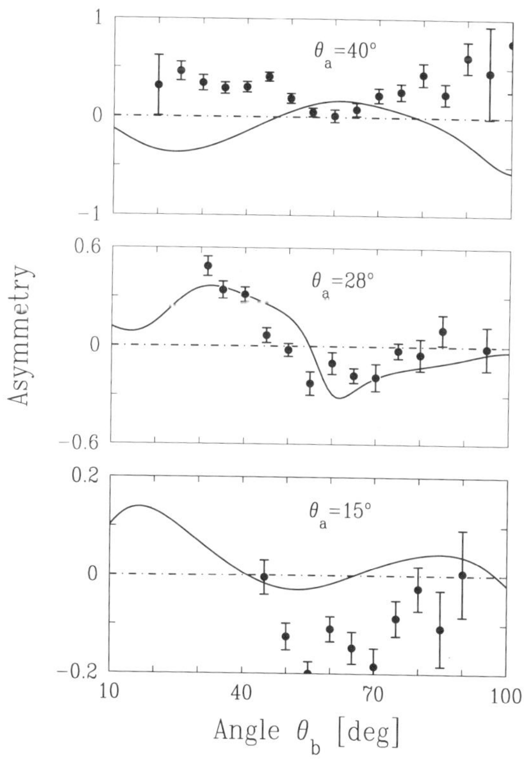

Figure 5. Spin up-down asymmetry for 147 eV spin-polarized electron impact ionization of xenon. The curves and experimental data are the same as described in Fig. 4 with the exception that these results are independent of normalization.

In other words, the question concerning whether or not there is any new physics to be found with spin-polarized beams can be answered in the affirmative.

The obvious next question concerns the missing physics in the present calculations. The obvious answer is relativistic effects on the projectile which have been omitted here. This possibility has been investigated by Mazevet and McCarthy[17] and is discussed elsewhere in this book. When Relativistic and non-relativistic calculations were compared, it was found that relativistic effects on the projectile are not very important and do not explain the significant differences between theory and experiment seen in Fig. 5.

Another Non-Relativistic Exchange Effect

The non-relativistic spin effects that have been included in this work are manifested through the exchange amplitude (recall that the asymmetry is zero if the exchange amplitude is ignored). There is another type of exchange effect which has not been included in the theory. Figure 6 represents a generic scattering collision between an incident projectile and an arbitrary charge distribution ρ.

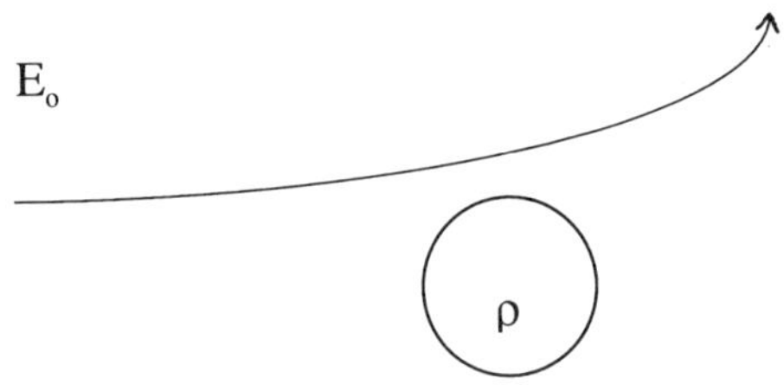

Figure 6. Schematic representation of electrons scattering from a charge density ρ.

If the projectile is an electron and the charge cloud are also electrons, one should take into account the indistinguishability of particles in the calculation of the projectile wavefunction. Consequently, instead of solving the homogeneous Eq. (8), one should solve an inhomogeneous equation with an exchange potential

$$(T_0 + U_i - \varepsilon_i)\chi_i^+(r) = \int V_{ex}(r,r')\,\chi_i^+(r')\,dr' \tag{12}$$

where V_{ex} is the non-local exchange potential. To determine the importance of this effect, we used the Furness-McCarthy[18] local approximation for exchange and replace Eq. (12) with

$$(T_0 + U_i - \varepsilon_i)\chi_i^+(r) = U_{ex}(r)\,\chi_i^+(r) \tag{13}$$

where

$$U_{ex} = -\tfrac{1}{2}\left\{\sqrt{(\varepsilon_i - U_i)^2 + 4\pi\rho} - (\varepsilon_i - U_i)\right\} \tag{14}$$

For the two final state electrons, the charge density for the final state ion is used.

The effect of the exchange potential on the asymmetry parameter for J = 1/2 scattering is shown in Fig. 7. It is seen that the exchange potential has almost the effect of a sign reversal on the θ_a = 40° case which brings theory into reasonable qualitative agreement with experiment. For the θ_a = 28° case, the exchange potential had little effect and for θ_a = 15°, there was a small change in the direction of the data. It should be noted, however, that the scales on the three curves are different with the 15° case being the most magnified. From this limited data set, it would appear that the previous disagreement between experiment and theory can be attributed to neglecting exchange distortion in the calculation of the continuum electron wavefunctions.

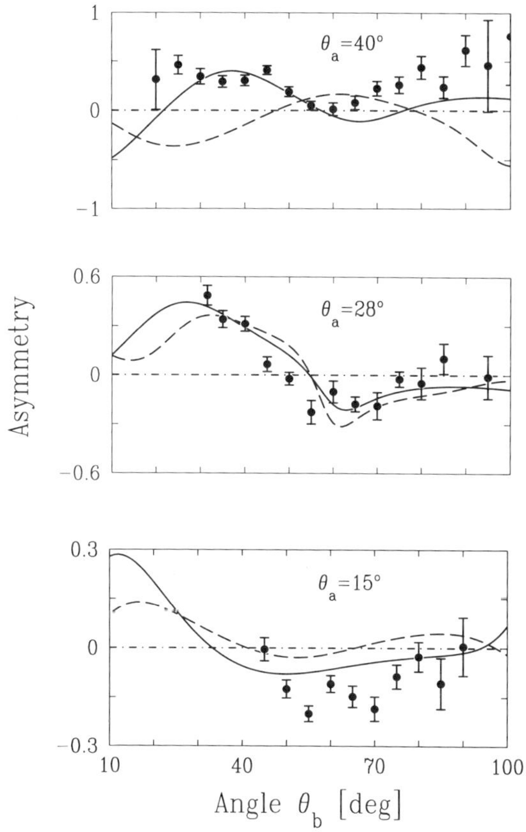

Figure 7. Effect of exchange distortion on the spin up-down asymmetry for 147 eV spin-polarized electron impact ionization of xenon with the ion being left in the J=3/2 state. The experimental data are the same as described in Fig. 5. The theoretical curves are: dashed - DWBA results neglecting exchange distortion; solid - DWBA results including exchange distortion.

SUMMARY

The field of (e,2e) is now over two decades old and during that time significant progress has been made. Experiments are now capable of measuring absolute DCS both in an out-of the scattering plane. The recent experimental advance of using spin-polarized electrons for the (e,2e) work has opened new opportunities for probing new physical effects. As a result of the work which has been performed so far, it is clear that large spin-dependent effects may be expected even for those cases in which relativistic spin-flip forces are small. It is also clear that the spin-dependent experiments provide a more sensitive probe than unpolarized beams of electrons. For example, it was shown here that a theory which provided very good agreement with experimental spin-averaged DCS was in very bad agreement with spin-asymmetry measurements. The deficiency, which was not observable using unpolarized electrons, was traced to the neglect of exchange distortion in the calculation of the continuum electron wavefunctions. The experiments which have been performed so far indicate that this new work should provide a sensitive mechanism for investigating the effects of exchange scattering. We expect that the role and importance of exchange will become clearer as additional experimental data become available.

REFERENCES

1. M. Duemmler, G. F. Hanne, and J. Kessler, in *Proceedings of the XVIII International Conference on the Physics of Electronic and Atomic Collisions, Abstracts of Contributed Papers*, edited by T Andersed, B. Fastrup, F. Folkmann, and H. Knudsen (Aarhus University, Aarhus, Denmark), p. 144 (1993).
2. G. F. Hanne, Phys. Rep.95:95 (1983).
3. K. Bartschat and D. H. Madison, J. Phys. B: At. Mol. Opt. Phys. 20:5839 (1987).
4. K. Bartschat and D. H. Madison, J. Phys. B: At. Mol. Opt. Phys. 21:2621 (1988).
5. N. Anderson, J. W. Gallagher, and I. V. Hertel, Phys. Rep. 165:1 (1988).
6. N. T. Padial, G. D. Menezes, F. J. da Paixao, G. Csanak, and D. C. Cartwright, Phys. Rev. A 23:2194 (1990).
7. G. F. Hanne in *Correlations and Polarization in Electronic and Atomic Collisions and (e,2e) Reactions* (Institute of Physics Conference Series Number 122, P. J. O. Teubner and E. Weigold, Eds.) p. 15 (1991)
8. T. Simon, M. Laumeyer, G. F. Hanne, and J. Kessler, Contribution to 7th International Symposium on Polarization and Correlation in Electronic and Atomic Collisions, Universität Bielefeld, Bielefeld, Germany, 29-31 July 1993 (unpublished).
9. G. Baum, W. Blask, P. Freienstein, L. Frost, S. Hesse, W. Raith, P. Rappolt, and M. Streun, Phys. Rev. Lett. 69:3037 (1992).
10. D.H. Madison, R.V. Calhoun and W.N. Shelton, Phys. Rev. A 16:552 (1977).
11. S. Jones, D. H. Madison, and M. K. Srivastava, J. Phys. B: At. Mol. Opt. Phys. 25:1899 (1992).
12. T. Rösel, J. Röder, L. Frost, K. Jung, H. Ehrhardt, S. Jones, and D. H. Madison, Phys. Rev. A 46:2539 (1992).
13. S. Jones, D. H. Madison, and G. F. Hanne, Phys. Rev. Lett. 72:2554 (1994).
14. D.H. Madison, V.D. Kravtsov, S. Jones and R.P. McEachran, Phys. Rev. A 53:2399 (1996)
15. I. P. Grant, B. J. McKenzie, P. H. Norrington, D. F. Mayers, and N. C. Piper, Comput. Phys. Commun. 21:207 (1980).
16. B. Granitza, X. Guo, J. Hurn, J. Lower, S. Mazevet, I.E. McCarthy, Y. Shen, and E. Weigold, Aust. J. Phys. 49:383 (1996).
17. S. Mazevet and I.E. McCarthy, private communication.
18. J.B. Furness and I.E. McCarthy, J. Phys. B: At. Mol. Opt. Phys. 6:2280 (1973).

ELECTRON-MOMENTUM SPECTROSCOPY OF SOLIDS BY THE *(e,2e)* REACTION

A. S. Kheifets[1], M. Vos[1], S. A. Canney[1], X. Guo[1], I. E. McCarthy[1], and E. Weigold[2]

[1]Electronic Structure of Materials Centre, Flinders University of South Australia, Adelaide 5001, Australia
[2]Research School of Physical Sciences and Engineering, Australian National University, Canberra, ACT 0200, Australia.

INTRODUCTION

The *(e,2e)* reaction has been used for many years as a probe of target electronic structure of atoms and molecules [1]. The notation *(e,2e)* refers to a process in which an incident electron (energy $E_\circ$, momentum $\mathbf{p}_\circ$) knocks out a target electron, with subsequent detection of both outgoing electrons in time coincidence. After detection of the *(e,2e)* event, the energies (E_f and E_s) and momenta ($\mathbf{p}_f$ and $\mathbf{p}_s$) of both outgoing electrons are determined. Here indices f and s specify the faster and the slower of the two outgoing electrons, respectively. The binding energy ε and momentum $\mathbf{q}$ of the target electron *before* the collision are given by the following conservation laws:

$$\varepsilon = E_\circ - E_s - E_f \; , \qquad \mathbf{q} = \mathbf{p}_s + \mathbf{p}_f - \mathbf{p}_\circ \tag{1}$$

Under conditions of a high incident energy and a large momentum transfer

$$E_\circ >> \varepsilon \; , \qquad |\mathbf{p}_\circ - \mathbf{p}_f| > 1 \; , \tag{2}$$

the mechanism of an *(e,2e)* reaction is particularly simple: it is a free electron binary collision between the projectile and the target electron. We use atomic units of momentum 1 a.u. = $a_\circ^{-1}$ in Eq. (2) and throughout the paper. Although the target electron is bound, it can be assigned the attributes of a free electron, i.e. the energy ε and momentum $\mathbf{q}$, with a certain probability $\rho(\varepsilon, \mathbf{q})$. Under this binary collision assumption the *(e,2e)* cross-section is given by a simple formula:

$$\frac{d\sigma}{d\mathbf{p}_s d\mathbf{p}_f} = (2\pi)^4 N \, \frac{p_s p_f}{p_\circ} \, f_{ee} \, \rho(\varepsilon, \mathbf{q}) \; , \tag{3}$$

Coincidence Studies of Electron and Photon Impact Ionization
Edited by Whelan and Walters, Plenum Press, New York, 1997

where the electron-electron factor f_{ee} is the Mott cross-section for the free electron collision, and N is the number of atoms in the target[2]. The spectral, or energy resolved, momentum density $\rho(\varepsilon, \mathbf{q})$ gives the probability of finding an electron with a given energy ε and momentum $\mathbf{q}$. It is obvious therefore that the *(e,2e)* technique directly determines the electron distribution in momentum space in the system of interest, and hence is commonly referred to as electron momentum spectroscopy (EMS). In solids it allows the determination of the dispersion relation between the electron energy and momentum $\varepsilon = E_j(\mathbf{q})$ in a particular energy band j.

Angle-resolved photoelectron spectroscopy (ARPES) is another experimental technique which is able to map the energy-momentum relation in solids. Although this technique has been very successful and is widely used, the theoretical understanding of the intensities of the observed peaks is far from complete. This is because the interaction processes are very complicated, involving a many-body problem in the initial and final states. Hence, it is not straightforward to measure the momentum densities of electrons using ARPES.

A technique that does enable the measurement of momentum densities is Compton scattering. This technique probes the target electrons with high energy photons. A measurement of the total momentum density of all target electrons integrated over a plane perpendicular to the scattering vector is obtained. Thus, only a projection of the momentum is obtained and information about the energy is not resolved.

The application of EMS to solids has been restricted primarily by the complications arising from their high atomic density. Both the incoming and outgoing electrons can undergo additional scattering other than the *(e,2e)* event itself. For solids one usually has to use a transmission mode in order to be able to access all possible values of the momentum $\mathbf{q}$. To obtain detailed information on the electronic structure of the target, high coincidence count rates and good energy and momentum resolution are required. The early *(e,2e)* experiments on solids [2–4] were unable to resolve any valence band structure because of poor energy resolution. It was not until the experiments of Ritter *et al.* [5], where the energy resolution had improved to 6 eV, that structure in the valence bands could be resolved. However, these experiments were still very restricted by coincidence rates and energy and momentum resolution.

Significant progress in *(e,2e)* experiments on solids has been achieved by the Flinders University group. The major difficulties involved in the measurements of solids were successfully overcome by utilising multi-parameter detection techniques and an improvement in the energy resolution [6]. A parallel development in theory based on the density-functional approach permitted calculations of energy-resolved electron momentum densities for a large number of crystalline and disordered materials[7,8]. This served as a basis for understanding and interpreting a vast amount of experimental data [9–12] (see also Dennison and Ritter [13] for a review).

Very recently a further improvement in energy resolution down to less than 1 eV was achieved by the Flinders group [14]. Target preparation techniques were also developed to a stage where monocrystalline targets became available for *(e,2e)* transmission experiments. On the theoretical side, a much better understanding of multiple scattering was achieved by Monte Carlo simulations of the experimental process which accounted for additional electron collisions other than the *(e,2e)* collision [15]. In addition, the finite life-time effects due to many-electron correlations in the ionized system were also included in the model. As a result of these developments a quantitative EMS spectroscopy of solids became possible for the first time. In a very recent EMS study on aluminum complete agreement has been achieved between the experimental results and theoretical calculations for the entire electronic structure (energy, momentum and intensity) [16].

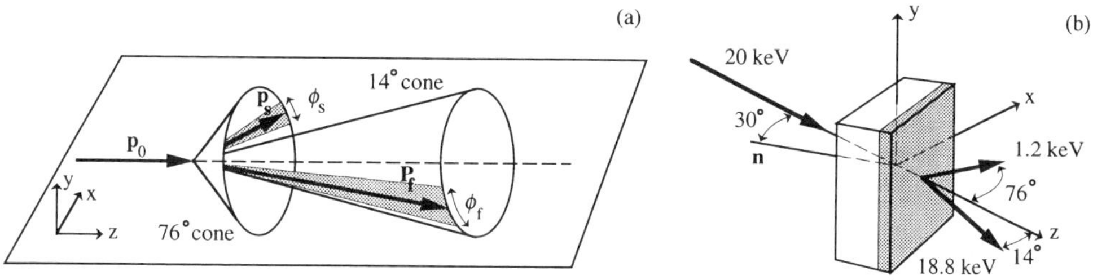

Figure 1: (a) Geometry of a transmission *(e,2e)* experiment. A range of outgoing azimuthal angles are highlighted to show how different target electron momenta can be measured. (b) An illustration of the surface that is studied in a transmission *(e,2e)* measurement on a thin film. The depth that is probed depends on the energy of the electrons and the angle of the target. This depth is dominated by the energy of the slowest electron that is detected.

The purpose of this paper is to give an overview of the latest achievements with the experiments and the theory of *(e,2e)* on solids. We concentrate primarily on the results of the Flinders group. Recent results on the reflection type *(e,2e)* done elsewhere are reported in other sections of the present volume.

EXPERIMENTAL TECHNIQUE

The asymmetric non-coplanar *(e,2e)* reaction is utilized for EMS in transmission mode. During a typical *(e,2e)* experiment on valence electrons the incident energy is 20 keV, with the energy of the fast electron nominally 18.8 keV and the energy of the slow electron about 1.2 keV. Constant polar angles of 14° and 76° degrees are used for the detection of the fast and slow electron respectively; the geometry is shown in Fig. 1.

With this geometry the measured momentum $\mathbf{q}$ for a coplanar event, i.e. one with all three momentum vectors in the same plane, is zero. A range of azimuthal (out of plane) angles are measured simultaneously by both detectors. The fast electron is detected in a hemispherical analyser whilst a toroidal analyser detects the slow electron. Each electron analyser has a two-dimensional position-sensitive detector mounted at its exit, enabling a range of target electron binding energies ε and momenta $\mathbf{q}$ to be measured simultaneously. This is a major advantage as it greatly improves the *(e,2e)* coincidence rate.

THEORETICAL MODEL

The spectral momentum density (SMD) which enters into Eq. (3) is defined according to Neudachin *et al.* [2] as

$$\rho(\varepsilon, \mathbf{q}) = \sum_{\alpha} |\langle \mathbf{q}\,(N-1, \alpha)|N, 0\rangle|^2\, \delta(\varepsilon - E_{N-1,\alpha} + E_N)\ . \tag{4}$$

Here $|N, 0\rangle$ and $|N-1, \alpha\rangle$ are the many-electron wave functions of the ground and ionized states of the crystal, respectively, and $|\mathbf{q}\rangle$ is the normalised plane wave. The summation is over all the final states α of the ionized system.

In the independent-particle approximation Eq. (4) simplifies to

$$\rho_j(\varepsilon, \mathbf{q}) = |\langle \mathbf{q}|\Psi_{j\mathbf{k}}\rangle|^2\, \delta(\varepsilon - E_j(\mathbf{k}))\ , \tag{5}$$

where we specify a one-electron state $\Psi_{j\mathbf{k}}$ by the band index j and crystal momentum $\mathbf{k}$. The spatial integration in Eq. (5) is performed over the crystal volume. Using Bloch's theorem it can be reduced to the unit cell Ω transforming Eq. (4) into

$$\rho_j(\epsilon, \mathbf{q}) = (2\pi)^{-3} \sum_{\mathbf{G},\mathbf{k}\in 1^{st}\ \mathrm{BZ}} \left| \int_\Omega d^3r\, \psi_{j\mathbf{k}}(\mathbf{r}) e^{-i\mathbf{q}\cdot\mathbf{r}} \right|^2 \delta_{\mathbf{q},\mathbf{k}+\mathbf{G}}\, \delta(\epsilon - E_j(\mathbf{k})) \, . \tag{6}$$

Here the electron momentum $\mathbf{q}$ is translated to the first Brillouin zone by the reciprocal lattice vector $\mathbf{G}$.

To calculate the one-electron wavefunctions $\psi_{j\mathbf{k}}(\mathbf{r})$ and one-electron energies $E_j(\mathbf{k})$ one has to perform a band-structure calculation on the material of interest. The calculations reported in [7–12] relied on the linear muffin-tin orbital (LMTO) method as described by Skriver [17]. It is convenient to make an atomic-sphere approximation and substitute the unit cell with a number of space filling atomic spheres each of which represents a non-equivalent atomic position. The electron potential is spherically symmetric within the spheres. The tails of the LMTO orbitals outside the spheres are chosen to have zero kinetic energy. The Bloch sum of the tails is canceled within the spheres. This requirement serves as a boundary condition on the spherically symmetric Schroöedinger equation within each sphere. The one-electron wave function within any particular sphere can be written in the form of the central-field expansion and hence the SMD can be readily evaluated.

The finite experimental energy resolution can be easily incorporated into the model by substituting the δ-function of energy entering Eqs. (6) by a normalised Gaussian function $\delta\epsilon^{-1}\Gamma\left[(\epsilon - E_j(\mathbf{k})/\delta\epsilon\right]$, where $\delta\varepsilon = (8\ln 2)^{-1/2} \cdot \mathrm{FWHM}$.

In practice it is of interest to calculate the SMD either in a particular direction ($\mathbf{q} = q\mathbf{e}$, $\mathbf{e}$ is the unit vector) or as a spherical average ($|\mathbf{q}| = q$) over the irreducible wedge of the Brillouin zone. These two types of calculations refer to either a crystalline oriented target, or a polycrystalline disordered target, respectively. In both cases the momentum dependence is reduced to a single scalar variable q and the momentum broadening can be included by a Gaussian convolution with the experimental momentum width.

Besides the instrumental effects, an additional energy broadening can be caused by the finite life-time of the hole created by the *(e,2e)* event. This effect can be described if instead of Eq. (5) we use a more general expression:

$$\rho_j(\varepsilon, \mathbf{q}) = |\langle \mathbf{q} | \Psi_{j\mathbf{k}} \rangle|^2 \, Im\; G_{j\mathbf{k}}(\varepsilon) \, , \tag{7}$$

where the retarded, or one-hole, Green's function $G_{j\mathbf{k}}(\varepsilon)$ describes the final state of the ionized system after removing an electron with the crystal momentum $\mathbf{k}$ from the band j. The Green's function can be presented as

$$G_{j\mathbf{k}}(\varepsilon) \; = (\varepsilon - E_j(\mathbf{k}) - i\gamma(\varepsilon))^{-1} \tag{8}$$

where $\gamma(\varepsilon)$ is inverse lifetime of the hole. To calculate γ from first principles one has to use a realistic model for the screening and relaxation caused by ionization. Alternatively, one can use Eq. (8) with an empirical dependence $\gamma(\varepsilon)$ obtained from independent experiments, i.e. from angle-resolved photoemission.

The results of the previously described LMTO calculations would describe the experimental data if all *(e,2e)* events were "clean" i.e. further elastic or inelastic scattering of either the incoming or both outgoing electrons does not occur. The Monte Carlo simulation gives an estimate of the rate at which certain momentum transfer (elastic scattering) and energy loss (inelastic scattering) combinations occur and how this in turn affects our measured spectra. For a detailed description of the Monte Carlo procedure as previously applied to *(e,2e)* spectroscopy of solids see the paper by Vos and Bottema [15]. The simulation involves the inclusion of both elastic and inelastic scattering events for any of the incident and two outgoing electrons. For elastic scattering the cross-sections are calculated in the Born approximation from actual Hartree-Fock wavefunctions of the free atoms constituting the solid. For inelastic scattering only bulk plasmon excitations are considered.

RESULTS

Aluminum

Aluminum being a nearly free-electron metal, has a band structure that is well understood. Hence the agreement reached between the experiment and theory provides an excellent insight into the current understanding of *(e,2e)* in solids [16] .

The energy-resolved momentum density for a polycrystalline aluminum film is shown in Fig. 2 (a). In this plot the dispersion curve of the valence band clearly stands out above the background. The densities are represented on a linear grey-scale with white corresponding to the highest intensity. The dispersion curve looks very similar to a free-electron parabola as is expected for metallic aluminum. The parabola extends from 4 eV binding energy at the top, down to 16 eV at the bottom. It should be noted that the experimental binding energies are referenced to the vacuum level. The Fermi level corresponds to 4 ± 0.5 eV binding energy in the experiment. The measured bandwidth of 12 eV is close to that calculated for aluminum using the LMTO method (11.5 eV).

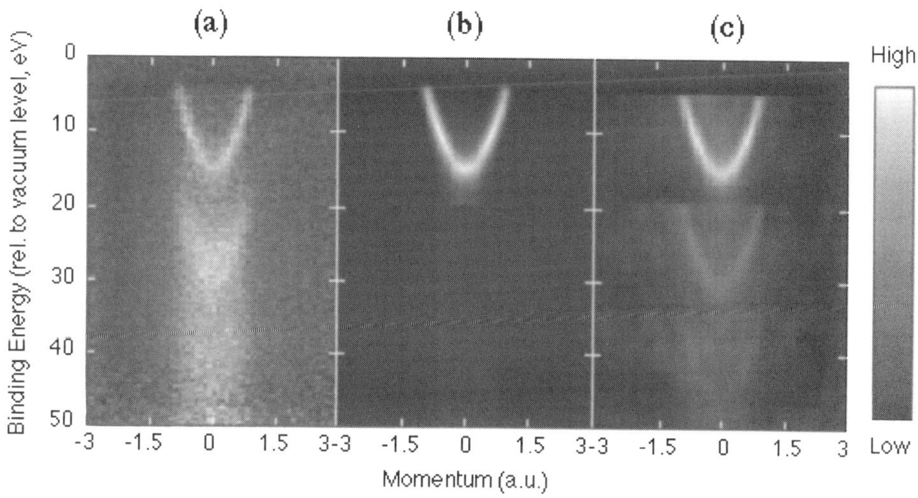

Figure 2: (a) Energy resolved momentum density from *(e,2e)* measurements on polycrystalline aluminum film, (b) as calculated using the LMTO method convoluted with Gaussians representing the finite experimental energy and momentum resolution and a Lorentzian taking into account energy broadening due to finite life-time effects, (c) simulated using the Monte Carlo procedure taking the LMTO momentum density as an input.

The energy-resolved electron momentum density of the LMTO calculation is presented in Fig. 2 (b). This calculation has been convoluted with an instrumental width of 1.0 eV to match the measured energy resolution of the *(e,2e)* experiment. The experiment also had a finite momentum resolution which was determined by the angular resolution of each of the three electron beams and was estimated to be 0.1 a.u. The additional energy broadening caused by the finite lifetime of the hole left behind by the ejected electron has also to be taken into account. Levinson *et al.* [18] measured the inverse lifetimes in their angle-resolved photoemission experiments on aluminum.

These data have been used to obtain the function $\gamma(\varepsilon)$ which appears in Eq. (8) and produces a Lorentzian-type broadening.

The effects of elastic and inelastic scattering on the energy resolved momentum density were investigated using the Monte Carlo procedure. The Monte Carlo results are shown in Fig. 2 (c). The binding energy scale of the plots of Fig. 2 ranges from 0 to 50 eV so that a discussion on the multiple scattering events can also be made. Qualitative improvement of the agreement between theory and experiment is obtained after inclusion of the Monte Carlo simulation. The valence band of the measurement is

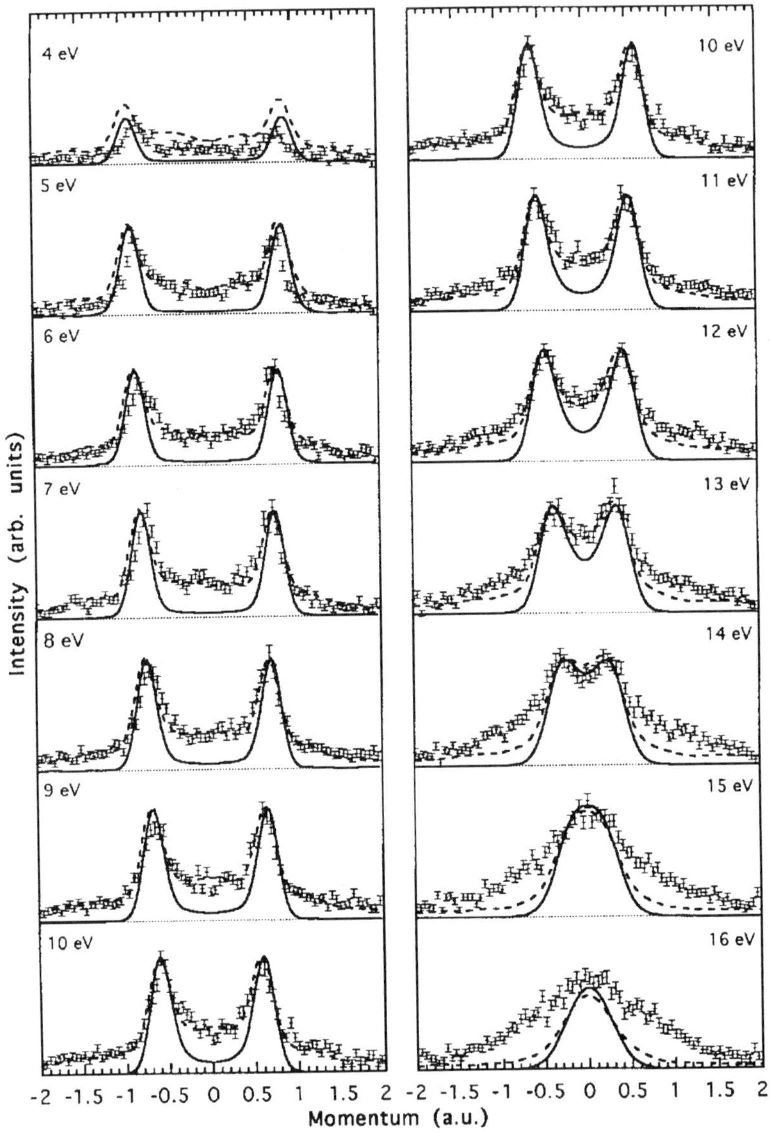

Figure 3: The momentum profiles of the measured aluminum sample for a series of energies are shown with error bars in each panel. Also shown is the LMTO calculation (solid lines) with a Gaussian convolution with 1 eV energy resolution, 0.1 a.u. momentum resolution, and a Lorentzian broadening for the finite lifetime. The dashed lines show the above LMTO calculation with the Monte Carlo simulation

simulated very well. There is extra intensity inside and outside the valence band from the Monte Carlo simulation that is not seen in the LMTO calculation. In the *(e;2e)* measurement the plasmon dispersion of aluminum can clearly be seen. The plasmon dispersion is a replica of the aluminum valence band shifted to 15 eV lower binding energy, i.e. inelastic scattering resulting in energy loss. The Monte Carlo simulation also predicts this plasmon dispersion, however, the intensity of the plasmon distribution is not as high as the measurement. Also the plasmon distribution from the experiment appears to be "filled in" and this is not seen in the Monte Carlo simulation. The Monte Carlo simulation does predict a plasmon distribution of double excitation, i.e. 30 eV energy loss.

For a quantitative study, that is the determination of the energy–momentum position of the electrons and the relative intensities, a more thorough analysis is required. For this purpose a series of momentum profiles is constructed by taking slices through the valence band region (width of slices are 1 eV) with the centre of those slices being the binding energy indicated. The experimental momentum profiles are shown (with error bars) in Fig. 3. From these plots one can see two distinct peaks at the top of the band (4 eV) which are symmetric around zero momentum. As the energy increases these peaks disperse inwards, i.e. towards zero momentum, finally forming one peak centered at zero momentum at the bottom of the band (16 eV).

The momentum profiles for the LMTO calculation are shown as solid lines in Fig. 3. The agreement between the experiment and the LMTO calculation is excellent. The Monte Carlo simulation is shown in Fig. 3 by dashed lines. It is clear that the Monte Carlo simulation has significantly increased the intensity between the two peaks, i.e. where the LMTO predicted zero intensity. The mechanism responsible for this is elastic scattering. Up until 12 eV the final agreement is excellent. Above 12 eV the peak intensities themselves are in very good agreement although the measurement shows intensity at higher momentum that is not predicted in either of the theoretical models. This excess intensity is most likely a result of inelastic processes that are not accounted for in the Monte Carlo procedure. These processes include surface plasmon excitation and the effects of momentum transfer during plasmon excitation and single electron excitations.

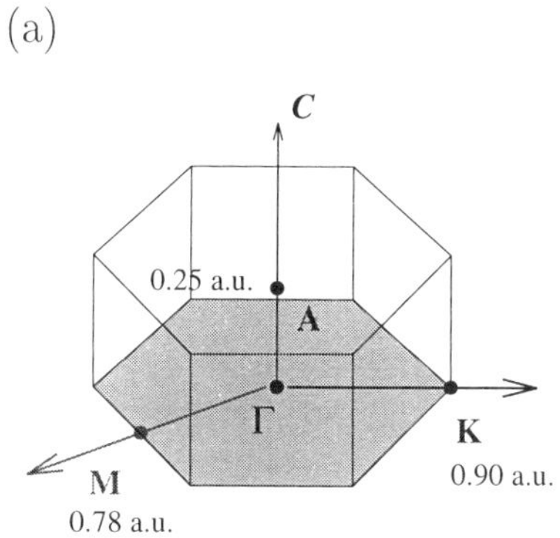

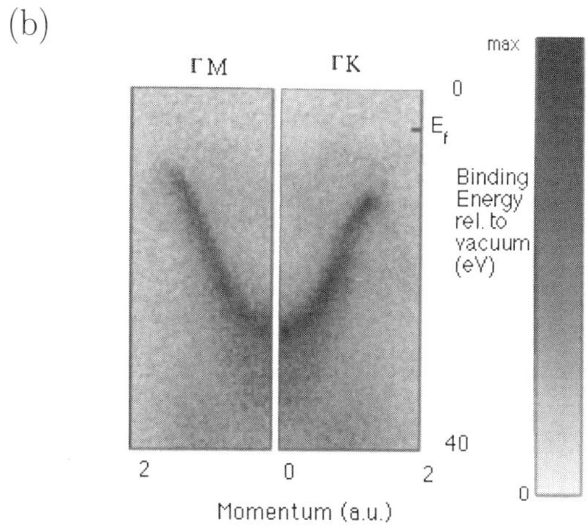

Figure 4: (a) Brillouin zone of graphite (upper-half). Distances from the zone centre Γ are indicated in atomic units. (b) Experimental spectral momentum density in ΓM and ΓK directions

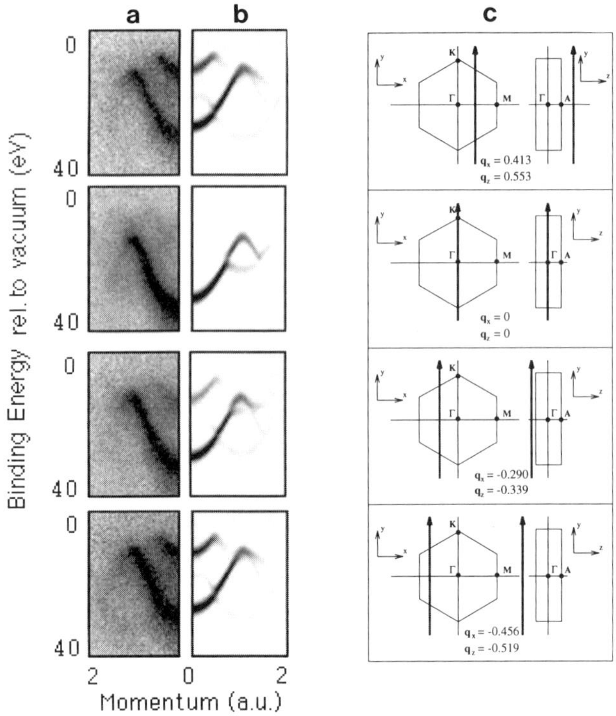

Figure 5: Spectral momentum density of natural graphite outside the basal plane. Experiment (a), LMTO calculation (b) and the Brillouin zone scheme (c). The measurement probes the electron momentum along the direction indicated as a solid arrow. The component q_y is varying in the experiment whereas q_x and q_z are fixed.

Natural graphite

The *(e,2e)* experiment on natural graphite[19] is the first reported EMS measurement on a single crystal target with a thoroughly controlled orientation and fairly well known electronic structure. Hence, it can be used to create a 3D momentum-density map complementary to the charge density distribution in real space.

Graphite is a layered compound with strong covalent bonding in plane and weak van der Waals bonding in between the planes. The large distance between the planes results in a quasi two-dimensional energy band structure with three of the carbon valence electrons in dispersing σ bands and one electron occupying a weakly dispersing π band. The orientation of the π bonding is such that its electrons have maximum momentum density along the *c*-axis and zero momentum density in the basal plane. The σ bands have non-zero densities in the basal plane.

The energy-resolved momentum density of graphite in the basal plane along the two high-symmetry directions ΓK and ΓM, indicated in Fig. 4 (a), is represented in Fig. 4 (b). As expected, only the σ-band contribution can be seen in these directions. Near the bottom of the valence band dispersion is parabolic. At larger momenta the anisotropy is clearly seen as predicted by the LMTO calculation[5] .

The out-of-plane SMD of graphite is shown in several panels of Fig. 5. The kinematic arrangement of the experiment is such that one component of the valence electron momentum q_y can be varied continuously whereas the others q_x and q_z are locked by the position of the slow electron detector. Several panels of Fig. 5 corresponds to various combinations of q_x, q_z indicated on the Brillouin zone scheme (c). When $q_z = 0$ (second top panel) the contribution from the π band is absent as expected. As $|q_z|$ increases the π band intensity increases too. This behaviour is in very good agreement with the LMTO calculation shown in Fig. 5(b). Somewhat larger energy broadening in the experiment, especially near the bottom of the valence band, can be attributed to the finite lifetime effects which are not included in the calculation.

Aluminum oxide

Aluminum oxide is the most complicated target studied so far by EMS [20]. In its trigonal form $\alpha{-}Al_2O_3$ has a unit cell with 10 non-equivalent atomic positions which results in an unusually complex pattern of the valence energy bands.

Figure 6 represents the measured spectral momentum density for the aluminum oxide polycrystalline sample (left panel) and the LMTO calculation of a spherically averaged $\alpha-$ Al_2O_3 (right panel), respectively. Aluminum oxide has two distinct features in its valence band structure namely an upper valence band and a lower valence band. A previous (e, 2e) experiment[21] on an aluminum–aluminum oxide thin foil with an energy resolution of 4.5 eV reported dispersionless structures in both the upper and lower valence band. This is manifestly not the case in the present experiment. The present measured spectral momentum density plot shows a "dual parabola" dispersion pattern spanning 8 eV in the upper valence band (Fig. 6 left panel). The LMTO calculation shows a similar "dual parabola" dispersion pattern with nearly the same energy span. Near zero momentum both the EMS measurement and the LMTO calculation exhibit lower intensity throughout the upper valence band.

There is a gap between the upper valence band and lower valence band. The EMS measurement and the LMTO calculation give nearly the same gap. In the lower valence band, the LMTO calculation indicates a dispersion of about 5 eV. The EMS measurement shows a similar "bowl" shape in the spectral momentum density (the full width of momentum distribution becoming narrower with increasing binding energy in the lower valence band) but it shows less dispersion. Nevertheless, one can see that

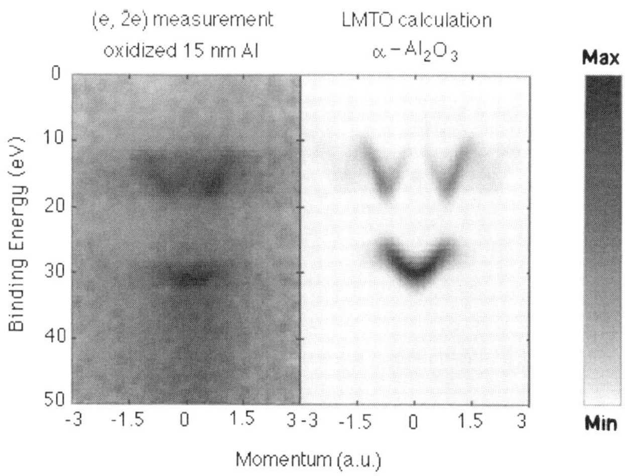

Figure 6: The spectral momentum density plots as measured for the aluminum oxide sample (left panel) and calculated using the LMTO method of spherically averaged $\alpha{-}Al_2O_3$ (right panel), respectively. The binding energy is relative to the vacuum level. The intensity scale is shown on the right hand side.

the measured major features in the valence band of aluminum oxide are qualitatively reproduced by the LMTO calculation.

CONCLUSION AND FUTURE DIRECTIONS

We have shown that EMS of solids is past its demonstration phase and is capable of producing quantitative data on energy-resolved momentum densities of solids. A considerable amount of research has still to be done and it is likely that this technique will make significant and interesting contributions for many years to come. As an immediate task we consider the 3D electron momentum mapping in single crystals (graphite, silicon). We also plan experiments on polycrystalline films of $3d$ transition metals (Ni, Fe, Cu) and their oxides which show strong electron-correlation effects. Understanding and interpreting these experiments will require further development of theory beyond the independent-particle approximation.

REFERENCES

1. I.E. McCarthy and E. Weigold, Electron momentum spectroscopy of atoms and molecules, *Rep. Prog. Phys.* 54:789 (1991).
2. N. M. Persiantseva, N. A. Krasil'nikova, and V. G. Neudachin, Investigation of the collective properties of a degenerate electron gas in aluminum by the (e,2e) method, *Sov. Phys. JETP* 49:530 (1979).
3. U. Amaldi Jr., A. Egidi, R. Marconero, and G. Pizzella, Use of a two channeltron coincidence in a new line of research in atomic physics, *Rev. Sci. Instrum.* 40:1001 (1969).
4. R. Camilloni, A. Guardini-Guidoni, R. Tiribelli, and G. Stefani, Coincidence measurement of quasifree scattering of 9-keV electrons on K and L shells of carbon, *Phys. Rev. Lett.* 29:618 (1972).
5. A.L. Ritter, J.R. Dennison, and R. Jones, Spectral momentum density of amorphous carbon from (e,2e) spectroscopy, *Phys. Rev. Lett.* 53:2054 (1984).
6. P. Storer, R.S. Caprari, S.A.C. Clark, M. Vos, and E. Weigold, An improved condensed matter spectrometer with parallel detection in energy and momentum, *Rev. Sci. Instrum.* 65:2214 (1994).
7. A. S. Kheifets and Y. Q. Cai., Energy-resolved electron momentum densities of diamond-structure semiconductors, *J. Phys.: Condens. Matter* 7:1821 (1995).
8. A. S. Kheifets and M. Vos, Spectral momentum density calculation in graphite *J. Phys.: Condens. Matter* 7:3895 (1995).
9. M. Vos, P. Storer, S.A. Canney, A. S. Kheifets, I. E. McCarthy, and E. Weigold, Energy resolved electron momentum densities of graphite films, *Phys. Rev. B* 50:5635 (1994).
10. M. Vos, P. Storer, Y. Q. Cai, A. S. Kheifets, I. E. McCarthy, and E. Weigold, Electron structure of amorphous Si measured by (e,2e) spectroscopy, *J. Phys.: Condens. Matter* 7:279 (1995).
11. Y. Q. Cai, M. Vos, P. Storer, A. S. Kheifets, I. E. McCarthy, and E. Weigold, The valence electronic structure of polycrystalline SiC as observed by (e,2e) spectroscopy, *Phys. Rev. B* 51:3449 (1995).
12. Y. Q. Cai, P. Storer, A. S. Kheifets, I. E. McCarthy, and E. Weigold, Energy resolved momentum densities of amorphous germanium and the effect of hydrogen absorption by (e,2e) spectroscopy, *Surface Science* 334:276 (1995).
13. J.R. Dennison and A.L. Ritter, Application of (e,2e) spectroscopy to the electronic structure of valence electrons in crystalline and amorphous solids, *J. Electron Spectr. and Rel. Phen.* 77:99 (1996).
14. S.A. Canney, M.J. Brunger, I.E. McCarthy, P.J. Storer, S. Utteridge, M. Vos, and E. Weigold, (e,2e) spectroscopy of solids with improved energy resolution, *J. Elect. Spectr. and Rel. Phen.* (1996, in press).

15. M. Vos and M. Bottema, Monte Carlo simulations of (e,2e) experiments on solids, *Phys. Rev. B* 54:5946 (1996)
16. S. A. Canney, M. Vos, A. S. Kheifets, N. Clisby and I. E. McCarthy, Measured energy-momentum densities of the valence band of aluminium, *J. Phys.: Condens. Matter* (1996, in press).
17. H. L. Skriver. *The LMTO method.* Springer-Verlag, Berlin, 1984.
18. H.J. Levinson, F. Greuter, and E.W. Plummer, Experimental band structure of aluminium *Phys. Rev. B* 27:727 (1983).
19. M. Vos, Z. Fang, S. Canney, A. S. Kheifets, I.E. McCarthy and E. Weigold, Energy-momentum density of graphite by electron-momentum spectroscopy, *Phys. Rev. Lett.* (1996, to be published).
20. X. Guo, M. Vos, A. S. Kheifets, Z. Fang, S. Canny, S. Utteridge, I. E. McCarthy and E. Weigold, Electronic structure investigation of an oxidized aluminum film with electron momentum spectroscopy, *Phys. Rev. B* (1996, in press).
21. P. Hayes, M. A. Bennett, J. Flexman and J. F. Williams, Symmetric transmission electron momentum (e,2e) spectroscopy of an aluminum-aluminum oxide thin foil, *Phys. Rev. B* 38:13371 (1988).

VALENCE BAND RECONSTRUCTION IN GRAPHITE BY GRAZING INCIDENCE (e,2e) SPECTROSCOPY

S. Iacobucci[1], S. Rioual[1, #], D. Neri[2], R. Camilloni[1], P. Luches[3], S. Nannarone[3], G. Stefani[4]

[1] IMAI-CNR Area Ricerca Roma, Via Salaria km 29.300 c.p. 10
00016 - Monterotondo, Italy
[2] Universita' degli Studi di Roma La Sapienza, Piazzale A.Moro, 2
00187 - Roma, Italy
[3] Unita' INFM Modena e Dip. Fisica Univ. Modena, Via G.Campi 213/A
41100-Modena, Italy
[4] Unita' INFM Roma3 and Dip. Fisica Univ. Roma Tre, Via Vasca Navale 84
00146-Roma, Italy

INTRODUCTION

Today most calculations of the electronic structure of solids are performed within the density functional theory. The approach relies on the existence of a model system of N-non interacting fermions, which has the same charge density as the actual interacting system.

An effective one particle potential exists, which generates the charge density, and it contains an exchange-correlation term which is known only approximately. One of the most relevant features for a homogeneous gas of interacting electrons is that the occupation number is not any more a step function, because in the presence of correlation there is a finite probability that the momentum states outside the Fermi level are occupied. Hence, the momentum density expectation value is thought to be more sensitive than charge density or total energy to the presence of correlations in the ground state of the system. A momentum density spectroscopy should then be able to provide a careful test for the aforementioned effective potential. This is just one example for being interested in measuring momentum density in solids and surfaces.

A variety of experimental methods have been already developed to investigate energy dispersion and density of occupied and empty states in solids[1]. In contrast, only a limited number of methods measure, more or less directly, wave function related quantities as real space electron density (x-ray form factor) or momentum density. Relevance of the momentum density in understanding the electronic structure has been well established for atoms and molecules[2]. Although very sensitive to fine details of the wave function, the

[#] Present Address: Lab. Coll. Electr. et Atom., Université de Bretagne Occidentale, Brest (F)

Coincidence Studies of Electron and Photon Impact Ionization
Edited by Whelan and Walters, Plenum Press, New York, 1997

electron momentum density is not as widely investigated as other observable quantities, because to measure it, ionisation experiments with completely determined kinematics are needed. Coincidence experiments of this kind, in particular (e,2e)[3] and coincidence Compton scattering experiments[4], are the only available spectroscopies capable of providing both binding energies and momentum densities of the bound states of the quantum system under study.

To perform an (e,2e) experiment, both free electrons produced by an electron impact ionisation are to be analysed in momentum and detected coincident in time, thus allowing a complete balance of energy and momentum. When applied to solids, the (e,2e) technique measures the momentum distribution of an energy selected state, as well as its energy versus momentum dispersion in a zone of real momenta that extends all the way down to zero value. This unique capability was realised[5] even before the first experiment had been performed and is not by chance that the first momentum distribution of a single orbital was measured on a carbon thin film[6]. Technical difficulties, however, have hampered till very recently the application of (e,2e) spectroscopy to solids. It was not before the end of the eighties that the (e,2e) has been largely applied to investigate bulk properties of amorphous and crystalline solids[7].

REFLECTION (e,2e)

Even though the first application to surfaces was envisaged[8] shortly after the first (e,2e) experiment[6], it has been only recently that the feasibility of such an experiment has been demonstrated. Experiments performed in transmission mode and high energy asymmetrical kinematics must rely on the shortness of the mean free path of the low energy ejected electron in order to achieve surface sensitivity for a technique which is not inherently surface sensitive[9]. Experiments performed in grazing angle reflection geometry[8,10,11] were suggested to be surface sensitive in as much as the penetration depth of the electrons does not exceed the first few layers of the sample. Reflection (e,2e) measurements from surfaces have been unsuccessfully tried in a few laboratories since the late '70s. Finally, the existence of pairs of correlated electrons, escaping from surfaces under bombardment of low energy electrons at normal incidence, was demonstrated by Kirschner et al.[12]. Iacobucci et al.[13] have shown the feasibility of the (e,2e) experiment in grazing angle geometry and intermediate energy. These two pioneering experiments investigate quite different kinematic regions of the (e,2e) process, the latter being suitable for momentum spectroscopy.

The reflection (e,2e) technique consists of measuring simultaneously the energy E_0 of the incident electron, the energies E_e and E_s of the two final electrons, and the probability of them being emitted into solid angles $d\Omega_e$ and $d\Omega_s$ oriented along the directions θ_e and θ_s both on the same side of the surface as the incoming beam. At intermediate and high energies, depending on the amount of momentum transfer $\vec{\mathbf{K}} = \vec{\mathbf{K}}_0 - \vec{\mathbf{K}}_s$, two approximations can be used to describe the (e,2e) process. In the dipolar limit (i.e. vanishing momentum transfer) the (e,2e) mechanism is equivalent to photoionisation[14] and binding energy ($\varepsilon(\vec{\mathbf{q}})$) spectroscopy is possible. The dipolar (e,2e) differs from photoionization in as much as the momentum associated with the transition, $\vec{\mathbf{q}} = \vec{\mathbf{K}}_s + \vec{\mathbf{K}}_e - \vec{\mathbf{K}}_0$, is not uniquely determined and can be changed by changing the geometry of the process while keeping fixed the energy balance $\varepsilon(\vec{\mathbf{q}}) = E_0 - E_s - E_e$. In the impulsive (binary) limit, i.e. momentum transfer roughly equal to the ejected electron momentum, the spectral momentum density of the electrons bound in the target can also be measured[7]. Several experiments performed on gaseous targets[15] and thin films have shown that the impulsive condition can be satisfied both in symmetric ($E_s = E_e$) and asymmetric conditions; hence both kinematics permit measurement of the momentum distribution.

In the Plane Wave Impulse Approximation[7] and within the non interacting bound particle models, it is easily verified that the five fold differential cross section can be written (with implicit antisymmetrisation)[13] as:

$$\frac{d^5\sigma}{d\Omega_s d\Omega_e dE} \propto \frac{K_s K_e}{K_o} \times \sum_t \left| \left\langle \vec{K}_e \cdot \vec{K}_s \cdot \prod_{j=1, j\neq t}^{N-1} \Psi_q(\vec{r}_j) \middle| v_{o,t} \middle| \vec{K}_o \cdot \prod_{j=1}^{N} \Psi_q(\vec{r}_j) \right\rangle \right|^2$$

where E satisfies the energy balance, t and j are valence electrons indexes, $v_{o,t}$ is the Coulomb interaction between the incoming electron and the bound electron, $\Psi_q(\vec{r}_j)$ is a one electron Bloch function with energy $\varepsilon(\vec{q})$ and crystal momentum $\vec{q}' = \vec{q} + \vec{G}$, $\vec{G}$ being a reciprocal lattice vector.

The main features of the five fold differential cross section are readily understood when a simplified description of the bound states is adopted. In the quasi-free electron model, $\Psi_q(\vec{r}_j)$ can be written as linear combination of plane waves. Furthermore, in the frozen core assumption the cross section becomes

$$\frac{d^5\sigma}{d\Omega_s d\Omega_e dE} \propto \frac{K_s K_e}{K_o} \cdot \frac{1}{K^4} \cdot \sum_{\vec{G}} \left| C_{\vec{q}'-\vec{G}} \right|^2 \qquad (1)$$

where the C_i's are coefficients of the ground state wave function of the linear expansion. Therefore, the five fold differential cross section factorizes in a kinematical factor depending only upon the kinematics of the collision (first two terms of (1)), times a form factor which is solely determined by the target electronic structure.

EXPERIMENTAL

The experimental equipment is based on an Ultra High Vacuum (UHV) chamber containing an Electron Gun (EG), two electron analyzers - a Hemispherical Deflector Analyzer (HDA) and a Cylindrical Mirror Analyzer (CMA) -, and a sample holder mounted on a five-degrees of freedom manipulator. The apparatus permits to work in reflection geometry and at small grazing angle ($\theta_O = 0^\circ - 17^\circ$).

The EG is a commercial gun (Leybold EQ11/35) that has been modified in order to automatically scan the electron beam energy in the range 0-5000 eV under computer control. The sample holder is mounted on a XYZ translator allowing a ±180° rotation around the Z-axis and a ±10° tilt of the sample to adjust the grazing angle θ_O. The target temperature is controlled from -120 °C to 900 °C, thus allowing to condense molecular specimen on the target surface.

The HDA consists of a dispersive hemispherical element, equipped with a three-element electrostatic lens and it can rotate in the scattering plane ($\theta_S = 0^\circ - 12^\circ$). The analyzer has been optically aligned in air by using a laser beam and achieving an angular accuracy and repeatability of ±0.2°. Working in the constant pass energy mode the HDA features an energy resolution of 0.4 eV at 300 eV. It is used to analyze the fast scattered electrons. Their energy distribution is obtained by scanning the electron beam energy while keeping the HDA tuned at a fixed energy.

The CMA is a commercial (RIBER) single-pass, modified to optimize the luminosity. It is mounted with its main axis normal to target surface and it is used to detect Auger and low energy emitted electrons working in ΔE/E constant mode. Equipped with an exit slit of

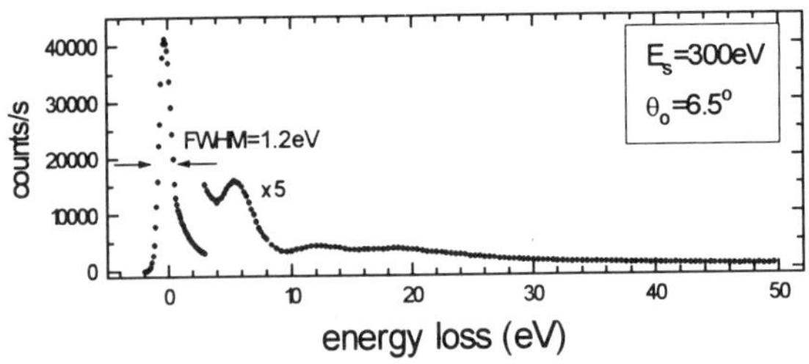

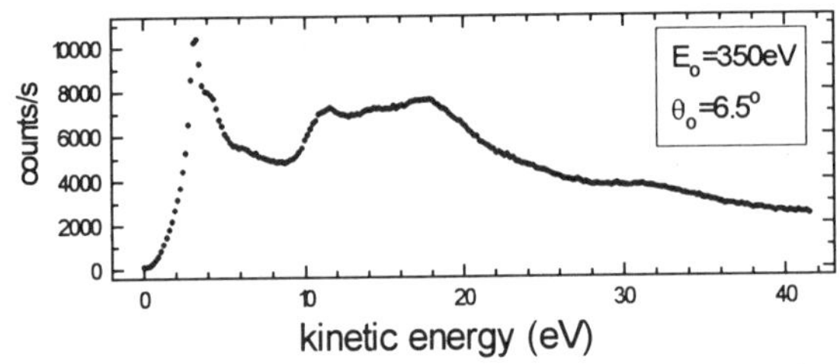

Fig. 1. (Top) A typical AREEL spectrum as measured with the HDA during a coincidence scan on HOPG. (Bottom) An example of secondary electron spectrum measured with the CMA on the same sample.

1.5mm diameter it features an energy resolution of 2% (FWHM), an angular acceptance of 3° and a field of view of about ±1mm. An example of e secondary electron spectrum, as measured on highly oriented pyrolitic graphite (HOPG) is shown in fig.1 (bottom).

The coincidence spectrometer has been characterized using gaseous targets by measuring (e,2e) spectra relative to the He 1S ionization. An (e,e'Auger) spectrum has also been measured by detecting the scattered electron after ionizing the Cl 2p core level in coincidence with the associated $L_{2,3}VV$ Auger electron in gaseous CCl_4 molecule[16].

Energy and angular resolutions of the analyzers have been sacrified in order to obtain an optimization of the areas individually imaged by the two analyzers with that excited by the EG. This is of crucial relevance in order to maximize the overall luminosity of the coincidence spectrometer. In particular, for solid targets the HDA is operated at low-energy resolution (resolving power $\cong 3\ 10^2$). The best overall energy resolution achieved is 1.2 eV and mostly limited by the HDA resolution. A typical angular resolved electron energy loss (AREEL) spectrum as measured during a coincidence scan on HOPG is reported in fig. 1 (top).

The coincidence electronic chain is a conventional one[17]. The fast electronic pulses coming from the two single channel multipliers are fed through two fast amplifiers to constant fraction discriminators. The shaped pulses are then sent directly to the scalers to measure "single" count rates and, to establish time correlation between them, to a time multichannel analyzer (Lecroy 3001QVT) which yields the probability distribution of delay time between pulses (time spectrum). The time resolution of the spectrometer is about 12ns and it is largely due to the time spread of the trajectories in the two analysers. The *true* coincidence rate I_t must be discriminate from the nearly flat background of the uncorrelated events (*accidental* coincidence) which occurs at a rate I_a. The optimal acquisition time is achieved when the ratio I_t / I_a, which depends on the incident current I_0, is about one[17]. In the best case we measured $I_t \cong 0.05$Hz with a ratio $I_t / I_a \cong 0.5$ for $I_0 \cong 1$nA.

The HOPG sample was prepared according to UHV standard procedures by peeling in air and annealing at about 700°C at residual pressure of $1 \cdot 10^{-10}$. mbar. Working with 300 eV primary electrons at a typical grazing angle value of 5° , a probing depth less than 1 Å is expected. Cleanliness and orientation of the surface were checked by Auger electron spectroscopy and AREEL measurements. The best monitor for surface cleanliness was the persistence of the sharp $\pi \rightarrow \pi^*$ transition at 6.2 eV in the AREEL spectrum and the presence of a narrow angular distribution (FWHM=1.7°) for the peak of specularly reflected elastic

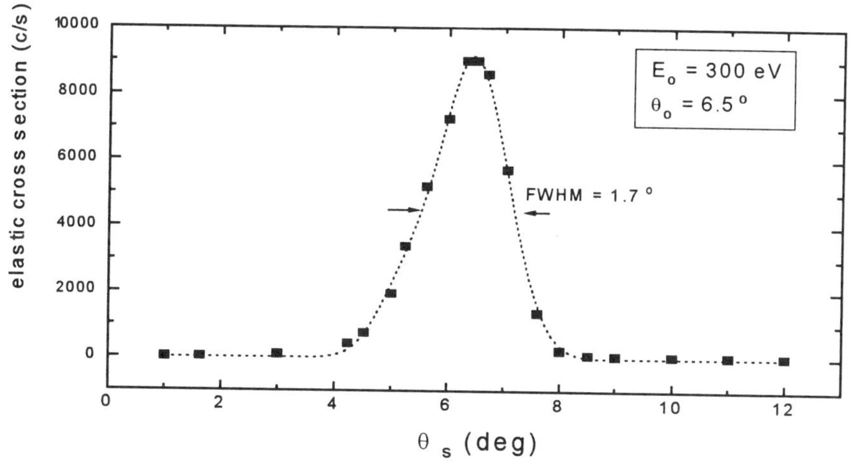

Fig. 2. Elastic cross section angular distribution as measured on HOPG (squares) and its best fit (dotted line).

electrons (see fig. 2). Reproducibility of these measurements over a period of six weeks has been assumed as a guarantee for the stability of the surface conditions and of the energy calibration during the long acquisition time needed for coincidence measurements.

GRAZING ANGLE INTERACTION MECHANISM

The cross section (1) has been computed assuming that the (e,2e) event is generated by a single impulsive collision where the momentum perpendicular to the solid surface necessary to reflect the final electrons in the backward hemisphere is provided by an elastic reflection of the incoming (scattered) electron with the crystal surface. In this case, and within a zero order approximation, the momentum transfer in (1) is to be replaced by a "reflected momentum transfer" $\vec{\mathbf{K}}' = \vec{\mathbf{K}}_0' - \vec{\mathbf{K}}_s$, where $\vec{\mathbf{K}}_0'$ is the specularly reflected plane wave (double step model).

Theoretical fundaments to this double collision model can be found in Mills' calculation for the electron scattering cross-section at surfaces[18]. Experimental support comes from angle resolved electron energy loss (AREEL) measurements that have shown this mechanism to be the dominant one for valence[19] and core[20] ionization.

To understand whether or not this model is applicable to (e,2e) events, the symmetry of the cross section (1) is a key feature.

The angular distribution of the scattered electron has been measured tuning the spectrometer at a fixed binding energy (9 eV). The measurement has been done at two grazing angles $\theta_0 = 4.7^\circ$ and 6.7°, and for the scattered angle variable around the specular reflection. The measure at the smaller grazing angle is shown in fig. 3 and it is evident the symmetry of the five fold differential cross section around the specular reflection direction. Similar measurements at 6.7° grazing angle, not shown in the figure, display similar symmetry around the elastically reflected beam. Regardless of the collisional model adopted, because of the CMA large accepted solid angle, as a first approximation the form factor of (1) doesn't depend upon θ_S.

On the other hand, the θ_S dependence of the kinematical factor drastically changes if a single or a double collision model is adopted. It can be readily seen if one explicitly looks at the momentum transfer dependence on the scattering angle:
$K = [K_o^2 + K_s^2 - 2 K_o K_s \cos(\theta_O + \theta_S)]^{1/2}$ for the single-event model, which bears a symmetry

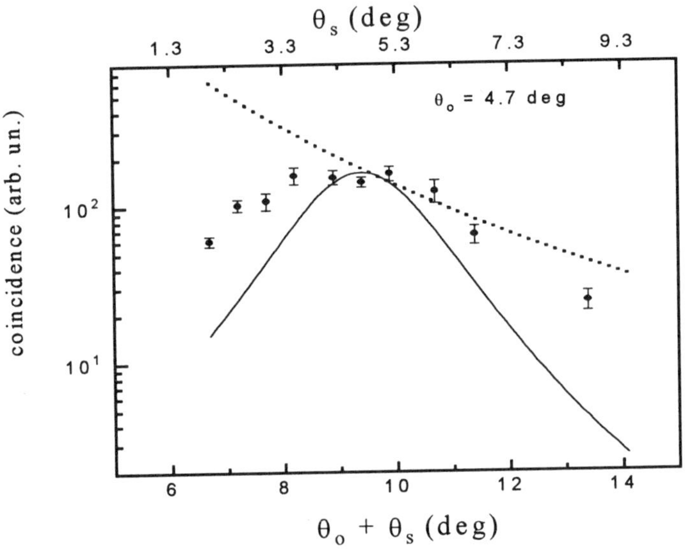

Fig. 3. Angular distribution of the (e,2e) cross section for θ_o= 4.7° and fixed binding energy (9eV). The take off angle is θ_s while the scattering angle is $\theta_o + \theta_s$. Continuous and dashed lines are the kinematical factors for the double and the single collision model respectively.

for the cross section around the $\vec{K}$ direction (i.e. forward direction), and:
$K' = [K_o^2 + K_s^2 - 2 K_o K_s \cos(\theta_O - \theta_S)]^{1/2}$ for the double step model, which bears a symmetry around the $\theta_O = \theta_S$ direction, (i.e. specular reflection direction).

The shape of the measured cross section (fig. 3) clearly supports the double collision hypothesis.

It seems therefore plausible that at high-intermediate energies, the first order collision model conjugated with the Plane Wave Impulse Approximation[7] can be applied to reflection (e,2e) experiments to measure the energy dispersion bands $\varepsilon(\vec{q})$ and the momentum distribution of surface states.

RESULTS

HOPG binding energy spectra have been measured under strict reflection geometry ($\theta_0 = \theta_s$) at various reflection angles and for a selection of ejected energies. Two spectra measured for ejected electron energies of 3.7 eV and 14.2 eV are reported in fig.4 together with their best fits (full line).

The row spectra are affected by multiple loss effects. To remove them, the Van Cittert method has been applied [21]. Details of the procedure will be given in a forthcoming publication. The result of the deconvolution is shown by the dotted line in fig. 4. It clearly appears that multiple scattering largely affects the transition intensities, but it does not change substantially the energy positions. It has to be noted that, although sizeable, the multiple scattering effect is less important in grazing angle geometry than in transmission.

The vacuum level is the origin for the binding energy scale which has been derived from the energy conservation law. The Fermi level ε_F, as derived from an independent measurement, is also reported. The overall energy resolution of the experiment was 1.3 eV.

Under the assumptions used to derive the cross section in equation (1), the features in the binding energy spectrum are to be interpreted on the basis of graphite band structure. It might be speculated that the kinetic energy of the ejected electron is too low for it to be

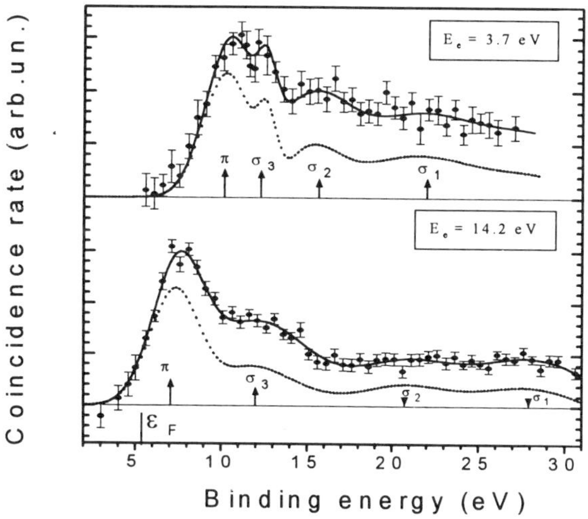

Fig. 4. (e,2e) spectra as measured on HOPG together with their best fits (continuous line) and after multiple scattering correction (dotted line)

considered a plane wave. Hence, the (e,2e) cross section should be determined by the joint density of states in the volume of momentum space selected by the experiment. Though this is true in principle, in practice the experiment was performed selecting a given final state and keeping it unchanged while scanning the binding energies of the initial state. This is similar to a constant final state photoemission experiment [22], hence the cross section behaviour is dominated by bound state characters. The momentum $\vec{\mathbf{q}}$ reconstructed by the momentum conservation in the kinematics of the experiment has non vanishing components both in the parallel ($q_{//}$) and in the perpendicular ($q_{\perp}$) directions to the HOPG $\hat{\mathbf{c}}$ axis. Taking into account the finite analyser angular acceptances, the spectrometer work function and the refraction across the surface potential barrier of the sample [13], the reconstructed values of $q_{\perp}$ range from nearly the middle to the boundaries of the first Brillouin zone in the ΓMK plane (from 0.65 to 1.47 $Å^{-1}$). The $q_{//}$ component ranges between 1.67 and 2.08 $Å^{-1}$, i.e. from the bottom to the middle of the third Brillouin zone along the ΓA direction. According to the HOPG momentum density [23], in the volume of real momenta sampled by the experiment the four main bands of graphite display similar values of the momentum probability; hence, they should yield similar amplitudes for the (e,2e) cross section. Consequently, based on the calculated dispersion bands [24], four main components to the (e,2e) binding energy spectrum have to be expected respectively for the π , σ_3 , σ_2 and σ_1 bands. The agreement between energy positions is good over the momentum range which has been measured. The σ_2 peak is broader than the others, which is to be expected as the band σ_2 is affected by larger energy dispersion in the region of momentum sampled in fig. 4. The energy position of the peak corresponding to the band σ_1 is deeper than theoretically expected by about 1 eV. This discrepancy is overcome if it is assumed for its dispersion curve the one previously measured in a transmission (e,2e) experiment [23], instead of the calculated one [24]. Nonetheless, an intrinsic loss contribution cannot be excluded as the origin of this features [25]. Different binding energy spectra of the same kind of fig. 4, measured for fixed ejected electron energies and fixed mirror geometry's, allow the reconstruction of HOPG valence band structure in the reconstructed momentum range.

SUMMARY AND CONCLUSIONS

The feasibility of (e,2e) experiments in reflection geometry has been established for energies of the primary electrons of 300 eV and for several incidence angles and ejected electron energies..

The mechanism that leads to formation of correlated electron pairs appears to be the same irrespective of the energy and of the kinematics chosen: namely, an elastic reflection by the crystal lattice plus an inelastic collision with the valence band electrons.

A simple impulsive approximation is sufficient to explain the main features of the measured cross sections.

Finally, it has been demonstrated that grazing angle (e,2e) can be used for band mapping.

ACKNOWLEDGMENTS

This work has been made possible by the financial support of EEC Human Capital and Mobility, Contract No. ERBCHRXCT930359 and Progetto Finalizzato Chimica Fine del CNR.

REFERENCES

1. H. Lüth, *Surfaces and interfaces in solids*, Springer-Verlag, Heidelberg, 1993.
2. M. A. Coplan, J.H. Moore, J.P. Doering, *Rev. Mod. Phys*. 66: 98 (1994).
3. G. Stefani, L. Avaldi, R. Camilloni, *J. de Phys. IV, Colloq. C* 6/3: 1 (1993).
4. J. R. Schneider, F. Bell, Th. Tschentschez, A.J. Rollason, *Rev. Sci. Instrum*. 63: 1119 (1992).
5 V. G. Neudachin, G.A. Novoskoltseva, Yu.F. Smirnov, *Z. Exsper. Teo. Fiz*. 55: 1039. (1968); *Soviet. Phys. J. Exper. Ther. Phys*. 28: 540 (1969).
6. R. Camilloni, A. Giardini Guidoni, R. Tiribelli and G.Stefani, *Phys. Rev. Lett*. 29: 618 (1972).
7. I. E. Mc Carthy, E. Weigold, *Rep. Prog. Phys*. 54: 789 (1991).
8. A. D'Andrea and R. Del Sole, *Surf. Sci*. 71: 306 (1978).
9. Y. Q. Cai, M. Vos, P. Storer, A.S. Kheifets, I.E. McCarthy, E. Weigold, *Phys. Rev. B* 51: 3449 (1995).
10. V. G. Levin, V. G. Neudachin and Yu. F. Smirnov, *Phys. Stat. Sol*. (b) 49: 489 (1972).
11. O. M. Artamonov, *Zh. Tekh. Fiz*. 55: 1190 (1985); *Sov. Phys. J. Tech. Phys*. 30: 681 (1985).
12. J. Kirschner, O. M. Artamov and A. N. Terekhov, *Phys. Rev.Lett*. 69: 1711 (1992).
13. S. Iacobucci, L. Marassi, R. Camilloni, S. Nannarone, G. Stefani, *Phys. Rev. B* 51: 10252 (1995).
14. A. Hamnett, W. Stoll, G. Branton, C. E. Brion and M. J. Van der Wiel, *J. Phys. B* 9:_ 945 (1976).
15. L. Avaldi, R. Camilloni, E. Fainelli and G. Stefani, *J. Phys.B: At. Mol. Phys*. 20: 4163 (1987).
16. R. Camilloni, S. Iacobucci, L. Marassi, S. Nannarone, G. Stefani, *Vuoto* XXIV: 64 (1995).
17. G. Stefani, L. Avaldi, R. Camilloni, New Directions in: *Research with Third Generation Soft X-ray Synchrotron Radiation Sources*, NATO ASI E Vol. 254, Edited by A.F. Schlachter and F.J. Wuilleumier, Kluwer Academic Publisher 1994, p. 161-190 and references therein.
18. D. L. Mills, Surf. Sci., 48: 59 (1975).

19. S. Iacobucci, P. Letardi, M. Montagnoli, P.Nataletti and G. Stefani, *J. Electron Spectrosc. Relat. Phenom*. 67: 479 (1994).
20. R. Gunnella, I. Davoli, R. Bernardini and M. De Crescenzi, *Phys. Rev. B* 52: 17091 (1995).
21. P. H. van Cittert, *Z.Phys*. 69: 298 (1931).
22. A. R. Law, M. T. Johnson, H. P. Hughes, H. A. Padmore, *J. Phys. C: Solid State Phys*. 18: L297 (1985).
23. M. Vos, P. Storer, S. A. Canney, A. S. Kheifets, I. E. Mc Carthy, E. Weigold, *Phys. Rev. B* 50: 5635 (1994).
24. A. Kheifets, *J. Phys. C* 7: 3895 (1995).
25. N. A. Krasil'Nikova and N. M. Persiantseva, *Phys. Rev. Lett*. 69 A: 287 (1978).

ENERGY DISTRIBUTION OF CORRELATED ELECTRON PAIRS EXCITED BY LOW ENERGY ELECTRONS FROM W(001) MEASURED BY A TIME-OF-FLIGHT (e,2e) SPECTROMETER

S.N. Samarin, O.M. Artamonov, H. Schwabe and J. Kirschner

Max-Planck-Institut für Mikrostrukturphysik, Weinberg 2, D-06120 Halle, Germany.

INTRODUCTION

Over the last twenty years the (e,2e) spectroscopy was evolved as a tool for studying the electronic structure of solids[1,2,3] and scattering dynamics of fast electrons on solid targets [4,5,6]. All these works were done in the transmission mode using high energy incident electrons. The fast electron knocks out a target electron, then scattered and ejected electrons are detected in coincidence. In this geometry the scattering problem is kinematically complete: energy E_o and momentum $\mathbf{q}_o$ of the incident electron are known and energies E_1, E_2 and momenta $\mathbf{q}_1$, $\mathbf{q}_2$ of both time correlated electrons are measured. Using energy and momentum conservation laws one can obtain the precollision binding energy ε and momentum $\mathbf{q}$ of the target electron:

$$\varepsilon = E_o - E_1 - E_2 \tag{1}$$

$$\mathbf{q} = \mathbf{q}_1 + \mathbf{q}_2 - \mathbf{q}_o \tag{2}$$

It can be assumed and experimentally verified[3] that at sufficiently high incident electron energy and momentum transfer the precollision target electron momentum $\mathbf{q}$ is equal and opposite to the measured recoil momentum $\mathbf{q}_o - \mathbf{q}_1 - \mathbf{q}_2$. The data obtained with this mode of (e,2e) experiments are related to the bulk electronic structure of a solid.

In order to apply this technique to the solid surface analysis and for studying the scattering dynamics of low energy electrons the (e,2e) experiments in reflection geometry were designed[7,8,9,10]. In this mode an incident electron impinges onto the sample surface and two outgoing electrons are detected in coincidence. Two different modes in reflection geometry experiments were used: 1) with normal incidence and low electron energy[7,8,9], 2) with grazing angle of incidence and medium primary electron energy[10]. We will focus our attention on experiments with the normal and small off-normal angle of incidence and low energy primary electrons. In this case the momentum $\mathbf{q}_o$ of incident electron is pointed from the vacuum to the solid and the total momentum $\mathbf{k}_t$ of two outgoing electrons is pointed from the solid to the vacuum. Thus to describe the ionization in reflection geometry further interaction, besides the electron-electron collision, must be considered.

Coincidence Studies of Electron and Photon Impact Ionization
Edited by Whelan and Walters, Plenum Press, New York, 1997

There are at least three models involving additional elastic scattering to describe (e,2e) experiment in reflection geometry. In a first model the momentum perpendicular to the surface is provided by the crystal that reflects the incident electron along the specular direction. This momentum is necessary to satisfy the conservation of the full momentum of the system under consideration. In principle not only the (0,0) specular beam but all diffracted beams in the backward hemi-sphere should be considered. Their contribution to the number of observable (e,2e) events in reflection geometry depends on the arrangement of the experiment. A second model can be considered where the incident electrons are isotropically scattered by the ion cores and then some of them create a correlated electron pair with a total momentum pointing toward the vacuum. A third model involves an elastic scattering of two correlated electrons on the ion core potential that provides the momenta necessary for electrons to escape from the surface. This third process is much less probable than the first and the second ones because it calls for the elastic scattering of two electrons on the ion core potential.

In the present work we studied the role of elastic scattering of the incident electron in the correlated electron pair creation and estimated the contribution of the multi-step electron-electron scattering. For this purpose we compared the energy distribution of correlated electron pairs generated by the normal incident electron beam and the energy distribution for off-normal incidence.

EXPERIMENT

Experiments were carried out using a μ-metal chamber with a vacuum in the range of 10^{-10} - 10^{-11} Torr. The W(001) crystal was mounted on a rotatable holder. The sample cleaning procedure included the oxidation of the sample followed by high temperature

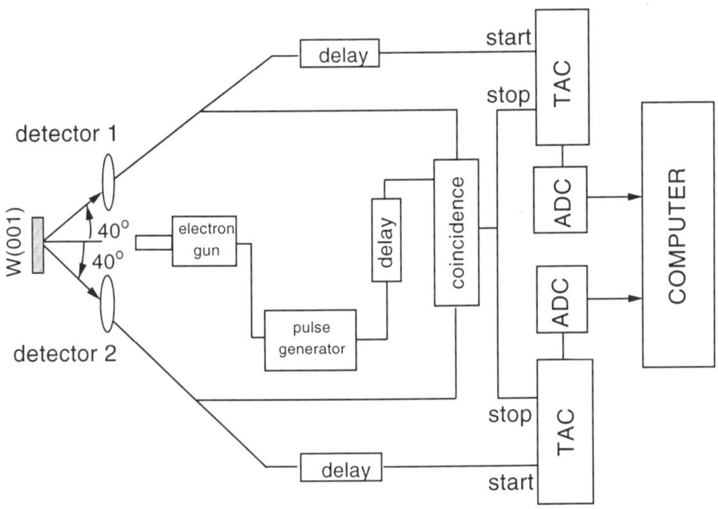

Figure 1. Experimental set-up of the (e,2e) time-of-flight spectrometer. ADC is Analog-to-Digital Converter., TAC is Time-to-Amplitude Converter.

flashes to remove carbon from the sample surface. The cleanliness of the surface was monitored by Auger spectroscopy. We used two 75 mm MCP-based position sensitive electron detectors. The detectors and the electron gun were in the same plane containing the sample surface normal. Detectors were placed in a symmetric position with respect to the electron gun axis at ± 40° with respect to this axis. The distance L between the sample and detectors can be varied in the limits of 130 - 263 mm. The position sensitive detectors allowed us to control the electron beam position on the sample observing the specular beam as well as diffracted beams. In order to measure the energies of both correlated electrons generated by one incident electron, we combined the coincidence technique with a time-of-flight electron energy measurement.

Our experimental set-up is shown in fig. 1. The incident electron current of the order of 10^{-14} A, on average, was pulsed with a width of about 1 ns and a repetition rate of 2.5×10^6 pulses per second, the electron beam diameter on the sample being not more than 1 mm. The trigger pulse from the generator was used as a reference point on the time of flight scale. When the incident electron generates a correlated pair of electrons and they are detected by the detectors, two pulses start two time-to-amplitude converters (TAC). A stop pulse to both TAC comes from a logic unit that delivers a stop pulse only when two delayed and shaped pulses from detectors and delayed trigger pulse from generator coincide within a time window of 200 ns. A valid event represents a point in a two-dimensional time-of-flight coordinate system. One of the coordinates of this point defines the time of flight of the first electron of the pair and the second coordinate defines the time of flight of the second correlated electron of the pair. Once a time T an electron takes to pass the distance L between the sample and detector is measured, one can calculate the kinetic energy of this electron:

$$E = \frac{m}{2}\left(\frac{L}{T}\right)^2 , \tag{3}$$

where m is the rest electron mass (the energies are very low).

The energy resolution of the time-of-flight electron energy measurement can be obtained from equation (3):

$$\Delta E = 2\left(\frac{m}{2}\right)^{-\frac{1}{2}} L^{-1} E^{\frac{3}{2}} \Delta T \tag{4}$$

It is clear from (4) that the energy resolution of the time-of-flight electron energy measurement depends on the flight distance L, the energy E to be measured and the time resolution ΔT. On the other hand the time resolution depends mostly on the modulating pulse width. If we take our values: ΔT = 1 ns, L = 263 mm, then ΔE = 0.05 eV for E = 5 eV. Experimental estimation of the energy resolution ΔE in the energy range 4 - 5 eV gives for ΔE the value of about 0.4 eV that is approximately equal to the half-width of the electron energy distribution in the incident electron beam. This means that in our case the energy resolution is not limited by the time resolution but by the energy spread of the incident electron beam.

EXPERIMENTAL RESULTS AND DISCUSSION

An example of the two-dimensional time-of-flight distribution of correlated electron pairs (panel a) is shown in fig. 2. The coincidence events are displayed on a two-dimensional gray scale plot, white being the highest number of events. A ridge-like maximum on this picture represents the most probable combinations of electron energies within time correlated electron pairs. This distribution looks very symmetric with respect to the diagonal (white line) of the frame. Points on this line represent correlated pairs with equal times of flight of both electrons of a pair.

The time-of-flight distribution of correlated electrons was converted to the energy distribution of pairs (panel b of fig.2). The ridge-like maximum of the time-of-flight distribution is transformed to the band along line $E_1+E_2 = 14$ eV in the energy distribution.

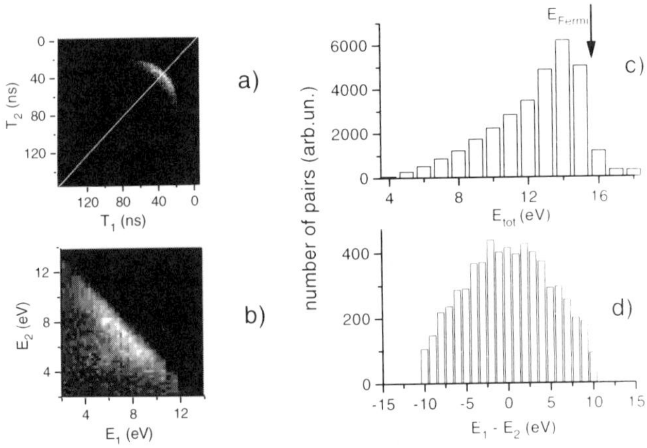

Figure 2. Time-of-flight and energy distributions of correlated electron pairs for the normal incidence, $E_p = 20$ eV. a) Two-dimensional time-of-flight distribution of correlated electrons. The white point in the upper right corner corresponds to the accidental coincidence of two elastically reflected electrons. b) Two dimensional energy distribution of correlated electrons. c) Histogram of the correlated pairs distribution as a function of the total energy of a pair. The height of a column represents the number of events within the total energy band $E_{tot} \pm 0.5$ eV. d) Energy sharing distribution of correlated pairs. The height of a column represents the number of pairs with the energy difference $E_1 - E_2$ within the total energy band $E_{tot} = 14 \pm 0.5$ eV.

If we take the width of this band equal to 1 eV, then within this band the sum of energies of two correlated electrons (we call it "total energy" of a pair) is equal to 14±0.5 eV. Once

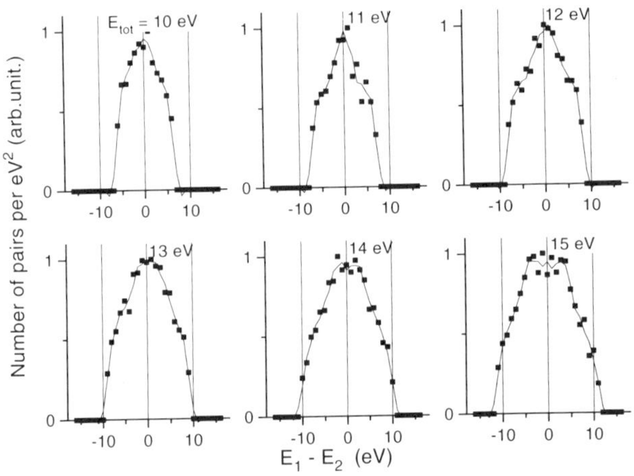

Figure 3. Energy sharing distributions of correlated pairs for different total energies of pairs. Normal incidence, $E_p = 20$ eV.

we know the total energy of the pair E_{tot}, with the incident electron energy $E_p = 20$ eV, we can calculate the binding energy E_b of the target electron: $E_b = E_p - E_{tot} = 20 - 14 = 6$ eV.

Taking into account that the work-function of W(001) is 4.6 eV, one sees that most likely correlated pairs are created when the target electron is excited from the energy levels just below Fermi level. Further analysis of the energy distribution is made by calculating the number of events in a total energy band $E_{tot} \pm 0.5$ eV for different E_{tot}. The distribution of the number of correlated electron pairs as a function of the total energy of pairs ("total energy distribution") is shown in the panel c of fig.2. The maximum of this distribution corresponds to the target electron excitation from energy levels just below the Fermi level. Correlated electron pairs with lower total energies may originate either in the excitation of bound electrons from deeper energy levels or in a multi-step electron-electron scattering.

The energy sharing distribution (panel d of fig. 2) represents the number of pairs with energy difference $\Delta = E_1 - E_2$ within the total energy band 14 ± 0.5 eV. Energy sharing distributions for different total energies of pairs are plotted in fig. 3. One can see that these distributions are symmetric with respect to the energy $\Delta E = 0$ where energies of both correlated electrons are equal.

We measured the energy distributions of correlated electron pairs for the different incident electron energies from 12 eV to 39 eV with an energy step of 1 eV. In the range from 12 eV to 24 eV the total energy distributions indicate that most likely target electrons are excited from the energy levels just below E_{Fermi}. For higher incident electron energies the maximum of the total energy distribution shifts to lower total energies of pairs (20 eV for $E_p = 39$ eV) and the relative number of pairs with lower total energy increases. That can be due to the multi-step electron-electron scattering. The energy sharing distributions are still symmetric for all incident energies.

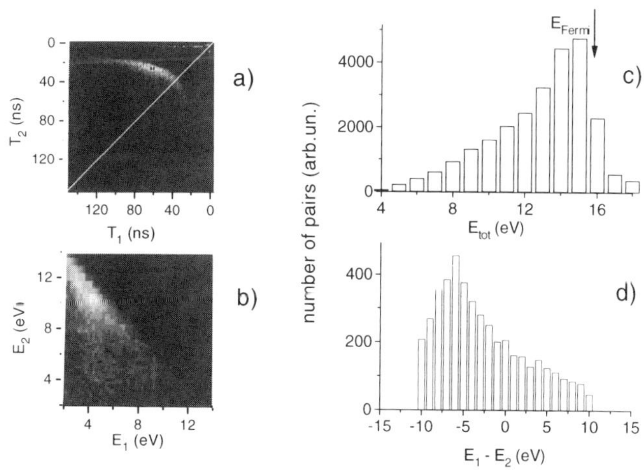

Figure 4. Time-of-flight and energy distributions of correlated electron pairs for off-normal incidence. E_p = 20 eV, angle of incidence $\alpha = 16^o$. a), b), c), and d) have the same meaning as in fig. 3. In contrast to fig. 3 time-of-flight distribution, two-dimensional energy distribution and energy sharing distribution are asymmetric.

To study the off-normal incidence excitation of correlated pairs the sample was rotated in the detector plane such that the incident electron beam made an angle $\alpha=16^o$ with the surface normal, primary electron energy was 20 eV. Experimental results for off-normal incidence are shown in fig. 4. Even on the time-of-flight scale (panel a) the distribution is quite asymmetric with respect to the diagonal (white line) of the picture. The energy distribution (panel b) clearly shows that one of the correlated electrons has most likely higher energy than the other one. The total energy distribution (panel c) has its maximum just below Fermi level. The energy sharing distribution for $E_{tot} = 14 \pm 0.5$ eV indicates that one of the two electrons most probably has 6 eV higher energy than the other one. We performed this experiment also for $\alpha = -16^{\circ}$ and obtained an asymmetry with reverse sign.

Fig. 5 shows energy sharing distributions for different total energies of pairs. For total energies higher than 11 eV these distributions are asymmetric with respect to the energy difference $\Delta = E_1 - E_2$. What could be reasons for this asymmetry?

As mentioned above Bragg scattered electrons (diffraction beams) can also take part in the correlated pair generation. However under present experimental conditions (low primary electron energy and geometry of experiment) they make too small angles with the sample surface to generate pairs that could be detected.

Correlated electron pairs can be excited either by specularly reflected incident electrons or by isotropically scattered ones. For normal incidence in both these cases the energy sharing distributions will be symmetric because the sample is a cubic crystal with a (001) face and the detectors are in symmetric positions with respect to the surface normal. That means in this geometry the two electron detectors are indistinguishable and this is the reason for the symmetric energy sharing distribution.

If the incident electron beam makes a small angle α with the normal to the sample surface there are in principle two ways (as well as for normal incidence) for incident electrons to be scattered. They can undergo isotropic elastic scattering by ion cores and then create correlated electron pairs. In this case the energy sharing distribution of correlated pairs will be again symmetric because isotropic scattering of incident electrons keeps the full symmetry of the (e,2e) process.

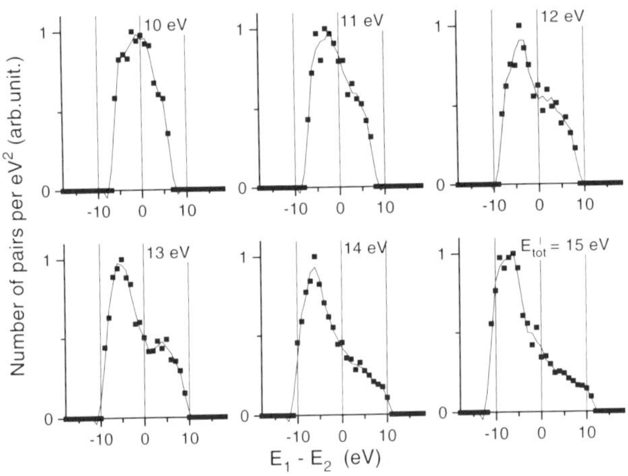

Figure 5. Energy sharing distributions of correlated pairs excited by the off normal incident electron beam for different total energies of pairs. Starting at $E_{tot} = 10$ eV they become asymmetric with respect to the energy difference $E_1 - E_2 = 0$.

On the other hand the incident electron beam can undergo specular reflection and then specularly reflected electrons can generate correlated electron pairs. In this case we can expect asymmetric energy sharing distribution of pairs because of the asymmetry of the kinematics of the (e,2e) process: one of the correlated electrons (with smaller scattering angle) will have most likely higher energy. Moreover this asymmetry will occur only for electrons generated in a single-step electron-electron collision. It means that as far as the energy sharing distribution for a total energy of pairs below cut-off is asymmetric, these pairs are created in single electron-electron collisions. By contrast the correlated electron pairs created in a multi-step process will have a more symmetric energy sharing distribution. This allows us to estimate the contribution of multi-step scattering to the creation of correlated pairs as well as the probing energy range of (e,2e) spectroscopy in reflection geometry in the low energy range.

The measured asymmetry in the energy sharing distribution within correlated pairs indicates that they are generated in asymmetric kinematics. One of the correlated electrons detected by detector called "number 2" has higher energy (smaller scattering angle). This is precisely the detector which makes a smaller angle with the specularly reflected beam. The analysis of this kinematics shows that correlated pairs are mostly created by specularly reflected electrons. It means that in the present geometry as well as for grazing incidence with 300 eV primary electrons[10], the specular reflection of incident electrons can be considered as a first step of correlated electron pair creation. Thus the momentum conservation law for the (e,2e) process in reflection kinematics must be rewritten as follows:

$$\mathbf{q} = \mathbf{q}_1 + \mathbf{q}_2 - \mathbf{K}_o, \qquad (5)$$

where $\mathbf{q}$ is the crystal momentum of the target electron, $\mathbf{q}_1$ and $\mathbf{q}_2$ are momenta of outgoing correlated electrons, $\mathbf{K}_o = \mathbf{q}_o - 2\mathbf{q}_{o\perp}$ is momentum of the specularly reflected incident electron, $\mathbf{q}_{o\perp}$ is the normal component of the incident electron momentum $\mathbf{q}_o$.

Correlated pairs with $E_{tot} = 15 \pm 0.5$ eV are created in a single electron-electron collision. The energy sharing distribution for this total energy (Fig. 5) indicates that they are generated in an asymmetric (e,2e) process. If we look at the energy sharing distributions for lower total energies of pairs, one can see that for $E_{tot} = 11$ eV and higher these distributions are also asymmetric. It means that correlated pairs at these total energies are mostly created in the (e,2e) process in the same kinematic conditions as considered above and shows a minor contribution of multi-step scattering in this energy range. It turns out that the energy range probed by the (e,2e) experiment in the present geometry is about 5 eV below the Fermi level.

CONCLUSIONS

In the present work we found that the energy distribution of correlated electron pairs excited by 20 eV primary electrons from W(001) depends on the kinematics of the (e,2e) process in reflection geometry. For normal incidence and symmetric detectors position the energy sharing distributions of correlated electrons are symmetric. On the other hand for off-normal incidence the energy sharing distributions are asymmetric for correlated pairs excited from energy levels between Fermi level and 5 eV below the Fermi level. It means that these pairs are mainly created in single electron-electron collisions. The present work clearly shows that under our experimental conditions correlated electron pairs are created mostly by specularly reflected incident electrons. This makes the (e,2e) experiment in reflection geometry kinematically better defined. The minor contribution of multi-step electron-electron collisions in the energy band of a few electron-volt below the Fermi edge and the energy resolution of 0.4 eV show (e,2e) experiments in reflection geometry to be a useful tool for studying the electronic structure of solids and the scattering dynamics of low energy electrons in the surface region.

REFERENCES

1. V.G. Levin, V.G. Neudachin, and Yu.F. Smirnov, On investigation of the structure of energy bands using quasielastic knock-out of an electron by an electron (e,2e), Phys. Stat. Sol. (b) 49:489 (1972).
2. M. Vos and I.E. McCarthy, Observing electron motion in solids, Rev. of Modern Phys. 67, N3:713 (1995).
3. Y.Q. Cai, M. Vos, P. Storer, A.S. Kheifets, I.E. McCarthy and E. Weigold, Direct imaging of the valence electronic structure of solid by (e,2e) spectroscopy, Solid State Communs. 95, N1:25 (1995).
4. Jeff Drucker, M.R. Scheinfein, J. Liu and J.K. Weiss, Electron coincidence spectroscopy studies of secondary and Auger electron generation mechanisms, J. Appl. Phys. 74(12):7329 (1993).
5. M.R. Scheinfein, Jeff Drucker and J.K. Weiss, Secondary-electron production pathways determined by coincidence electron spectroscopy, Phys. Rev. B47, N7:4068 (1993-I).
6. Jeff Drucker, M.R. Scheinfein, Delocalized secondary-electron generation studied by momentum-resolved coincidence-electron spectroscopy, Phys. Rev. B47, N23:15973 (1993-I).
7. J. Kirschner, O.M. Artamonov and A.N. Terekhov, Two-electron coincidence spectroscopy of scattering events at surfaces, Phys. Rev. Lett. 69:1711 (1992).
8. O.M. Artamonov, S.N. Samarin and J. Kirschner, Coincidence electron spectroscopy of W(100) in the threshold-energy region, Phys. Rev. B51:2491 (1995).
9. J. Kirschner, O.M. Artamonov and S.N. Samarin, Angle resolved energy correlated coincidence electron spectroscopy of solid surfaces Phys. Rev. Lett.75:2424(1995).
10. S. Iacobucci, L. Marassi, R. Camilloni, S. Nannarone, G. Stefani, Reflection (e,2e) spectroscopy on surfaces, Phys.Rev. B51, N15: 10252 (1995-I).

ELECTRON IMPACT IONIZATION OF HYDROGENIC IONS IN ASYMMETRIC GEOMETRIES

O.Rath Spivack [1], J.Rasch[1], Colm T.Whelan[1], R.J.Allan[2] and H.R.J.Walters[3]

[1] Department of Applied Mathematics and Theoretical Physics,
University of Cambridge, Cambridge CB3 9EW, U.K.

[2] CCLRC, Daresbury Laboratory, Warrington, WA4 4AD, U.K.

[2] Department of Applied Mathematics and Theoretical Physics,
The Queen's University, Belfast, BT7 1NN, U.K.

Introduction

Interest in collisional processes involving ions has recently increased, and more results, both theoretical and experimental, are becoming available. Some experimental measurements of total and single differential cross sections for electron impact ionization of positively charged ions have been obtained in the past few years [1–8], although measurements of triple differential cross sections (TDCS) are not yet available. The TDCS for electron impact ionization of He^+ and Li^+ has been calculated by Biswas and Sinha [9, 10] for very asymmetric geometry in the context of an eikonal approximation.

Here we present results for electron impact ionisation of a sequence of hydrogenic ions in coplanar asymmetric geometry. Our aim is to extend the work previously done for symmetric geometry [11, 12] and investigate the effects of different geometrical arrangements. The main trends observed for symmetric geometry also appear in this case, notably the enhancement of the recoil peak with increasing nuclear charge. In addition, new features dependent on the asymmetric geometry emerge. Their appearance and the way they vary with the scattering angle and/or the outgoing energies help our understanding of the mechanism of interaction in this process.

We also investigate the work by Biswas and Sinha [9] for electron impact ionisation of He^+ in asymmetric geometry and discuss the differences in results obtained with their method and ours.

Theory

In the Distorted Wave Born Approximation (DWBA) we obtain the following expressions for the the direct and exchange scattering amplitudes f and g[13]

$$f = < \chi^-(\mathbf{k}_f, \mathbf{r}_f)\chi^-(\mathbf{k}_s, \mathbf{r}_s) \mid \frac{1}{r_{sf}} \mid \chi^+(\mathbf{k}_0, \mathbf{r}_f)\psi_0(r_s) > \tag{1}$$

$$g = < \chi^-(\mathbf{k}_f, \mathbf{r}_s)\chi^-(\mathbf{k}_s, \mathbf{r}_f) \mid \frac{1}{r_{sf}} \mid \chi^+(\mathbf{k}_0, \mathbf{r}_f)\psi_0(r_s) > \ , \tag{2}$$

where $\chi^-(\mathbf{k}_f, \mathbf{r}_f)$ is the distorted wave for the outgoing fast electron, and $\chi^-(\mathbf{k}_s, \mathbf{r}_s)$ is the distorted wave for the outgoing slow electron, both calculated in the static potential of the final bare ion, $\chi^+(\mathbf{k}_0, \mathbf{r}_f)$ is the distorted wave for the incoming electron, calculated in the static exchange potential of the target ion, and $\psi_0(r_s)$ is the initial bound state hydrogenic function for an ion in its ground state, given by

$$\psi_0 = \phi_{1s} Y_{00} = Z^{3/2}\pi^{-\frac{1}{2}}e^{-Zr} \ . \tag{3}$$

The triple differential cross section in this approximation is then

$$\frac{d^3\sigma^{DWBA}}{d\Omega_f d\Omega_s dE} = (2\pi)^4 \frac{k_f k_s}{k_0}\{\frac{1}{4} \mid f^S + g^S \mid^2 + \frac{3}{4} \mid f^T - g^T \mid^2\} \ , \tag{4}$$

where the labels S and T refer to singlet or triplet case respectively (see[13]). Second and higher order effects in the interaction between the electrons and the target are all implicitely included. Electron-electron interaction in the final channel is estimated by introducing a Gamov factor

$$N_{ee} = \gamma/(e^\gamma - 1) \ , \tag{5}$$

where $\gamma = 2\pi/(\mid \mathbf{k}_f - \mathbf{k}_s \mid)$, so that the TDCS is given by

$$\frac{d^3\sigma}{d\Omega_f d\Omega_s dE} = N_{ee} \frac{d^3\sigma^{DWBA}}{d\Omega_f d\Omega_s dE} \ . \tag{6}$$

This factor and its use are discussed in[14–16]. Here we observe differences in the shape and magnitude of the cross section arising from estimating the post-collisional interaction (pci) with the above N_{ee} factor, or in the context of the eikonal approximation[9].

Results

We have applied the approximation described above to electron impact ionisation of hydrogenic ions with nuclear charge Z:

$$e^- + A^{(Z-1)+} \to e^- + e^- + A^{Z+}, \tag{7}$$

in coplanar asymmetric geometry. Results have been obtained for the hydrogenic sequence H, He^+, Li^{2+}, Be^{3+}, B^{4+}, C^{5+} at incident energies of 40eV and 100eV above threshold, following our previous work in symmetric geometry, and for He^+ at initial energies of 250eV, 500eV and 1000eV, in order to compare our results with those of Biswas and Sinha[9]. In the case of the hydrogenic sequence, we have also investigated the changes in the TDCS for different degrees of asymmetry, namely with a ratio of the energies of the outgoing electrons of 1.5 and 19.0, and for different scattering angles.

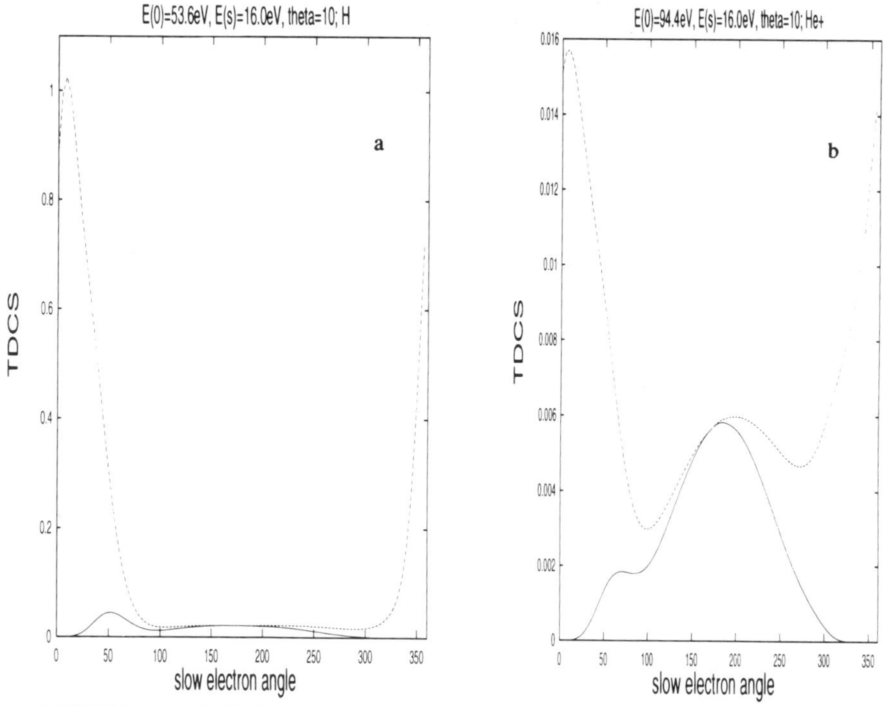

Figure 1. TDCS (in a.u.) for the ionisation of H and He^+ by electron impact in coplanar asymmetric geometry. The energy of the incident beam is 40eV above threshold and the exiting slow and fast electrons have a final energy of 16eV and 24eV respectively, i.e. R=1.5. The scattering angle is $\theta_f = 10^0$. Here - - - - is the TDCS obtained with the DWBA approximation described in the text; ——— includes the Nee correction.

The effects due to post collisional interaction, as taken into account by the N_{ee} factor, are clearly apparent if we compare, for example, the TDCS obtained with and without N_{ee} for H and He^+ at an incident energy of 40eV above threshold (Figure 1). We can see that the binary peak is greatly overestimated (and the recoil peak slightly underestimated) if we do not take pci effects into account. Also, the binary peak is shifted towards much larger angles by electron-electron repulsion in the final channel. All subsequent analysis refers therefore to results which include pci, i.e. the cross section is given by equation 6.

Many more parameters can be varied in asymmetric geometry, as opposed to the symmetric case, nevertheless the main result shown for the symmetric case[12] is mirrored by analogous results in this geometry. In parallel with the enhancement of the large angle peak in symmetric geometry as the nuclear charge Z increases, we observe a similar enhancement of the recoil peak in asymmetric geometry, independently of the scattering angle θ_f or of the ratio between the energies of the outgoing electrons (see Figures 2 and 3). The large angle peak is necessarily a result of electron-nucleus interaction. In asymmetric geometry the recoil peak need not reflect a direct collision. Both the target electron and the nucleus have a momentum about their centre of mass before the collision and this can be utilised to allow the electron to escape in the recoil direction[17] The dependence on the incident energy observed previously in the symmetric case is also apparent here: the relative magnitude of the recoil peak with respect to the binary one increases as the incident energy decreases, throughout the hydrogenic series. These results are summarised in Table 1, which gives the intensity ratio of the binary over the recoil peak. for various values of the parameters. For comparison, we give in Table 2 the intensity ratios of the binary versus large angle peak obtained in symmetric geometry. The main differences introduced by the asymmetry of the geometry are:

Table 1. Ratio of the binary versus recoil peak as a function of the nuclear charge Z for incident beam energies of 100eV and 40eV above threshold, and energy ratios of 1.5 and 19.0. The angle of the fast electron is fixed at 10^0 (4^0).

		$E_i - E_{ion} = 100eV$		$E_i - E_{ion} = 40eV$	
Z		R=1.5	R=19.0	R=1.5	R=19.0
1	H	9.440 (2.132)	6.106 (1.770)	1.902 (0.485)	1.900 (0.870)
2	He^+	1.486 (0.670)	0.770 (0.550)	0.251	0.221 (0.170)
3	Li^{++}	0.383	0.233	0.0834	0.0743
4	Be^{+++}	0.156	0.106	*	0.0402
5	B^{++++}	0.086	0.0616	*	0.0277
6	C^{+++++}	0.083	0.0422	*	0.0218

1. a less marked change from hydrogen to the rest of the series
2. much lower values of the intensity ratios.

We can see that the binary peak actually *disappears* in some cases , although not necessarily the most asymmetric ones (see for example Figure 3f and the ratios for Be^{3+}, B^{4+} and C^{5+} at 40eV above threshold and R=1.5 in Table 1).

Finally we derived the TDCS for the ionization of He^+ in coplanar asymmetric geometry at three different incident energies (250eV, 500eV and 1000eV), with fixed ejected electron energy 5eV, and fixed scattering angle 4^0, and at 1000eV incident energy for the same scattering angle but ejected electron energy 10eV. These are the same cases as considered by Biswas and Sinha[9]. We found that the differences between our results and theirs are mainly in the magnitude of the cross section, and can be ascribed to the different approximations used; but the main feature regarding the changes in the magnitude of the binary and recoil peaks as the parameters vary are in broad agreement. Let us summarise the main differences between our approximation and theirs. In our approximation we allow for elastic scattering of the incident electron in the field of the hydrogenic ion and for both exiting electrons in the field of the bare ion, electron-electron pci is accounted for via the N_{ee} factor. Biswas and Sinha use the *prior* form of the T-matrix:

$$T_{if}^{prior} = < \Psi_f^- \mid V_i \mid \phi_i > \ , \tag{8}$$

where the labels i and f refer to the initial and final channel respectively, and V_i is the perturbation interaction in the initial channel, given by

$$V_i = \frac{1}{r_{12}} - \frac{1}{r_1} \ , \tag{9}$$

where the labels 1 and 2 refer to the fast and slow electron respectively. For the prescription of the final state wave function Ψ_f^- they assume that the slow ejected electron is represented by a full Coulomb wave with charge Z, while the fast scattered electron is treated in the framework of the eikonal approximation. Their model is therefore only valid for strongly

Table 2. Ratio of the binary versus recoil peak as a function of the nuclear charge Z for incident beam energies of 100eV, 40eV and 20eV above threshold, in symmetric coplanar geometry.

Z		$E_i - E_{ion}$		
		100eV	40eV	20eV
1	H	9386.3	393.2	53.95
2	He^{+}	227.2	19.24	6.55
3	Li^{++}	19.47	3.52	2.11
4	Be^{+++}	5.02	1.67	1.35
5	B^{++++}	2.29	1.18	~ 1
6	C^{+++++}	1.38	1.05	~ 1

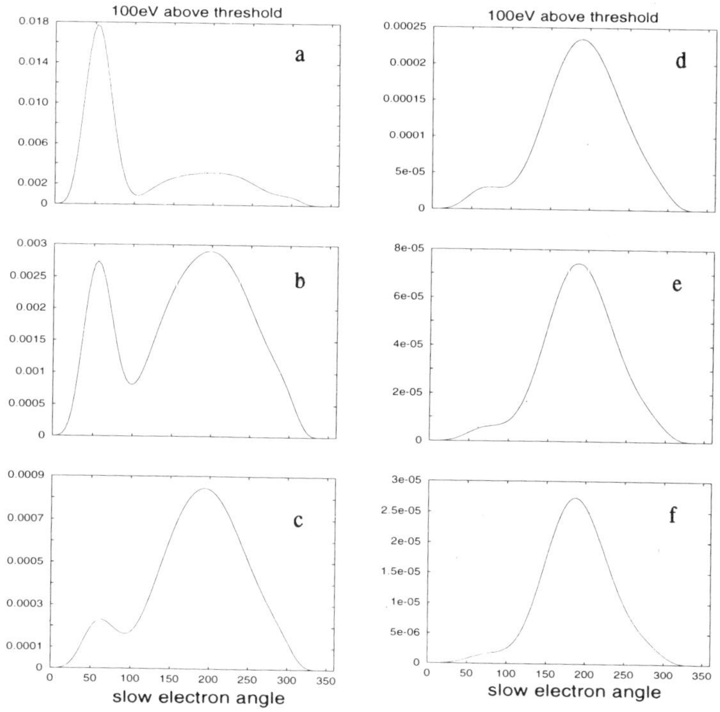

Figure 2. TDCS (in a.u.) for the ionisation of hydrogen and hydrogenic ions by electron impact in coplanar asymmetric geometry. The energy of the incident beam is 100eV above threshold and the exiting slow and fast electrons have a final energy of 40eV and 60eV respectively, i.e. R=1.5. The scattering angle is $\theta_f = 10^0$. Each frame corresponds to the following targets: (a)- H, (b)- He^+, (c)- Li^{2+}, (d)- Be^{3+}, (e)- B^{4+}, (f)- C^{5+}.

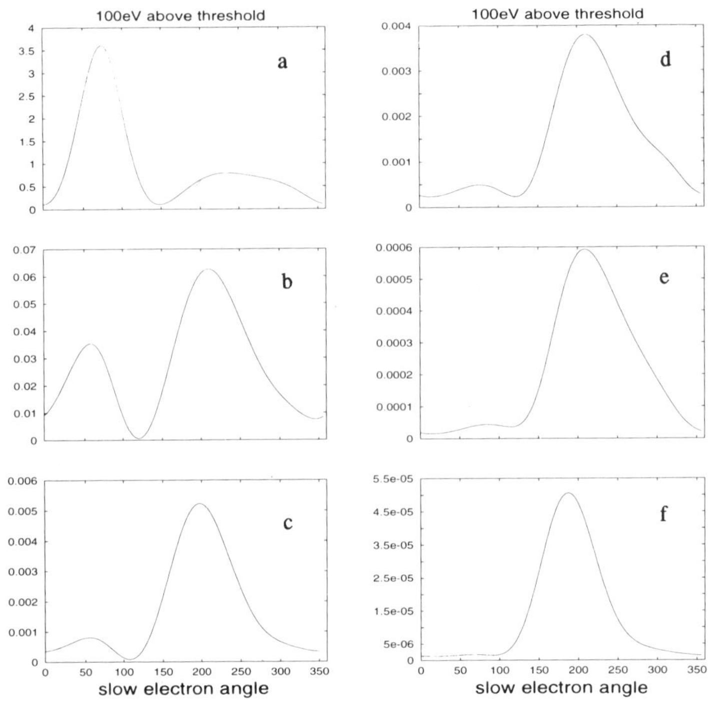

Figure 3. As Figure 2, but when the exiting slow and fast electrons have a final energy of 5eV and 95eV respectively, i.e. R=19.0.

asymmetric geometry, i.e. for small scattering angles and when the momentum of the fast scattered electron is much higher than the momentum of the slow electron. Within these limits, their approximation satisfies the three-body asymptotic boundary condition and takes therefore into account the electron-electron interaction in the final channel. We can expect our approximation to be better at reproducing the recoil peak, when is associated with a second order collision effect in the initial state, since in the approach adopted by Biswas and Sinha the emphasis is on the interaction in the final state while the interaction in the initial state is less fully taken into account.

Acknowledgments

We gratefully acknowledge financial support for this work from NATO (crg 950407) and the EC (HCM CHRX-CT93-0350), ORS is supported by a Daphne Jackson Research Fellowship at Lucy Cavendish College.

REFERENCES

1. K.T. Dolder and B. Peart, *Rep. Prog. Phys.* **39** 693 (1976).
2. D.H. Crandall, R.A. Phaneuf, B.E. Hasselquist and D.C. Gregory, Measured cross section for ionization of $C^{(3+)}$, $N^{(4+)}$, and $O^{(5+)}$ ions with contribution due to excitation-ionization, *J.Phys.* B **12** L249 (1979).
3. S.M. Younger, Electron impact ionization cross section for highly ionized hydrogen- and lithium-like atoms, *Phys. Rev.* A **22** 111 (1980).
4. Younger S M 1982 *Phys. Rev. Lett.* **26** 3177 (1982).
5. N.R. Badnell, D.C. Griffin and M.S. Pindzola, Electron impact ionization of $Ca^{(+)}$, *J.Phys.* B **24** L275 (1991).

6. S. Rachafi, D.S. Belić, M. Duponchelle, J. Jureta, M. Zambra, Z. Hui and P. Defrance, Absolute cross section measurements for electron impact ionization of $Ar^{(7+)}$, *J. Phys.* B **24** 1037 (1991).
7. D. Fang, W. Hu, J. Tang, Y. Wang and F. Yang, *Phys. Rev.* A **47** 1861 (1993).
8. A. Müller, G. Hoffmann, B. Weissbecker, M. Stenke, K. Tinschert, M. Wagner and E. Salzborn, *Phys. Rev Lett.* **63** 758 (1989).
9. R. Biswas and C. Sinha, Electron-impact ionization of a hydrogenic ion, *Phys. Rev.* A **50** 350 (1994).
10. R. Biswas and C. Sinha, Triple differential cross section for ionization of helium-like lithium by electron impact,*J.Phys.* B **28** 1311 (1995).
11. O. Rath Spivack, J. Rasch, C.T. Whelan, R.J. Allan and H.R.J. Walters, (e,2e) processes with positive ion targets, in: Proceedings of "Euroconference on Coincidence Studies", G. Stefani and L. Avaldi, eds., Trieste (1995).
12. O. Rath Spivack, J. Rasch, C.T. Whelan, R.J. Allan and H.R.J. Walters, (e,2e) processes with hydrogenic ion targets in coplanar symmetric geometry, submitted to J. Phys. B., (1996).
13. C.T. Whelan, R.J. Allan, H.R.J. Walters and X. Zhang, (e.2e), effective charges, distorted waves and all that!, in: (e,2e) and related processes, C.T. Whelan, H.R.J. Walters, A. Lahmam-Bennani and H. Ehrardt, eds., Kluwer Academic, (1993).
14. J. Botero and J.H. Macek, Threshold angular distribution of (e,2e) cross section of helium atoms, Phys. Rev. Lett. **68** 576 (1992).
15. C.T Whelan, R.J. Allan, J. Rasch, H.R.J. Walters, X. Zhang, J. Röder, K. Jung and H. Ehrhardt, Coulomb three-body effects in (e,2e) collisions: the ionization of H in coplanar symmetric geometry, *Phys. Rev.* A **50** 4394 (1994).
16. J. Röder, J. Rasch, K. Jung, C.T. Whelan, H. Ehrhardt, R.J. Allan and H.R.J. Walters, Coulomb three-body effects in low energy impact ionization of $H(1s)$, *Phys. Rev.* A **53** 225 (1996).
17. C.T. Whelan, H.R.J. Walters and R.M. Dreizler, *Aust. J. Phys.* **46** 241 (1993).

BETHE-RIDGE EFFECTS IN (e,2e) ON H^- (USING 6DIME)

S. P. Lucey, J. Rasch, Colm T. Whelan, R. J. Allan[1], H. R. J. Walters[2]

D. A. M. T. P., Silver Street, Cambridge, CB3 9EW England

[1]C. C. L. R. C. Daresbury Laboratory, Warrington, WA4 4AD England

[2] D. A. M. T. P., David Bates Building, Queen's University of Belfast, BT7 1NN Northern Ireland

INTRODUCTION

The study of (e,2e) processes on neutral targets such as Hydrogen and Helium has provided a measure of knowledge as to the importance of competing physical effects that govern the shape of experimental Triple Differential Cross Sections [2, 3, 4]. All of these studies emphasise that the calculated Triple Differential Cross Section is sensitive to the the form of both the initial and final states that one uses to approximate the problem. It is therefore reasonable, before embarking on a study of (e,2e) on H^- also to anticipate a great sensitivity to the form of approximation that one uses, and to expect competing physical processes to have different relative effects to (e,2e) on neutral targets. It is the purpose of this paper to enlarge upon the theory presented in [5] and to demonstrate the applicability of the 6-Dimensional Integral Method [6, 7] in the study of (e,2e) on H^-.

The physics of (e,2e) on a negative ion are fundamentally different from those of a neutral target. For instance, in the initial channel one must now describe the Coulombic repulsion between the incoming electron and the H^- target. This extra complication is, however, more than balanced out by the simplicity of the final state, in which the only Coulomb interaction is between the two escaping electrons. Thus, unlike (e,2e) on neutral targets, the need for solving a 3-body continuum problem is eradicated. This implies that the emphasis of (e,2e) on H^- is not on the final but on the initial state description and in particular the representation one employs for the wavefunction of H^-.

Atomic units are used throughout.

THEORETICAL FORMULATION

Suppose an initial electron, momentum k_0 and position $\mathbf{r}_f$, is incident upon a H^- target causing two electrons to be ejected in to the final state. These two final continuum electrons are labelled with respect to the detector in which they are found, s, indicating the 'slow' detector and , f, the 'fast' detector. Hence, the detected electrons are described by the momenta variables $\mathbf{k}_s$, $\mathbf{k}_f$ conjugate to the position variables $\mathbf{r}_s$, $\mathbf{r}_f$. In these coordinates, θ_s, will denote a clockwise rotation from the incident electron direction in the ionisation plane and, θ_f, an anticlockwise rotation. The remaining atomic bound electron is represented spatially by $\mathbf{r}_1$. The formula for the Triple Differential Cross Section of (e,2e) on any two electron system can be written,[8]

$$\frac{\mathrm{d}^3\sigma}{\mathrm{d}\Omega_\mathrm{f}\,\mathrm{d}\Omega_\mathrm{s}\,\mathrm{dE_s}} = (2\pi)^4 \frac{|\mathbf{k}_\mathrm{f}||\mathbf{k}_\mathrm{s}|}{|\mathbf{k}_0|} |\,\mathrm{T_{ba}}\,|^2, \tag{1}$$

$$|\,\mathrm{T_{ba}}\,|^2 \equiv 2\left(|\,\mathrm{f}\,|^2 + |\,\mathrm{g}\,|^2 + |\,\mathrm{h}\,|^2 - \Re(\mathrm{fg}^* + \mathrm{gh}^* + \mathrm{hf}^*)\right), \tag{2}$$

where f, g, h are the direct, exchange and capture ionisation amplitudes. These ionisation amplitudes can be written in a variety of ways, the most common and easily interpretable being the prior and post forms,[3],

$$\mathrm{f_{prior}} = \langle\, \Psi_\mathrm{b}^-(\mathbf{r}_\mathrm{f}, \mathbf{r}_\mathrm{s}, \mathbf{r}_1) \mid \mathrm{V_a} \mid \phi_\mathrm{a}(\mathbf{r}_\mathrm{f}, \mathbf{r}_\mathrm{s}, \mathbf{r}_1)\, \rangle, \tag{3}$$
$$\mathrm{f_{post}} = \langle\, \phi_\mathrm{b}\,(\mathbf{r}_\mathrm{f}, \mathbf{r}_\mathrm{s}, \mathbf{r}_1) \mid \mathrm{V_b} \mid \Psi_\mathrm{a}^+(\mathbf{r}_\mathrm{f}, \mathbf{r}_\mathrm{s}, \mathbf{r}_1)\, \rangle. \tag{4}$$

In equation (3) Ψ_a^+ is an eigenfunction of the total Hamiltonian, $\mathrm{H_T}$ with outgoing wave boundary conditions appropriate to the initial state of the system and ϕ_b is the final state asymptotic wavefunction, such that $\mathrm{H_b}\phi_\mathrm{b} = \mathrm{E_b}\phi_\mathrm{b}$. From $\mathrm{H_b}$ one can define $\mathrm{V_b}$ as that part of the total Hamiltonian undiagonalised by the final state, that is, $\mathrm{H_T} - \mathrm{H_b} = \mathrm{V_b}$. On the other hand, ϕ_a is the asymptotic initial state wavefunction, $\mathrm{H_a}\phi_\mathrm{a} = \mathrm{E_a}\phi_\mathrm{a}$, $\mathrm{H_T} - \mathrm{H_a} = \mathrm{V_a}$ and Ψ_b^- is the solution of $\mathrm{H_T}\Psi_\mathrm{b}^- = \mathrm{E_T}\Psi_\mathrm{b}^-$ with incoming wave boundary conditions appropriate to the final state of the system.

One therefore has a choice over which expression one adopts for the ionisation amplitudes, there are no hard and fast rules over which approach is the more appropriate. In general equations (3) and (4) are not equal, unless the eigenfunction Ψ of the total Hamiltonian is exact.

Final Channel Wavefunctions

The final state consists of two continuum electrons escaping in the field of a hydrogen atom. To a very good approximation the dominant interaction is the one between the two exiting electrons, that is, we shall assume that the motion of these two electrons is unaffected by the presence of the hydrogen atom. This allows us to make an ansatz for the total wavefunction of the system,

$$\Psi_\mathrm{b}^-(\mathbf{r}_\mathrm{f}, \mathbf{r}_\mathrm{s}, \mathbf{r}_1) \simeq \zeta(\mathbf{r}_\mathrm{f}, \mathbf{r}_\mathrm{s})\, \Phi_\mathrm{H}(\mathbf{r}_1), \tag{5}$$

where,

$$\zeta(\mathbf{r}_\mathrm{f}, \mathbf{r}_\mathrm{s}) = \mathrm{N_{out}}\, \mathrm{e}^{\mathrm{i}\mathbf{k}_\mathrm{f}.\mathbf{r}_\mathrm{f}}\, \mathrm{e}^{\mathrm{i}\mathbf{k}_\mathrm{s}.\mathbf{r}_\mathrm{s}}\, {}_1\mathrm{F}_1(\mathrm{i}\alpha_\mathrm{sf}, 1, -\mathrm{i}(|\mathbf{k}_\mathrm{sf}||\mathbf{r}_\mathrm{sf}| + \mathbf{k}_\mathrm{sf}.\mathbf{r}_\mathrm{sf})), \tag{6}$$

and

$$\mathrm{N_{out}} = \frac{1}{(2\pi)^3}\, \mathrm{e}^{-\pi\alpha_\mathrm{sf}/2}\, \Gamma(1 - \mathrm{i}\alpha_\mathrm{sf}), \quad \Phi_\mathrm{H}(\mathbf{r}_1) = \frac{\mathrm{e}^{-|\mathbf{r}_1|}}{\sqrt{\pi}}, \quad \alpha_\mathrm{sf} = \frac{1}{2|\mathbf{k}_\mathrm{sf}|}. \tag{7}$$

Interestingly though this approximation is equivalent to the final asymptotic state, in truth therefore, equation (5) is saying $\Psi_\mathrm{b}^-(\mathbf{r}_\mathrm{f}, \mathbf{r}_\mathrm{s}, \mathbf{r}_1) \simeq \phi_\mathrm{b}(\mathbf{r}_\mathrm{f}, \mathbf{r}_\mathrm{s}, \mathbf{r}_1)$.

Initial Channel Wavefunctions

The asymptotic initial channel wavefunction, $\phi_a(\mathbf{r}_f, \mathbf{r}_s, \mathbf{r}_1)$, defined in equation (3) must take into account the Coulombic repulsion between the incoming electron and the overall charge of the target. This is done by assuming the target to have an overall effective charge of -1. Hence, $\phi_a(\mathbf{r}_f, \mathbf{r}_s, \mathbf{r}_1) = \zeta(\mathbf{r}_f)\Phi(\mathbf{r}_s, \mathbf{r}_1)$, where

$$\phi_a(\mathbf{r}_f, \mathbf{r}_s, \mathbf{r}_1) = N_{in}\, e^{i\mathbf{k}_0.\mathbf{r}_f}\, {}_1F_1(i\alpha_0, 1, i(|\mathbf{k}_0||\mathbf{r}_f| - \mathbf{k}_0.\mathbf{r}_f))\, \Phi_{H^-}(\mathbf{r}_s, \mathbf{r}_1), \tag{8}$$

and

$$N_{in} = \frac{1}{(2\pi)^{3/2}}\, e^{-\pi\alpha_0/2}\, \Gamma(1 - i\alpha_0), \qquad \alpha_0 = \frac{1}{|\mathbf{k}_0|}. \tag{9}$$

The interaction between the incoming electron and the overall charge of the H^- ion is thus represented by a Coulomb wave, and the initial wavefunction of the target is represented by $\Phi_{H^-}(\mathbf{r}_s, \mathbf{r}_1)$.

If one wished to write down an approximation to the total wavefunction of the system in the initial state one must be very clear how one interprets the interaction between the incoming electron and the H^- target. If one assumes that the incoming electron interacts with the target "as a whole" then one can let $\Psi_a^+ \simeq \phi_a$, however, if one assumes that it interacts with one of the electrons in the target, for example the 's' electron then,

$$\Psi_a^+(\mathbf{r}_f, \mathbf{r}_s, \mathbf{r}_1) = N_{in}\, e^{i\mathbf{k}_0.\mathbf{r}_f}\, {}_1F_1(i\alpha_0, 1, i(|\mathbf{k}_0||\mathbf{r}_{sf}| - \mathbf{k}_0.\mathbf{r}_{sf}))\, \Phi_{H^-}(\mathbf{r}_s, \mathbf{r}_1), \tag{10}$$

which is clearly no longer of a separable form.

Before highlighting the relative advantages of each approach it is instructive to define the asymptotic Hamiltonians and the undiagonalised potentials. From equation (8) one may deduce that,

$$H_a = -\frac{1}{2}\nabla^2_{r_f} - \frac{1}{2}\nabla^2_{r_s} - \frac{1}{2}\nabla^2_{r_1} + \frac{1}{|\mathbf{r}_f|} - \frac{1}{|\mathbf{r}_s|} - \frac{1}{|\mathbf{r}_1|} + \frac{1}{|\mathbf{r}_1 - \mathbf{r}_s|}, \tag{11}$$

which implies that,

$$V_a = \frac{1}{|\mathbf{r}_f - \mathbf{r}_1|} - \frac{2}{|\mathbf{r}_f|} + \frac{1}{|\mathbf{r}_s - \mathbf{r}_f|}. \tag{12}$$

Similarly one can define H_b and V_b as,

$$H_b = -\frac{1}{2}\nabla^2_{r_f} - \frac{1}{2}\nabla^2_{r_s} - \frac{1}{2}\nabla^2_{r_1} + \frac{1}{|\mathbf{r}_s - \mathbf{r}_f|}, \tag{13}$$

$$V_b = -\frac{1}{|\mathbf{r}_f|} - \frac{1}{|\mathbf{r}_s|} - \frac{1}{|\mathbf{r}_1|} + \frac{1}{|\mathbf{r}_f - \mathbf{r}_1|} + \frac{1}{|\mathbf{r}_1 - \mathbf{r}_s|}. \tag{14}$$

The Post or the Prior Formalism

In the prior form of the direct ionisation amplitude one must calculate integrals of the form,

$$f_{prior} = < \Psi_b^-(\mathbf{r}_f, \mathbf{r}_s, \mathbf{r}_1)\ |\ V_a\ |\ \zeta(\mathbf{r}_f)\Phi_{H^-}(\mathbf{r}_s, \mathbf{r}_1) > . \tag{15}$$

In equation (15) the sole ambiguous term is the expression for the H^- wavefunction, hence this approach provides a good means of testing different H^- wavefunctions. If on the other hand one uses the post formalism,

$$f_{post} = < \zeta(\mathbf{r}_f, \mathbf{r}_s)\, \Phi_H(\mathbf{r}_1)\ |\ V_b\ |\ \Psi_a^+(\mathbf{r}_f, \mathbf{r}_s, \mathbf{r}_1)\ >, \tag{16}$$

there is a larger degree of freedom in the form of the initial channel, in which one could try and model the distortion/polarisation effects of the incoming electron on the target. Using this argument it would appear, that the post approach is the most satisfactory since it has greater generality. Unfortunately though there is a calculational consideration that must be taken in to account. This can be seen more clearly by rewriting equation (15) in integral notation,

$$f_{\text{prior}} = \int\int\int d^3|\mathbf{r}_f|\, d^3|\mathbf{r}_1|\, d^3|\mathbf{r}_s|\, e^{i\mathbf{q}.\mathbf{r}_f}\, e^{-i\mathbf{k}_s.\mathbf{r}_s}\, e^{-|\mathbf{r}_1|}\, {}_1F_1\Big(i\alpha_0, 1, i(|\mathbf{k}_0||\mathbf{r}_f| - \mathbf{k}_0.\mathbf{r}_f)\Big)$$
$$\Phi_{H^-}(\mathbf{r}_s, \mathbf{r}_1)\Big(\frac{1}{|\mathbf{r}_f - \mathbf{r}_1|} - \frac{2}{|\mathbf{r}_f|} + \frac{1}{|\mathbf{r}_s - \mathbf{r}_f|}\Big)\, {}_1F_1(-i\alpha_{sf}, 1, i(|\mathbf{k}_{sf}||\mathbf{r}_{sf}| + \mathbf{k}_{sf}.\mathbf{r}_{sf})). \quad (17)$$

Solving equations such as (17) rely on the ability to separate one of the integration variables out of the calculation thus leaving one with at most a 6-dimensional integration, which can be solved numerically. However, if $\Phi_{H^-}(\mathbf{r}_s, \mathbf{r}_1)$ is of a more complex form than a combination of exponential terms, it is not possible to separate this expression. For instance, if one used a Hylleras wave function to model the H^- wavefunction. One would need to solve the integral,

$$\int d^3|\mathbf{r}_1|\, \frac{e^{-\omega|\mathbf{r}_1|}\, e^{-\Omega|\mathbf{r}_1 - \mathbf{r}_s|}}{|\mathbf{r}_f - \mathbf{r}_1|}, \quad (18)$$

analytically in order to complete the calculation. Unfortunately analytical solutions to integrals such as equation (18) do not exist. Hence, if one could carry out a 9-dimensional numerical integration this full calculation would be feasible. Unfortunately such calculations are beyond the scope of computers at the present time. This restriction quashes the extra flexibility introduced in the post approach, hence the prior approach will be used.

(e,2e) on H^- Using the Prior Approach

As a starting point for this study, two contrasting representations of $\Phi_{H^-}(\mathbf{r}_s, \mathbf{r}_1)$ have been taken within the prior formalism. The first wavefunction is a Slater wavefunction, $\Phi^S_{H^-}$,

$$\Phi^S_{H^-}(\mathbf{r}_s, \mathbf{r}_1) = \frac{(Z^*)^{3/2}}{\sqrt{\pi}} e^{-Z^*(|\mathbf{r}_s| + |\mathbf{r}_1|)}, \quad (19)$$

where $Z^* = 11/16$. It is important to note at this stage that equation (19) treats both electrons symmetrically, though includes no radial or angular correlation in its description. The second wavefunction has a different philosophical bent, Φ^C_{H-},

$$\Phi^C_{H^-}(\mathbf{r}_s, \mathbf{r}_1) = \frac{m}{4\pi}(e^{-\alpha r_s - \beta r_1} + e^{-\beta r_s - \alpha r_1}), \quad (20)$$

where $m = 0.3948$, $\alpha = 1.039$, $\beta = 0.283$. Equation (20) shall be referred to as the Chandreskar representation [9]. Even though equation (20) is symmetrised, classically one can picture the Chandreskar representation as treating both electrons in a very asymmetric manner, (this wavefunction is radially correlated though contains no angular correlation).

The main defect in these two representations is their inability to account for angular correlation in a H^- target. However, this does not imply that the Slater and Chandreskar wavefunctions when used in an (e,2e) calculation will necessarily produce erroneous results. The interesting point in this calculation is, whether or not the wavefunctions predict different results and if so, why are they different?

Results for (e,2e) on H^-

In [5] the behaviour of the Triple Differential Cross Section with respect to the detected coplanar angles was discussed in detail and the above two wavefunctions were found to predict very different trends in behaviour as one varied the fast electron detector angle in asymmetric geometries. In this analysis it was reasoned that this difference in behaviour must be due to the representation of the H^- target, as this was the only difference between the two calculations. Working on this inference, a plane-wave approximation was constructed that ignored all interactions between the colliding particles.

It was shown that if one writes the Chandreskar wavefunction in full, one can express the plane wave direct ionisation amplitude as the sum of two components,

$$f^{pw} = f^{pw}(\alpha, \beta) + f^{pw}(\beta, \alpha), \tag{21}$$

where,

$$f^{pw}(\alpha, \beta) = \frac{M}{|\mathbf{q}|^2}\left[\frac{1}{(\alpha+1)^3}\frac{1}{(|\mathbf{Q}|^2+\beta^2)^2} - \frac{1}{(\alpha+1)^3}\frac{1}{(|\mathbf{k}_s|^2+\beta^2)^2} + \frac{1}{(|\mathbf{k}_s|^2+\beta^2)^2}\frac{(\alpha+1)}{(|\mathbf{q}|^2+(\alpha+1)^2)^2}\right], \tag{22}$$

M is constant, $\mathbf{q} = \mathbf{k}_0 - \mathbf{k}_f$ and $\mathbf{Q} = \mathbf{q} - \mathbf{k}_s$.* The second term in equation (22) is constant for any chosen slow electron detector energy and hence cannot be the cause of the different behaviour for the two wavefunctions. The third term in equation (22) is constant for any specific geometry and is a monotonically decreasing function of momentum transfer, $\mathbf{q}$. Hence, the first term of equation (22) is the only possible source of different behaviour for the two H^- wavefunctions. The first term of equation (22) can be re-expressed as, $e(\mathbf{q}, \mathbf{Q})$,

$$e(\mathbf{q}, \mathbf{Q}) \propto \frac{1}{|\mathbf{q}|^2}\frac{1}{(|\mathbf{Q}|^2+\beta^2)^2}. \tag{23}$$

There are several points to understand concerning equation (23),

1. Firstly, if $|\mathbf{Q}| \gg 0$ then the behaviour of $e(\mathbf{q}, \mathbf{Q})$ is dominated by the first term of equation (23), that is, $e(\mathbf{q}, \mathbf{Q}) \propto 1/|\mathbf{q}|^2$.

2. Secondly, if β is small, in the Bethe ridge region where $|\mathbf{Q}| \approx 0$ the properties of, $e(\mathbf{q}, \mathbf{Q})$, are governed by the second term of equation (23), that is, $e(\mathbf{q}, \mathbf{Q}) \propto 1/(|\mathbf{Q}|^2 + \beta^2)^2$. However, if $\beta \gg 0$ then the behaviour of $|\mathbf{Q}|$ is unimportant and $e(\mathbf{q}, \mathbf{Q}) \propto 1/|\mathbf{q}|^2$. In the case of the Chandreskar wavefunction, $\beta^2 = 0.0801$, whereas for the Slater wavefunction, $\beta^2 = 0.4727$. Thus one can infer from the above, that the behaviour of $|\mathbf{Q}|$ will have a far greater effect upon the first term in equation (22) and in turn upon the actual Triple Differential Cross Section when using the Chandreskar wavefunction as compared to the Slater wavefunction.

From point number 2, the existence of a divergence from the normal monotonically decreasing behaviour of the Triple Differential Cross Section has been demonstrated for small β. The question that still remains is, why should the maximum occur in the position that it does?

---

*In the case of a very asymmetric geometry one can neglect the influence of the exchange ionisation amplitude and assume that $|T^{pw}_{ba}| \propto |f^{pw}|^2$. This fact can easily be deduced from equation (22), from which one would obtain the exchange ionisation amplitude if one swapped the momenta $\mathbf{k}_s$ and $\mathbf{k}_f$. The prefactor term in equation (22) is far smaller for the exchange case.

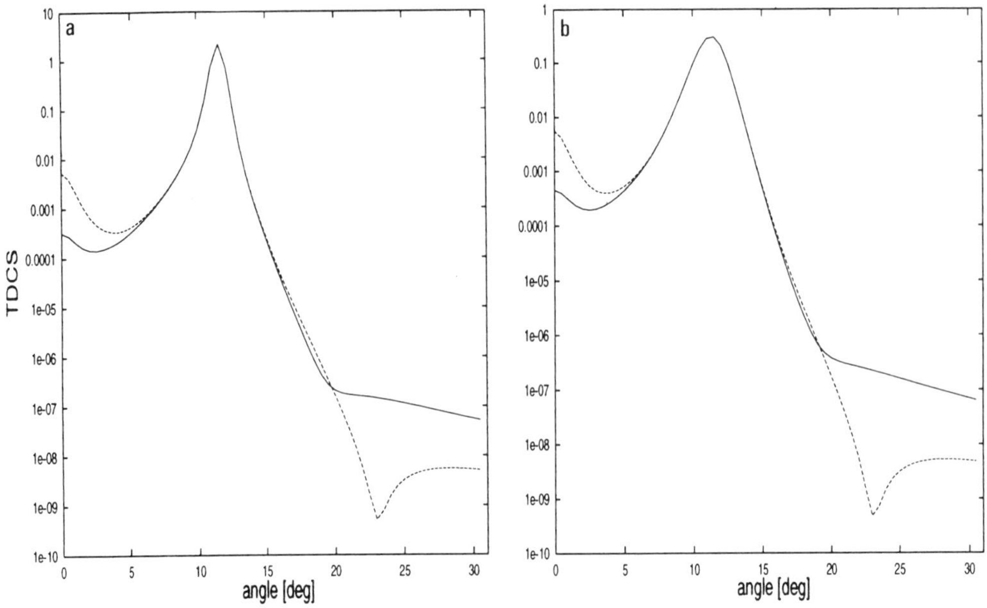

Figure 1. Variation of binary maxima with θ_f, for $E_0 = 5000$eV, $E_s = 200$eV, using a) the Chandreskar wavefunction and b) the Slater wavefunction. The bold curve is the full calculation and the broken line the plane-wave approach.

When Can $|\mathbf{Q}| = 0$?

Clearly $|\mathbf{Q}| = 0$ iff the components of $\mathbf{Q}$, Q_x, Q_y and Q_z simultaneously are zero. Hence, one can write,

$$Q_x = 0 \quad \Rightarrow \quad |\mathbf{k}_f| \sin\theta_f = |\mathbf{k}_s| \sin\theta_s, \tag{24}$$

$$Q_z = 0 \quad \Rightarrow \quad |\mathbf{k}_0| - |\mathbf{k}_f| \cos\theta_f - |\mathbf{k}_s| \cos\theta_s = 0, \tag{25}$$

the component $Q_y = 0$, $\forall\theta_f$. If one rewrites equation (24) in terms of $\cos(\theta_s)$,

$$|\mathbf{k}_s|^2 \cos(\theta_s)^2 = |\mathbf{k}_s|^2 - |\mathbf{k}_f|^2 + |\mathbf{k}_f|^2 \cos(\theta_f)^2, \tag{26}$$

and places this in to the square of equation (25) then one finds that,

$$\cos\theta_f = \frac{|\mathbf{k}_0|^2 + |\mathbf{k}_f|^2 - |\mathbf{k}_s|^2}{2|\mathbf{k}_f||\mathbf{k}_0|}, \tag{27}$$

which using the conservation of energy can be written,

$$\cos\theta_f = \frac{|\mathbf{k}_f|^2 - \epsilon}{|\mathbf{k}_f||\mathbf{k}_0|}, \tag{28}$$

where ϵ is the ground state energy of H^-. Hence, if the energy of the fast electron detector is fixed, there exist at most two values of θ_f for which $|\mathbf{Q}| = 0$, which explains why one observes a maximum in the trend of binary peaks in the Triple Differential Cross Sections for the Chandreskar representation.

If one studies high-intermediate incident energies then the prominence of the binary peak is even greater, than that demonstrated in [5], for example, in Figure (1), even the Slater wavefunction demonstrates a sensitivity to the value of $|\mathbf{Q}|$. From equation (22) one can

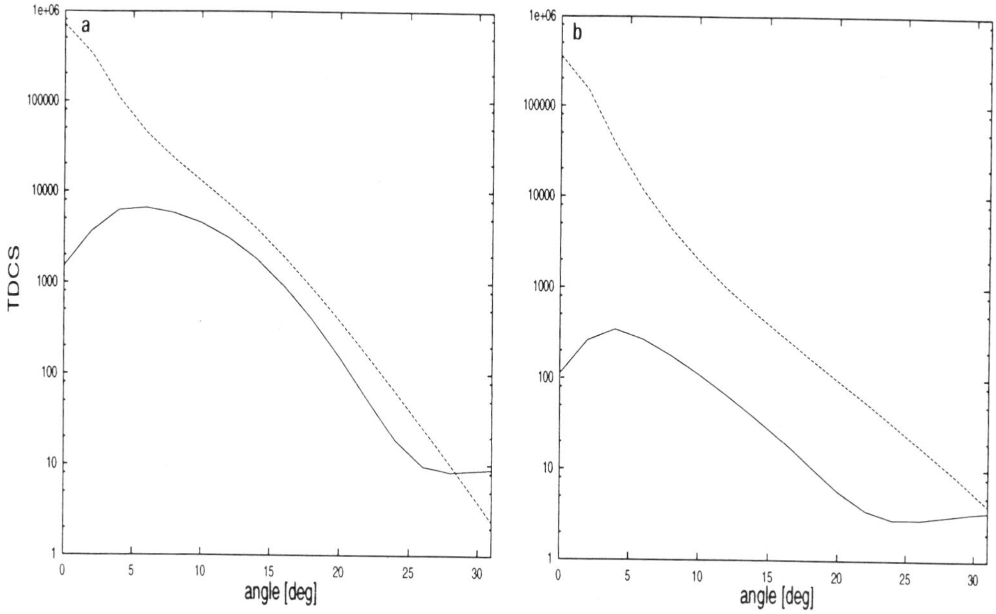

Figure 2. Variation of binary maxima with θ_f, for $E_0 = 20$eV, $E_s = 1$eV, using a) the Chandreskar wavefunction and b) the Slater wavefunction. The bold curve is the full calculation and the broken line the plane-wave approach.

see that this extra sensitivity is due to two causes. Firstly, as $|\mathbf{k}_s|^2$ is now large the second and third terms are very small in comparison to the first term, and secondly, the larger the momentum transfered to the target the smaller the $1/|\mathbf{q}|^2$ term in equation (23) and hence the greater the sensitivity of the behaviour of the Triple Differential Cross Section to $|\mathbf{Q}|$. Note, however, that the Chandreskar peak is an order of magnitude higher than the Slater peak even at these high energies. Whereas at low energies this behaviour disappears altogether, for both wavefunctions the Triple Differential Cross Section is dominated by the Momentum transfer components in equation (22), see Figure (2). A word of caution must be sounded though when analysing the results presented in Figure (2). The plane-wave calculations decrease steadily in size apart from a minor blip in the Chandreskar curve at $\approx 15°$, however, the full calculation on both graphs provide a maximum at $\approx 5°$. This feature is not due to Bethe ridge effects but is a consequence of the normalisation of the confluent hypergeometric function in equation (6). This problem is linked to a similar problem in calculations using 3-Coulomb wavefunctions and will be explained in a further work.

Improved Calculations

It is reasonable to expect the Chandreskar calculation to present a more accurate portrait of the behaviour of H^- in (e,2e) processes, intuitively, in classical terms, it is a more appealing picture of H^-. Furthermore in previous studies of electron loss from H^- when travelling through Noble gases, [10] a representation of the H^- wavefunction in which one electron was loosely bound and the other more tightly bound was found to more accurately describe experiment.

It is possible to reinforce the above statement by carrying out an (e,2e) calculation on H^- using a very accurate ground state wavefunction due to [11], of the form given below,

$$\Phi_{H^-}^{Tw}(\mathbf{r}_s, \mathbf{r}_1) = \sum_l P_l(\cos(\theta_{s1}))\, S_l(|\mathbf{r}_s|, |\mathbf{r}_1|), \tag{29}$$

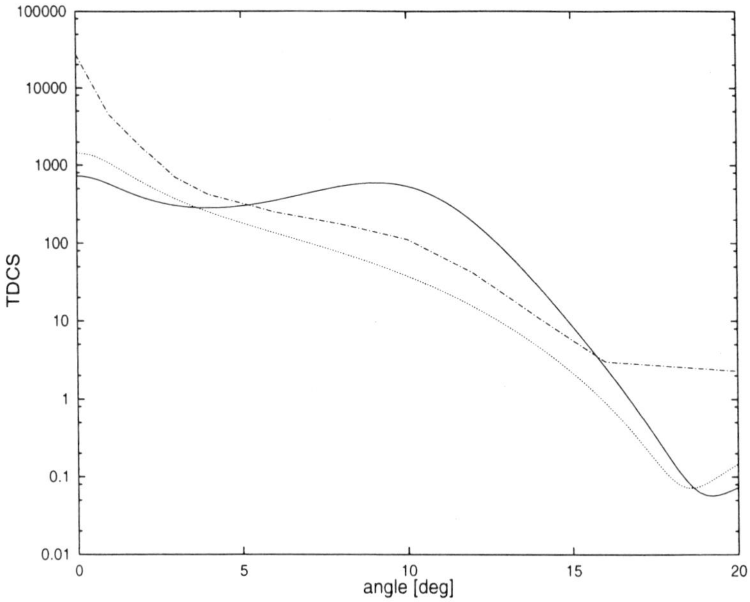

Figure 3. Variation of binary maxima with θ_f, for $E_0 = 150\text{eV}$, $E_s = 5\text{eV}$. The bold continuous curve is the Chandreskar, the dot-dash curve is the Slater and the dotted curve is the Tweed 21-parameter representation on the wavefunction of H^- .

form which one can expand the Legendre Polynomial in terms of spherical harmonics,

$$P_l(\cos(\theta_{s1})) = \sum_{lm} \frac{4\pi}{(2l+1)} (-1)^m \, Y_{lm}(\hat{\mathbf{r}}_s) \, Y_{l-m}(\hat{\mathbf{r}}_1), \tag{30}$$

and

$$S_l(|\mathbf{r}_s|, |\mathbf{r}_1|) = \sum_{kn} A_{lkn}(|\mathbf{r}_s|^k |\mathbf{r}_1|^n + |\mathbf{r}_s|^n |\mathbf{r}_1|^k) \; \exp\left(-\frac{p}{2}(|\mathbf{r}_s| + |\mathbf{r}_1|)\right), \tag{31}$$

where, $0 \leq k \leq n \leq 3$; $l, k, n \in \mathbb{N}$ and $A_{lkn}, p \in \mathbb{R}$. In order to calculate a Triple Differential Cross Section using wavefunction (29), it is necessary to separate out the $\mathbf{r}_1$ integration variable, thus leaving one with a series of products of 3-dimensional and 6-dimensional integrals. In this case, all the 3-dimensional integrals can be calculated analytically and the 6-dimensional integrals numerically[7]. By re-arranging these terms, it is possible to reduce the above calculation to one numerical 6-dimensional calculation.[†] From Figure (3) one can see that the in the Bethe ridge region, ($\theta_f \approx 10°$) Tweed's 21-parameter wavefunction demonstrates an intermediate behaviour. The general trend of Tweed appears to be closer to that of Slater rather than Chandreskar. Though Tweed does not exhibit a maximum at this energy, the Bethe ridge effects can be seen to portray a slightly larger effect than in the Slater case.

CONCLUSION

In the above it has been shown that the (e,2e) on H^- is very sensitive to the form of the H^- wavefunction. In addition an extra calculation has been performed using a wavefunction

[†] In order to test the accuracy of the above calculations a code similar to [12] has been constructed. The results of Figures (1, 2) were found to agree within 1% using the methods of [7] and [12].

due to Tweed that predicts a very accurate ground state energy for H^- , (Slater $= -0.027$, Chandreskar $= 0.014$, Tweed $= 0.028$, Experiment $= 0.028$). Even though all three wavefunctions predict very different Triple Differential Cross Section the trend in behaviour of the Tweed wavefunction in the Bethe region is seen to be closer to that of the symmetrical Slater wavefunction rather than the asymmetrical representation of Chandreskar.

ACKNOWLEDGEMENTS

We gratefully acknowledge support from the Royal Society, the EC(HCM CHCRX CT93 0350), NATO(CRG 950 407) and P.P.A.R.C. studentships.

REFERENCES

1. H. A. Bethe and E. E. Salpeter. "Quantum Mechanics of One- and Two-Electron Atoms," Plenum Publishing Corporation, New York (1977).
2. Colm T. Whelan, R. J. Allan and H. R. J. Walters. "PCI, Polarisation and Exchange Effects in (e,2e) Collisions," *Journal de Physique*. 3C6:39 (1993).
3. Colm T. Whelan, H. R. J. Walters and X. Zhang. "(e,2e) and Related Processes," Kluwer academic publishers, Netherlands (1993).
4. J. Rasch. "(e,2e) Processes with Neutral Atom Targets," Ph. D. Thesis, Cambridge University, (1996).
5. S. P. Lucey, Colm T. Whelan, R. J. Allan and H. R. J. Walters. "(e,2e) on Hydrogen Minus, H^-," *J. Phys. B.* 29:L489 (1996).
6. J. Rasch, S. P. Lucey and Colm T. Whelan. "6-Dimensional Integral Method," *Computer Physics Communications*. Accepted for publication (1996).
7. S. P. Lucey, J. Rasch, Colm T. Whelan and H. R. J. Walters "Aspects of (e,2e) : 6-Dimensional Integral Method," Proceedings of Denton conference (1996).
8. C. J. Joachain "Quantum Collision Theory," North Holland, Amsterdam (1983).
9. B. H. Bransden and C. J. Joachain "Physics of Atoms and Molecules," Longman, London and New York (1983).
10. D. P. Dewangan and H. R. J. Walters. "Electron Loss from H(2s), H(1s), He(1s), He(2s), Li(2s), H– and He– Projectiles Passing through Inert Gases- the Free Collision Model, " *J. Phys. B.* 11:3983 (1978).
11. R. J. Tweed. "Correlated Wavefunctions for Helium-like Atomic Systems," *J. Phys. B.* 5:810 (1972).
12. M.Brauner, J. S. Briggs and H. Klar. "Triply differential cross sections for ionisation of Hydrogen atoms by electrons and positrons," *J. Phys. B.* 22:2265 (1989).

The Ehrhardt Prizes 1996

Prof Dr rer.nat. Helmut Ehrhardt

These Prizes have been created to mark the retirement of Professor Helmut Ehrhardt from his Chair at the University of Kaiserslautern and his position as Deputy President of the Deutsche Forschungsgemeinschaft. Ehrhardt is known both as one of Europe's leading experimental physicists and also as a major figure in the German scientific establishment, he was the founding President of the University of Kaiserslautern and was responsible for the reconstruction of East German science from 1991 until his retirement. As a physicist he is, perhaps, most noted for his contribution to coincidence studies of atomic systems-including the first (e,2e) experiment in 1969- and many developments in Surface Science. Professor Ehrhardt has had a seminal influence on the development of Atomic Physics in Europe, both through the quality of his own scientific work and his unparalleled organizational skills. It is a testament to the latter that the entire experimental (e,2e) apparatus, from Kaiserslautern, has been transferred to Brest and that it is already producing significant results in its new site.

Two Ehrhardt prizes are to be awarded each year to "two physicists, in the early stages of their research career, who have made the most significant contribution to the understanding of atomic collision processes and coincidence studies." As this was the first year of the prizes it was thought appropriate that Professor Ehrhardt nominate the winners. There is no restriction on nationality and the prizes can be awarded to either theoreticians or experimentalists. The winners were Steve Jones of the University of Missouri Rolla in the US and Jens Rasch of the University of Cambridge. The work for which they were honoured is described in the following two papers.

THREE-BODY MODELS OF ELECTRON-HYDROGEN IONIZATION

S. Jones and D. H. Madison

Physics Department
University of Missouri-Rolla
Rolla, MO 65401 USA

INTRODUCTION

Although the importance of electron-electron correlations in (e,2e) reactions has long been realized, untill recently only little theoretical progress had been made including these correlations. Theoretical studies using "three-body" models (models with explicit dependence on the electron-electron coordinate) are hampered by the difficult six-dimensional integral for the scattering amplitude. "Two-body" models, on the other hand, retain only electron-ion interactions and the standard techniques of partial-wave analysis greatly simplify the integral. In contrast, the wave functions used in three-body models give rise to two-center integrals, which are *not* efficiently evaluated using partial-wave methods. Therefore, a different method was sought for. We have found that direct six-dimensional numerical quadrature is efficient (about five minutes on a workstation to compute one scattering amplitude). As a result, we now have the capability of calculating (e,2e) triply-differential cross sections using three-body wave functions of arbitrary complexity. This allows for a severe test of three-body models. The utility of our method is demonstrated by using two recently reported analytical three-body wave functions in an (e,2e) calculation.

Atomic units are used throughout this work.

THEORY

Here we consider only asymmetric collisions where the contribution from electron exchange is small. The triply-differential cross section (TDCS) for ionization of a hydrogen atom neglecting exchange is given by

$$\frac{d^3\sigma}{d\Omega_a d\Omega_b dE_b} = \frac{k_a k_b}{k_i}|f|^2. \tag{1}$$

Here $\mathbf{k}_i$, $\mathbf{k}_a$, and $\mathbf{k}_b$ are wave vectors for the incident, scattered, and ejected electrons,

Coincidence Studies of Electron and Photon Impact Ionization
Edited by Whelan and Walters, Plenum Press, New York, 1997

respectively, and

$$f = -(2\pi)^{-5/2} \left\langle \Psi_f^-(\mathbf{r}_a, \mathbf{r}_b) \left| -\frac{1}{r_a} + \frac{1}{|\mathbf{r}_a - \mathbf{r}_b|} \right| e^{i\mathbf{k}_i \cdot \mathbf{r}_a} \psi_i(\mathbf{r}_b) \right\rangle \tag{2}$$

is the direct amplitude, with ψ_i the wave function for the ground-state hydrogen atom and Ψ_f^- the exact final-state wave function.

To obtain the exact wave function, Ψ_f^-, we would need to solve

$$(H - E)\Psi_f^- = 0, \tag{3}$$

where

$$H = -\frac{1}{2}\nabla^2_{\mathbf{r}_a} - \frac{1}{2}\nabla^2_{\mathbf{r}_b} - \frac{1}{r_a} - \frac{1}{r_b} + \frac{1}{|\mathbf{r}_a - \mathbf{r}_b|} \tag{4}$$

is the full Hamiltonian and E is the total energy. Since there is no known solution for the three-body Hamiltonian, one must approximate Ψ_f^-. In this work, we have used two recently reported analytical three-body wave functions as approximations for Ψ_f^-. In a first model we use a wave function due to Alt and Mukhamedzhanov,[1] and in a second model we use the wave function of Berakdar.[2] Our results from these two models are compared with the published results from the earlier three-body model of Brauner, Briggs, and Klar[3] and with experiment.[4] All three models are discussed in detail below.

3C Model

Brauner, Briggs, and Klar[3] performed calculations for electron-hydrogen ionization using the "3C" wave function,

$$\Psi_{f,3C}^- = e^{i(\mathbf{k}_a \cdot \mathbf{r}_a + \mathbf{k}_b \cdot \mathbf{r}_b)} C(-1/k_a, \mathbf{k}_a, \mathbf{r}_a) C(-1/k_b, \mathbf{k}_b, \mathbf{r}_b) C(\mu/k_{ab}, \mathbf{k}_{ab}, \mathbf{r}_{ab}). \tag{5}$$

Here $\mathbf{r}_{ab} = \mathbf{r}_a - \mathbf{r}_b$, $\mathbf{k}_{ab} = \mu[\mathbf{k}_a - \mathbf{k}_b]$ with $\mu = 1/2$ the reduced mass of two electrons, and

$$C(\alpha, \mathbf{k}, \mathbf{r}) = \Gamma(1 - i\alpha)e^{-\frac{\pi}{2}\alpha} F(i\alpha, 1; -i(kr + \mathbf{k} \cdot \mathbf{r})) \tag{6}$$

is a Coulombic distortion factor with Γ the gamma function and F the confluent hypergeometric function. The above wave function, $\Psi_{f,3C}^-$, is asymptotically correct in the limit that all interparticle separations tend to infinity, where it reduces to the well-known Redmond asymptotic form,

$$\tilde{\Psi}_f^- = e^{i(\mathbf{k}_a \cdot \mathbf{r}_a + \mathbf{k}_b \cdot \mathbf{r}_b)} e^{i\phi}, \tag{7}$$

where the phase ϕ is given by

$$\phi = \frac{1}{k_a} \ln(k_a r_a + \mathbf{k}_a \cdot \mathbf{r}_a) + \frac{1}{k_b} \ln(k_b r_b + \mathbf{k}_b \cdot \mathbf{r}_b) - \frac{\mu}{k_{ab}} \ln(k_{ab} r_{ab} + \mathbf{k}_{ab} \cdot \mathbf{r}_{ab}). \tag{8}$$

AM Model

Recently, Alt and Mukhamedzhanov,[1] hereafter referred to as AM, derived in closed form a wave function for three charged particles in the continuum that is asymptotically correct in *all* asymptotic domains; that is, a wave function valid provided at least one electron is far from the ion regardless of the separation between the two electrons. The regions of validity for the AM wave function *enclose* the scattering region (the closed and finite region that contributes to scattering), and thus the AM wave function represents the proper boundary condition for this three-body problem. The AM wave function is given by

$$\Psi^{-}_{f,\text{AM}} = e^{i(\mathbf{k}_a\cdot\mathbf{r}_a+\mathbf{k}_b\cdot\mathbf{r}_b)} C(-1/k'_a, \mathbf{k}'_a, \mathbf{r}_a) C(-1/k'_b, \mathbf{k}'_b, \mathbf{r}_b) C(\mu/k'_{ab}, \mathbf{k}'_{ab}, \mathbf{r}_{ab}). \tag{9}$$

The novel feature of this wave function is that the electrons do not have fixed wave vectors. Rather, in each Coulomb distortion factor, local wave vectors are required as follows.

$$\mathbf{k}'_a = \mathbf{k}_a + \mathbf{K}\left(\mu/k_{ba}, \mathbf{k}_{ba}, \mathbf{r}_b\right), \tag{10}$$

$$\mathbf{k}'_b = \mathbf{k}_b + \mathbf{K}\left(\mu/k_{ab}, \mathbf{k}_{ab}, \mathbf{r}_a\right), \tag{11}$$

$$\mathbf{k}'_{ab} = \mathbf{k}_{ab} + \mathbf{K}\left(-1/k_a, \mathbf{k}_a, \boldsymbol{\rho}\right) - \mathbf{K}\left(-1/k_b, \mathbf{k}_b, \boldsymbol{\rho}\right). \tag{12}$$

Here $\boldsymbol{\rho} = \mu(\mathbf{r}_a + \mathbf{r}_b)$ is the coordinate for the center of mass of the two electrons and

$$\mathbf{K}(\eta, \mathbf{k}, \mathbf{r}) = \frac{r}{R}\left[\frac{F(1+i\eta, 2; -i(kr+\mathbf{k}\cdot\mathbf{r}))}{-iF(i\eta, 1; -i(kr+\mathbf{k}\cdot\mathbf{r}))}\right](\hat{\mathbf{k}} + \hat{\mathbf{r}}), \tag{13}$$

with $R = r_a + r_b + r_{ab}$, is the local correction to the wave vector. Note that $\mathbf{K}$ is a *complex* vector function. Since only the real part of $\mathbf{K}$ is needed to satisfy the boundary condition, we have neglected the imaginary part of $\mathbf{K}$ in the present work.

DS3C Model

Berakdar[2] proposed the DS3C (dynamic screening 3C) wave function,

$$\Psi^{-}_{f,\text{DS3C}} = e^{i(\mathbf{k}_a\cdot\mathbf{r}_a+\mathbf{k}_b\cdot\mathbf{r}_b)} C(-z_a/k_a, \mathbf{k}_a, \mathbf{r}_a) C(-z_b/k_b, \mathbf{k}_b, \mathbf{r}_b) C(-z_{ab}\mu/k_{ab}, \mathbf{k}_{ab}, \mathbf{r}_{ab}), \tag{14}$$

with charges

$$z_a = 1 - Z(\mathbf{r}_a, \mathbf{r}_b), \tag{15}$$

$$z_b = 1 - Z(\mathbf{r}_b, \mathbf{r}_a), \tag{16}$$

$$z_{ab} = -1 + Z(\mathbf{r}_a, \mathbf{r}_b) r_{ab}/r_a + Z(\mathbf{r}_b, \mathbf{r}_a) r_{ab}/r_b, \tag{17}$$

as an approximate analytical solution of the Schrödinger equation. Here Z represents a local screening of the nuclear charge. It is readily seen that the above charges, regardless of the choice for Z, satisfy

$$-\frac{z_a}{r_a} - \frac{z_b}{r_b} - \frac{z_{ab}}{r_{ab}} = -\frac{1}{r_a} - \frac{1}{r_b} + \frac{1}{r_{ab}}, \tag{18}$$

and thus the Schrödinger equation is invariant under this transformation. Berakdar's ansatz for the charge Z,

$$Z(\mathbf{r}_a, \mathbf{r}_b) = \left[\frac{3+\cos^2(4\alpha)}{4}\right]^2 \frac{r_{ab} r_a^2}{(r_a + r_b)^3}, \tag{19}$$

with $\alpha = \cos^{-1}(r_b/\sqrt{r_a^2 + r_b^2})$, was made for the purpose of both satisfying boundary conditions as well as giving the proper behavior on the Wannier ridge ($\mathbf{r}_b = -\mathbf{r}_a$). As a result, this a wave function designed to incorporate some proper physics of the nonasymptotic scattering region.

We would note, however, that including the proper physics of the Wannier ridge comes at the expense of strictly satisfying the asymptotic boundary condition. This can be seen as follows. In the limit that all interparticle separations tend to infinity, $\Psi^-_{f,\mathrm{DS3C}}$, Eq. (14), has the form

$$\widetilde{\Psi}^-_{f,\mathrm{DS3C}} = e^{i(\mathbf{k}_a\cdot\mathbf{r}_a+\mathbf{k}_b\cdot\mathbf{r}_b)}e^{i\phi_{\mathrm{DS3C}}}. \tag{20}$$

For this to be an asymptotic solution when all interparticle separations tend to infinity,

$$\phi_{\mathrm{DS3C}} = \frac{z_a}{k_a}\ln(k_ar_a+\mathbf{k}_a\cdot\mathbf{r}_a)+\frac{z_b}{k_b}\ln(k_br_b+\mathbf{k}_b\cdot\mathbf{r}_b)+\frac{z_{ab}\mu}{k_{ab}}\ln(k_{ab}r_{ab}+\mathbf{k}_{ab}\cdot\mathbf{r}_{ab}) \tag{21}$$

must, to leading order, satisfy (see *e.g.*, Ref. 3)

$$L\phi-\frac{1}{r_a}-\frac{1}{r_b}+\frac{1}{r_{ab}}=0, \tag{22}$$

where the linear operator L is given by

$$L=\mathbf{k}_a\cdot\nabla_{\mathbf{r}_a}+\mathbf{k}_b\cdot\nabla_{\mathbf{r}_b}. \tag{23}$$

Substituting ϕ_{DS3C}, given by Eq. (21), for ϕ in Eq. (22), and imposing the condition (18), we obtain

$$L\phi_{\mathrm{DS3C}}-\frac{1}{r_a}-\frac{1}{r_b}+\frac{1}{r_{ab}}=\widetilde{W}^-_{f,\mathrm{DS3C}}, \tag{24}$$

where we have defined a perturbing energy

$$\widetilde{W}^-_{f,\mathrm{DS3C}}=\ln(k_ar_a+\mathbf{k}_a\cdot\mathbf{r}_a)L\frac{z_a}{k_a}+\ln(k_br_b+\mathbf{k}_b\cdot\mathbf{r}_b)L\frac{z_b}{k_b}+\ln(k_{ab}r_{ab}+\mathbf{k}_{ab}\cdot\mathbf{r}_{ab})L\frac{z_{ab}\mu}{k_{ab}}. \tag{25}$$

Clearly, the functional dependence of the charge Z, Eq. (19), is too strong to satisfy Eq. (22) to leading order. Indeed, Z, given by Eq. (19), introduces terms in the perturbing energy of order $O(\ln r_a/r_a, \ln r_b/r_b)$, which decrease more slowly than Coulomb potentials.

NUMERICAL METHOD

We have evaluated the scattering amplitude, Eq. (2), with Ψ^-_f approximated by $\Psi^-_{f,\mathrm{AM}}$, Eq. (9), or $\Psi^-_{f,\mathrm{DS3C}}$, Eq. (14), using six-dimensional numerical quadrature over $\mathbf{r}_a$ and $\mathbf{r}_b$. Although these final-state wave functions are quite complicated, our numerical method can be understood by considering just simple plane-waves for the final state:

$$\begin{aligned} f_0 &= \left\langle e^{i(\mathbf{k}_a\cdot\mathbf{r}_a+\mathbf{k}_b\cdot\mathbf{r}_b)}\left|-\frac{1}{r_a}+\frac{1}{|\mathbf{r}_a-\mathbf{r}_b|}\right|e^{i\mathbf{k}_i\cdot\mathbf{r}_a}\psi_i(\mathbf{r}_b)\right\rangle \\ &= \int \mathbf{dr}_a\, e^{i\mathbf{q}\cdot\mathbf{r}_a}F(\mathbf{r}_a), \end{aligned} \tag{26}$$

where

$$F(\mathbf{r}_a)=\int \mathbf{dr}_b\, e^{-i\mathbf{k}_b\cdot\mathbf{r}_b}\left(-\frac{1}{r_a}+\frac{1}{|\mathbf{r}_a-\mathbf{r}_b|}\right)\psi_i(\mathbf{r}_b) \tag{27}$$

is the result of a first integration over the coordinate of the atomic electron. The final integration over the coordinate of the projectile can be performed efficiently if the z axis is taken along $\mathbf{q}=\mathbf{k}_i-\mathbf{k}_a$ and cylindrical coordinates (z,ρ,ϕ) are used:

$$f_0=\int_0^{2\pi}d\phi\int_0^{\infty}d\rho\,\rho\int_{-\infty}^{\infty}dz\,e^{iqz}F(\mathbf{r}). \tag{28}$$

This choice reduces the computational effort by orders of magnitude compared to the choice of spherical coordinates with the z-axis along the beam direction, since the dominant feature of the $\mathbf{r}_a$ contribution to the scattering amplitude is $e^{i\mathbf{q}\cdot\mathbf{r}_a}$ and these three-dimensional oscillations are treated in one dimension with the above choice of coordinate system.

RESULTS

In Figure 1, we compare our results with the 3C results[3] and with the experimental data of Ehrhardt *et al.*[4] Results are shown for three energies (10 eV, 5 eV, 3 eV) for the ejected electron and three angles (4°, 10°, 16°) for the scattered electron. For the case where the ejected electron has an energy of 10 eV, the AM and DS3C models are

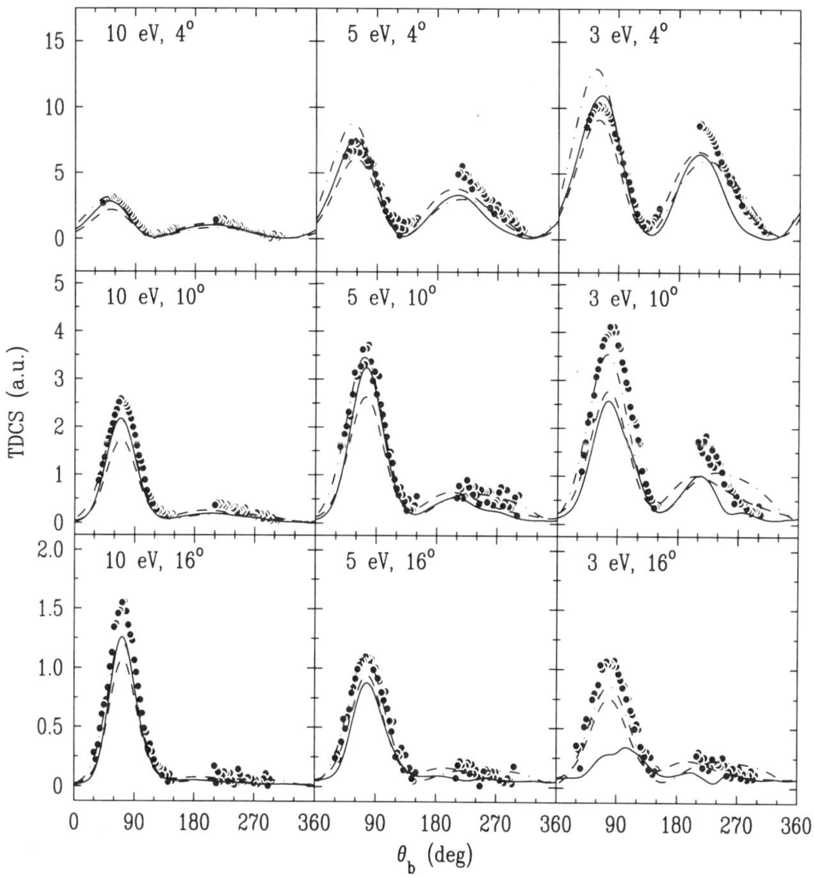

Figure 1. Triply-differential cross sections (TDCS) in the scattering plane for 150-eV electron-impact ionization of hydrogen vs. the angle θ_b of the ejected electron (measured clockwise from the forward beam direction), with the energy of the ejected electron and the angle of the scattered electron (measured counter-clockwise) fixed as indicated on the figure. Solid circles are the absolute (±15%) experimental data of Ehrhardt *et al.*[4] Theories: Solid line, AM; dot-dashed line, DS3C; dashed line, 3C.

in better agreement with experiment than the 3C model. For an ejected-electron energy of 5 eV, the DS3C is in excellent agreement with experiment for scattering angles of 10° and 16°, but does not do as well for the smallest momentum transfer case (4°). The AM model, on the other hand, does well for 4° and 10°, but not as well for the largest momentum transfer case (16°). This situation for the AM model becomes much worse for the lowest ejected-electron energy (3 eV), where the model fails completely for 16°. For hard collisons (*i.e.*, large scattering angles), a wave function that is accurate for small interparticle separations is needed, and therefore the above results indicate that the AM wave function is a poor approximation for small separations. In contrast, for an ejected-electron energy of 3 eV, the DS3C is again poor for the smallest momentum transfer case (4°). That the DS3C does poorly for 4° could conceivably be due to poor asymptotics, since for soft collisions (small scattering angles), significant contributions come from large electron-atom separations.

CONCLUSION

We have developed a computer program capable of calculating (e,2e) cross sections using three-body wave functions of arbitrary complexity. This allows for a severe test of three-body models. In particular, we found that the asymptotically correct wave function of Alt and Mukhamezhanov is a poor approximation for small interparticle separations. On the other hand, the fact the DS3C model does well for the larger momentum transfers (hard collisions) is a good indication that some proper physics for small separations has been accounted for. That the DS3C does not fare as well for the smallest momentum transfers (soft collisions) could be attributed to improper asymptotic behavior.

ACKNOWLEDGEMENTS

This worked was supported by the National Science Foundation.

REFERENCES

1. E.O. Alt and A.M. Mukhamedzhanov, Asymptotic solution of the Schrödinger equation for three charged particles, *Phys. Rev. A* 47:2004 (1993).
2. J. Berakdar, Approximate analytical solution of the quantum-mechanical three-body Coulomb contiuum problem, *Phys. Rev. A* 53:2314 (1996).
3. M. Brauner, J.S. Briggs, and H. Klar, Triply-differential cross sections for ionisation of hydrogen atoms by electrons and positrons, *J. Phys. B* 22:2265 (1989).
4. H. Ehrhardt, K. Jung, G. Knoth, and P. Schlemmer, Differential cross sections of direct single electron impact ionization, *Z. Phys. D* 1:3 (1986).

THE NORMALISATION OF THE EXPERIMENTAL TRIPLE DIFFERENTIAL CROSS SECTION OF NOBLE GAS ATOMS IN EXTREME ASYMMETRIC GEOMETRY

J. Rasch,[1] Colm T. Whelan,[1] R. J. Allan,[2] H. R. J. Walters[3]

[1]Department of Applied Mathematics and Theoretical Physics
University of Cambridge, Silver Street, Cambridge, CB3 9EW, England
[2]Daresbury Laboratory, Warrington, WA4 4AD, England
[3]Department of Applied Mathematics and Theoretical Physics
The Queen's University of Belfast, Belfast, BT7 1NN, England

INTRODUCTION AND EXPERIMENTAL METHOD

Electron-electron coincidence experiments, usually known as (e,2e) experiments, are the most powerful tool to study the process of the electron impact ionization[1,2]. Despite the extended activity in the field in the last years, the understanding of the process for targets heavier than H and He still presents many serious challenges for both, theoreticians and experimentalists alike. As far as the experiments are concerned, the challenge consists in (a) providing a large body of experimental data to explore the various zones of the Bethe surface[3], (b) determining the absolute scale of the triple differential cross-section (TDCS).

Recently (e,2e) measurements have been made on Argon 4p and Krypton 5p in very asymmetric geometry. In this paper we will consider how these might be put on an absolute scale. There are several ways of normalising the relative TDCS (for a review see Lahmam-Bennani et. al.[4]). A direct way to determine the absolute scale would be to use the relationship[4]

$$\frac{d^3\sigma}{d\Omega_1 d\Omega_2 dE} = \frac{I_c}{I_0} \frac{1}{nl(\epsilon_1 \Delta\Omega_1)(\epsilon_2 \Delta\Omega_2)\Delta E_c} \tag{1}$$

where I_c is the coincidence count rate, I_0 the incident electron rate, l the effective overlap length of the gas and electron beam, n the target density, $\Delta\Omega_1$ and $\Delta\Omega_2$ the acceptance solid angles of the analysers, ΔE_c the effective coincidence energy resolution and ϵ_1 and ϵ_2 the overall detection efficiencies of the analysers. Unfortunately, in the case of gas targets most of these quantities are rather difficult to determine or can only be measured in the experiment with a high relative error. One therefore has to resort to different techniques. One possibility would be to directly integrate the TDCS over the angular distribution and compare it with absolute double differential cross section (DDCS). However, this method also poses several difficulties (for a full discussion see Lahmam-Bennani et. al.[4]).

We will consider the normalisation procedure that was first used by the experimental group in Kaiserslautern[5] to put their Helium data taken at an impact energy of 600eV onto

Coincidence Studies of Electron and Photon Impact Ionization
Edited by Whelan and Walters, Plenum Press, New York, 1997

an absolute scale. It exploits the optical properties of the target and its cross sections for inelastic electron scattering and is one of the most widely used methods. In the next section we will outline the relation between the TDCS and its optical limit.

THE LIMITING PROCEDURE

The method of Jung et. al.[5] is based on work by Lassettre et. al.[11] and Inokuti[3] who found that in the limit of zero momentum transfer higher order effects tend to zero at any given energy so that a first Born treatment of the scattering problem becomes justified. In this case it is then possible to show that the double differential Generalised Oscillator Strength (GOS) is proportional to the Optical Oscillator Strength (OOS)

$$\int \left(\lim_{K \to 0} \frac{d\bar{f}(K)}{d\Omega_2 d\Omega_K dE} \right) d\Omega_K = \frac{d\bar{f}_{\text{opt}}}{dE} \tag{2}$$

where the double differential GOS and the OOS are defined as (see for example equation (4.111) in Bransden and Joachain[12])

$$\frac{d\bar{f}}{d\Omega_2 d\Omega_K dE} := 2\frac{E_2 - E_b}{K^2} \sum_{m_b} \left| \left\langle \chi_b^-(\boldsymbol{r}_2) \left| e^{i\boldsymbol{K}\boldsymbol{r}_2} \right| \phi(\boldsymbol{r}_2) \right\rangle \right|^2 \tag{3}$$

$$\frac{d\bar{f}_{\text{opt}}}{d\Omega_2 dE} := \frac{2}{3}(E_2 - E_b) \sum_{m_b} |\langle \chi_2(\boldsymbol{r}_2) \left| \boldsymbol{r}_2 \right| \phi \rangle|^2 \tag{4}$$

It should be noted however that in the case of Hydrogen and Helium we are ionising an s-state. This means that we do not have to deal with degenerate magnetic sublevels. In the Argon(4p) and Xenon(5p) experiment the gas target consisted of unpolarised atoms so that no distinction between different sublevels has been made. We therefore have to average over the magnetic sublevels m_b. Barred quantities will always denote such an average.

Making use of a 1st Born treatment the TDCS can be related to the GOS, equation (3), in the form

$$\frac{d^3\bar{\sigma}}{d\Omega_1 d\Omega_2 dE} = 2\frac{k_1 k_2}{k_0} \frac{1}{K^2} \frac{1}{(E_2 - E_b)} \frac{d\bar{f}}{d\Omega_2 d\Omega_K dE} \tag{5}$$

where $\boldsymbol{K} := \boldsymbol{k}_0 - \boldsymbol{k}_1$ denotes the vector of the momentum transfer. It can furthermore be shown that

$$\lim_{K \to 0} \frac{d\bar{f}}{d\Omega_2 d\Omega_K dE} = \frac{1}{8\pi^3 \alpha k_2} \frac{d\bar{\sigma}}{dE} \left(1 + \beta P_2(\hat{\boldsymbol{k}}_2 \hat{\boldsymbol{K}})\right) \tag{6}$$

so that we get for the limit of the TDCS

$$\lim_{K \to 0} K^2 \frac{d^3\bar{\sigma}}{d\Omega_1 d\Omega_2 dE} = \frac{1}{4\pi^3 \alpha} \frac{k_1}{k_0} \frac{1}{(E_2 - E_b)} \frac{d\bar{\sigma}}{dE} \left[1 + \beta P_2(\hat{\boldsymbol{k}}_2 \hat{\boldsymbol{K}})\right] \tag{7}$$

It should be noted that the value of β depends of course on the target subshell. For an s-state we get the well-known result $\beta = 2$. In the case of a p-state the result is more complicated[13].

The limit itself can physically not be reached, since conservation of energy implies

$$\frac{1}{2}k_0^2 = E_b + \frac{1}{2}k_1^2 + \frac{1}{2}k_2^2 \tag{8}$$

where E_b is the ionization energy of the target bound state and therefore $|\boldsymbol{k}_1 - \boldsymbol{k}_2| > 0$. In order to get this limit one has to vary the angle of the fast electron by measuring the yield of the slow electron parallel and antiparallel to the momentum transfer. One obtains two curves which have to join in the limit so that by extrapolating the two curves across the unphysical region we get the cross section for the point of zero momentum transfer. In the present work we have measured the relative TDCS at several ϑ_1's between 1° and 8°. The minimum momentum transfer K measured was 0.21 and 0.26 a.u. in Kr and Xe, respectively.

DISTORTED WAVE BORN APPROXIMATION (DWBA)

The TDCS for closed shell atoms is of the form[2,13]

$$\frac{d^3\sigma}{d\Omega_1 d\Omega_2 dE} = (2\pi)^4 \frac{k_1 k_2}{k_0} \left[|f - g|^2 + |g|^2 + |f|^2\right] \tag{9}$$

where

$$f = \left\langle \chi_a^-(\boldsymbol{k}_1, \boldsymbol{r}_1)\chi_b^-(\boldsymbol{k}_2, \boldsymbol{r}_2) \left| \frac{1}{|\boldsymbol{r}_1 - \boldsymbol{r}_2|} \right| \chi_i^+(\boldsymbol{k}_0, \boldsymbol{r}_1)\phi(\boldsymbol{r}_2) \right\rangle \tag{10}$$

$$g = \left\langle \chi_a^-(\boldsymbol{k}_1, \boldsymbol{r}_2)\chi_b^-(\boldsymbol{k}_2, \boldsymbol{r}_1) \left| \frac{1}{|\boldsymbol{r}_1 - \boldsymbol{r}_2|} \right| \chi_i^+(\boldsymbol{k}_0, \boldsymbol{r}_2)\phi(\boldsymbol{r}_1) \right\rangle \tag{11}$$

$\chi_i^+(\boldsymbol{k}_0, \boldsymbol{r}_1)$ denotes the distorted wave representing incoming electron calculated in the triplet local exchange potential of the atom whereas $\chi_a^-(\boldsymbol{k}_1, \boldsymbol{r}_1)$ and $\chi_b^-(\boldsymbol{k}_2, \boldsymbol{r}_2)$ are the distorted waves calculated in the triplet exchange potential of the atom for the fast electron (label 1) and ion for the slow outgoing electron (label 2).

The triplet local exchange potential is of the form[2,13]

$$V_{\text{loc}} = \frac{1}{2}\left[\left(\frac{1}{2}k_0^2 - V_{\text{static}}(\boldsymbol{r}_1)\right) - \sqrt{\left(\frac{1}{2}k_0^2 - V_{\text{static}}(\boldsymbol{r}_1)\right)^2 + 2\rho(\boldsymbol{r}_1)}\right] \tag{12}$$

The following two quantities enter the local exchange approximation

$$V_{\text{static}}(\boldsymbol{r}_1) := 2\sum_i \int \phi^i(\boldsymbol{r}_2)|\frac{1}{|\boldsymbol{r}_1 - \boldsymbol{r}_2|}\phi^i(\boldsymbol{r}_2)\, d^3r_2 + \int \varphi(\boldsymbol{r}_2)|\frac{1}{|\boldsymbol{r}_1 - \boldsymbol{r}_2|}\varphi(\boldsymbol{r}_2)\, d^3r_2 \tag{13}$$

$$\rho := 4\pi\left(\sum_i (\phi^i(\boldsymbol{r}_1))^*\phi^i(\boldsymbol{r}_1) \pm (\varphi(\boldsymbol{r}_1))^*\varphi(\boldsymbol{r}_1)\right) \tag{14}$$

and the sum is over all doubly occupied Slater-type orbitals ϕ^i. In the case of the slow outgoing electron equation 14 also contains the one singly occupied Slater-type orbital φ which is missing in the atom potential. Note that equation (12) reduces to the term $V_{exchange}$ as defined in Appendix D of Whelan et. al.[2] for Helium. Equation (12) is therefore a straightforward generalisation of the Helium case[13].

In the case considered here where E_2 the energy of the slow outgoing electron is comparable to the energy of the target bound state one should expect exchange to be important. In fact our calculations showed that the exchange potential is extremely important in this case[13]. It should however be noted that the exchange amplitude g which we include in our calculation should be negligible compared with the direct amplitude f. In fact our calculation showed that the exchange amplitude contributed on average not more than about 1-2% to the overall cross section.

In order to account for and test the significance of the electron-electron repulsion we multiplied the pure DWBA cross section equation (9) by an M_{ee} factor

$$\frac{d^3\sigma}{d\Omega_1 d\Omega_2 dE} = M_{ee}\frac{d^3\sigma^{\text{DWBA}}}{d\Omega_1 d\Omega_2 dE} \tag{15}$$

This factor was suggested by Ward and Macek[14] to model the repulsion effects of the two outgoing electrons (see also "An explanation of the structure observed in out-of-plane symmetric measurements on helium" in this volume). It retains the normalisation of the DWBA and exhibits the correct Wannier threshold law in the zero energy limit in contrast to the N_{ee} factor as used in Whelan et. al.[15] and Röder et. al.[16] which does not have this property but is very similar in shape. As expected the M_{ee} factor makes hardly any difference in this geometry (see Figures 1-4).

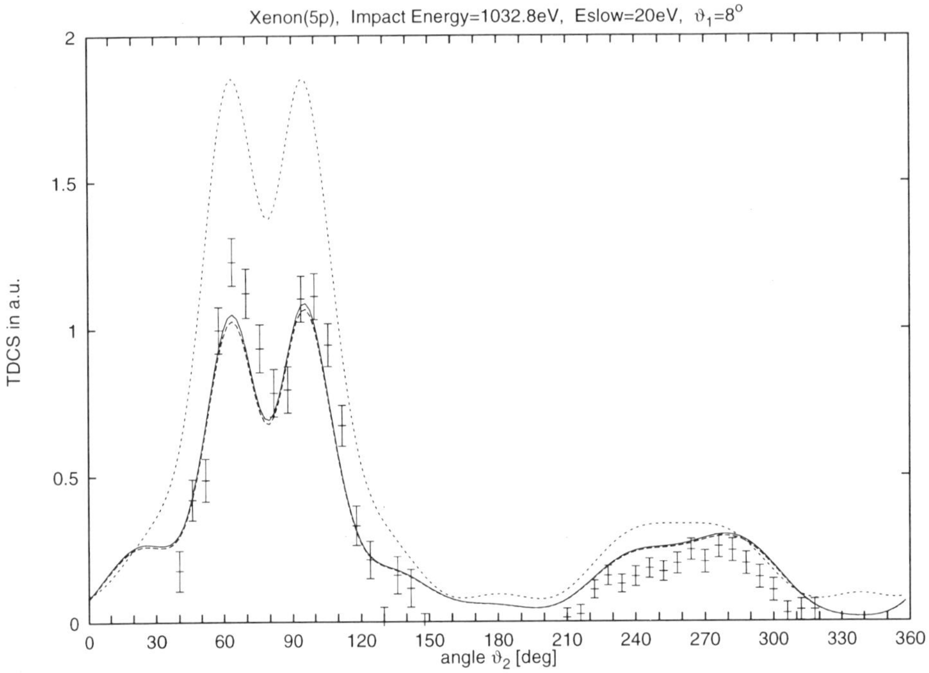

Figure 1. Xenon(5p) at an Impact Energy of 1000eV, energy of slow outgoing electron is 20eV, fast electron fixed at an angle of $\vartheta_1 = 8°$. Experimental data scaled to give the best fit to the theory. Shown are DWBA (solid line), DWBA multiplied with the M_{ee} factor (long dashed), 1st Born calculation (short dashed).

COMPARISON OF EXPERIMENT AND THEORY

In the Figures 1-4 we compare theoretical and experimental results. Shown are a 1st Born calculation, the DWBA and the DWBA multiplied with the M_{ee} factor as defined in equation (15). We have applied the standard normalisation method as given by equation (7) to put the experimental data onto an absolute scale. If we now compare experiment and theory we find that the overall shape agreement is good. Especially in the binary region and for larger scattering angles of the fast outgoing electron ($\vartheta_1 = 8°$ for Krypton and Xenon) we get very good shape agreement with experiment. This is also the region where the 1st Born does rather poorly. Only in the recoil direction we get some deviation.

However, we had to rescale the experimental data to get the best fit to the theory. In particular in Figure 1 and 2 the experimental data is down scaled by a factor of 3.5 whilst for Kr in Figure 3 and 4 it is down scaled by a factor of 6.6 and 4.3, respectively. Thus the question arises: what is going wrong? In the following we will elucidate this question by first concentrating on theoretical problems. We will then show that the normalisation works very well for Helium and then describe the problems that arise for Kr and Xe.

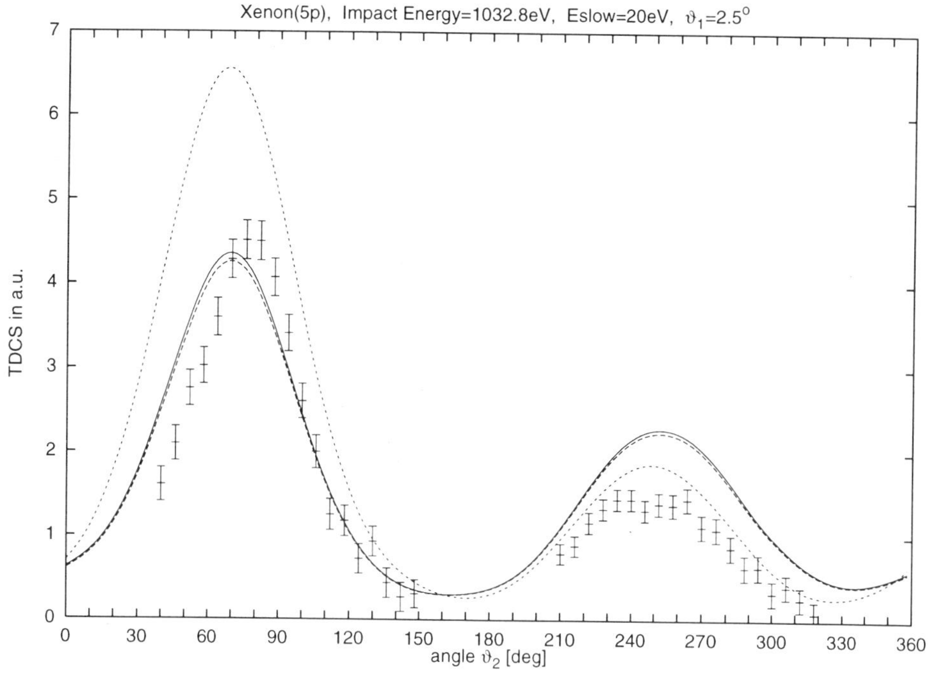

Figure 2. Xenon(5p) at an Impact Energy of 1000eV, energy of slow outgoing electron is 20eV, fast electron fixed at an angle of $\vartheta_1 = 2.5°$. Experimental data scaled to give the best fit to the theory. Shown are DWBA (solid line), DWBA multiplied with the M_{ee} factor (long dashed), 1st Born calculation (short dashed).

WHAT CAN GO WRONG THEORETICALLY?

On the theoretical side there are two main problems that can arise. In very asymmetric geometry the problem of getting the TDCS to convergence is a highly nontrivial problem. The mathematical problems encountered in the calculation and how they have been tackled is explained in great detail in 13. In the next section we will merely summarise the results.

CONVERGENCE OF THE DWBA CALCULATION

Whilst the very asymmetric geometry of the scattering event makes it possible to normalise the cross section it imposes a severe problem on the theory. Although higher order effects like post collisional interactions (e-e repulsion) and polarisation effects can be neglected we have to face a convergence problem introduced by the slowly vanishing overlap integrals. These overlap integrals between the fast incoming and fast outgoing electron tend to go to zero very slowly for high l-values. This can very easily be seen as follows. The radial Born integrals we have to solve are of the form

$$I_{\text{Born}}(l_1, l_2, \lambda, l_0, l_b) := \int_0^\infty \int_0^\infty u_{l_1}^a(k_1 r_1) u_{l_2}^b(k_2 r_2) \frac{r_<^\lambda}{r_>^{\lambda+1}} u_{l_0}^0(k_0 r_1) w_{l_b}(r_2) r_1^2 r_2^2 \, dr_1 \, dr_2 \quad (16)$$

For high l-states the fast incoming and outgoing electron no longer penetrates the atom and a Bethe approximation to the radial matrix element becomes justified. In this case the integral (16) reduces to

$$I(l_1, l_2, \lambda, l_0, l_b) = I_{0,R}(l_1, l_2, \lambda, l_0, l_b) + B_{0,\infty}(l_2, \lambda, l_b) J_{0,\infty}(l_1, \lambda, l_0) \quad (17)$$

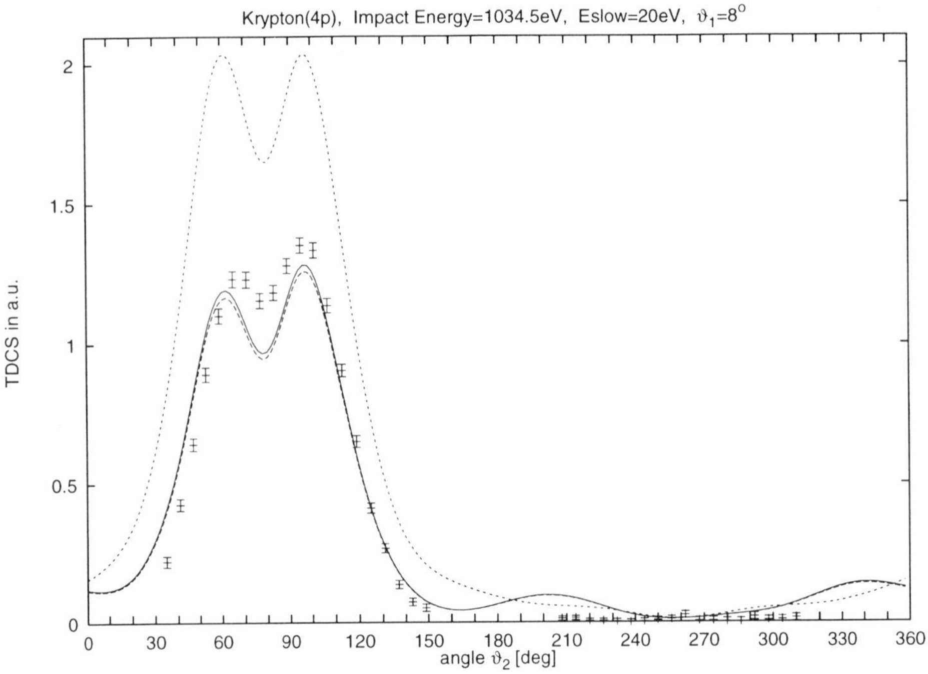

Figure 3. Krypton(4p) at an Impact Energy of 1000eV, energy of slow outgoing electron is 20eV, fast electron fixed at an angle of $\vartheta_1 = 8°$. Experimental data scaled to give the best fit to the theory. Shown are DWBA (solid line), DWBA multiplied with the M_{ee} factor (long dashed), 1st Born calculation (short dashed).

where we have defined

$$I_{0,R}(l_1, l_2, \lambda, l_0, l_b) := \int_0^R r_1^2 u_{l_1}^a(k_1 r_1) u_{l_0}^0(k_0 r_1)\, dr_1 \times \left[\frac{1}{r_1^{\lambda+1}} \int_0^{r_1} r_2^{\lambda+2} u_{l_2}^b(k_2 r_2) w_{l_b}(r_2)\, dr_2 + r_1^{\lambda} \int_{r_1}^{\infty} dr_2\, r_2^{-\lambda+1} u_{l_2}^b(k_2 r_2) w_{l_b}(r_2) \right] \tag{18}$$

$$J_{0,\infty}(k_1, l_1, k_0, l_0, \lambda) := \int_R^{\infty} r_1^{-\lambda+1} u_{l_1}^a(k_1 r_1) u_{l_0}^0(k_0 r_1)\, dr_1 \tag{19}$$

$$B_{0,\infty}(l_2, \lambda, l_b) := \int_0^{\infty} r_2^{\lambda+2} u_{l_2}^b(k_2 r_2) w_{l_b}(r_2)\, dr_2 \tag{20}$$

The parameter R is the outer limit to which the solution of the differential equation will be integrated. The paper 17 deals extensively with the evaluation of these integrals so that we merely summarise the results here. Both electrons only see an atom potential and are asymptotically essentially free. This means that the distorted waves $u_{l_1}^a(k_1 r_1)$ and $u_{l_0}^0(k_0 r_1)$ can be approximated by Bessel functions resembling this free behaviour for high l-values. The integral over these two Bessel functions can now be solved analytically due to the work by 18. The analytic solution is a finite sum over associated Legendre functions of the second kind with argument $\chi := (k_0^2 + k_1^2)/(2k_0 k_1)$. Because k_0 and k_1 are nearly equal in this geometry χ goes to 1 and we are approaching the pole of the Legendre function. This means that more and more l-states are contributing to the scattering event (in the unphysical limit χ would be equal to 1 and an infinite number of these l-states would contribute).

The worst case for evaluating these integrals would be if we replaced all distorted waves by plane waves. Getting the TDCS to convergence for this situation not only tests the accuracy

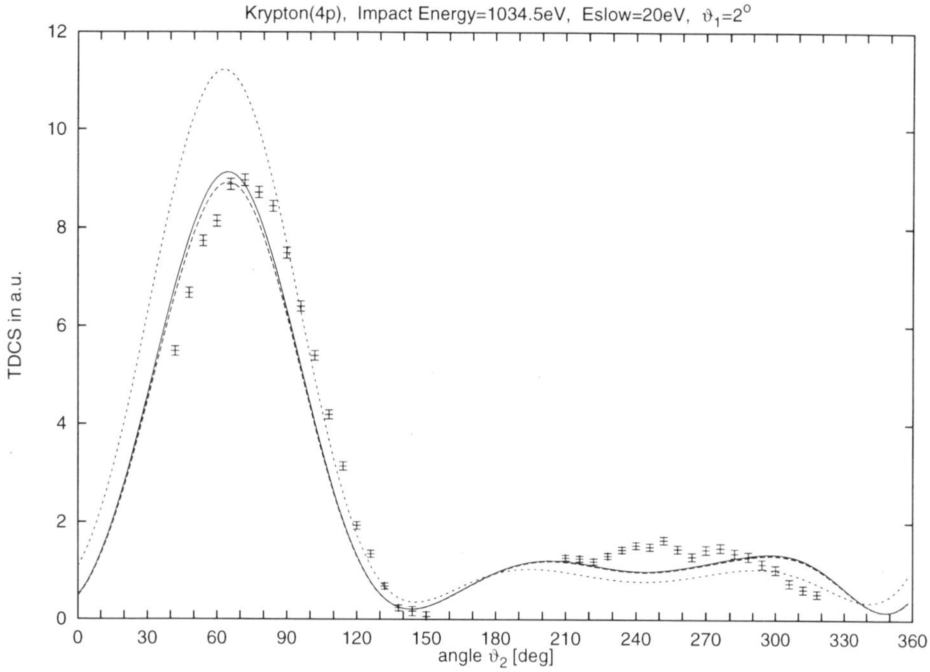

Figure 4. Krypton(4p) at an Impact Energy of 1000eV, energy of slow outgoing electron is 20eV, fast electron fixed at an angle of $\vartheta_1 = 2°$. Experimental data scaled to give the best fit to the theory. Shown are DWBA (solid line), DWBA multiplied with the M_{ee} factor (long dashed), 1st Born calculation (short dashed).

of the integration routine but also the overall accuracy of the angular momentum algebra and the differential equation solver. As described in 17 we were able to reproduce the analytic result available for Hydrogen for the case of $E_0 = 1000eV$, $E_1 = 985.4eV$, $E_2 = 1eV$. Note that these energies are even more extreme than the ones we have to consider here. Up to $l = 1000$ l-states where needed for the convergence. In the case considered here we achieved convergence for little more than $l = 400$. In Figure 5 we give a comparison for different converged results for Xenon(5p). Although merely $l = 10$ l-states were needed for the slow outgoing electron (we used $l = 15$ to be safe) we do not get a converged result for $l = 80$ for the fast incoming and outgoing electron both calculated in the triplet potential of the ion. At about $l = 80$ the numerically calculated phase shifts are zero within the numerical accuracy so that the Bethe approximation becomes fully justified. Within the Bethe approximation we have retained the monopole, dipole and quadrupole partial wave integral. Higher order integrals can be neglected. However we were surprised to see that especially for the Hydrogen case even the quadrupole term gave a significant contribution to the cross section 17. At about $l = 200$ for the fast incoming and outgoing electron the cross section is almost indistinguishable from a calculation using $l = 400$ l-states.

PHOTOIONIZATION CROSS SECTIONS

By relating the TDCS to photoionization cross sections we have to make sure that theoretical photoionization cross sections using Hartree-Fock based wave functions reproduce experimental measurements. This not only checks the program but also tests the quality of the wave function used. We remark that they were generated using a variational principle in order to minimise the energy of the atomic system[19], it is therefore not obvious how these wave functions can reproduce other properties of the system. This is especially important

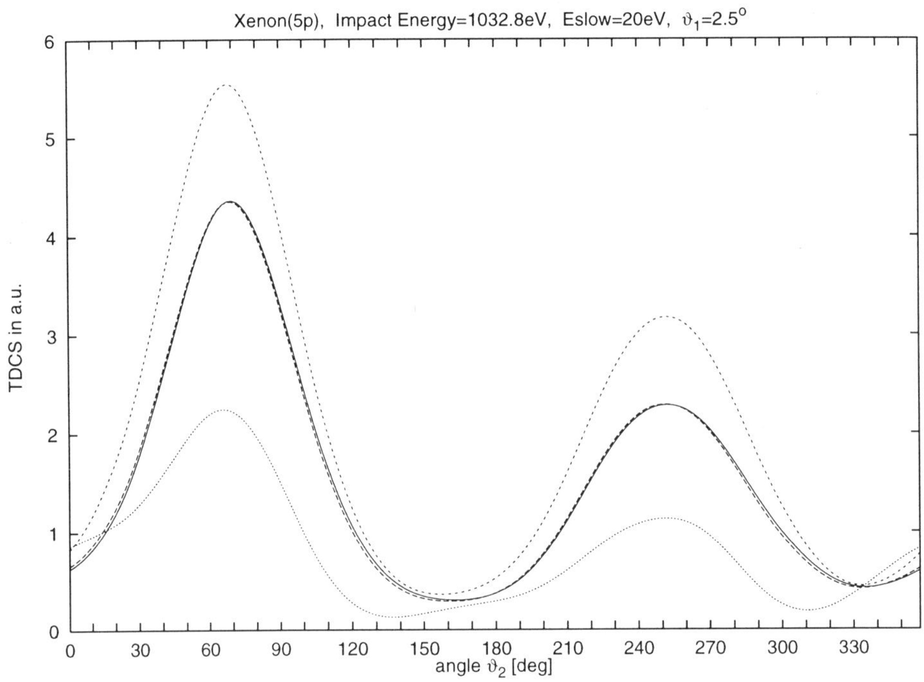

Figure 5. Xenon 5p at an Impact Energy of 1032.8eV, energy of slow outgoing electron is 20eV, fast electron fixed at an angle of $\vartheta = 2.5°$. Comparison of the different stages of the convergence is made. Shown is the DWBA using $l = 400$ l-states for fast incoming and outgoing electron (solid curve), using $l = 200$ (long dashed), using $l = 80$ (short dashed) and using only $l = 40$ (dotted). Number of l-states for the slow outgoing electron: $l = 15$.

in our case where the energy of the slow electron is very close to the ionization threshold and significant deviations from the photo cross sections can be expected. However, since the Hartree-Fock wave functions are convoluted with the distorted wave of the slow outgoing electron we can only indirectly test the bound state wave functions; we therefore test the quality of the distorted waves at the same time. The derivation of the photo cross sections for our Hartree-Fock wave functions can be found elsewhere[13].

We have compared the experimentally found values for β and the photo ionization cross section $\sigma := \frac{d\sigma}{dE}$ with the theoretical ones. We find very good agreement with the experimental β values for Krypton and Xenon. However there is a very significant difference for the σ values in the case of Krypton. Whilst the theoretical value of σ for the calculation for Xenon is to within 11% of the experiment we get more than a factor of 2 for Krypton. The experimental values of σ and β at the required energy loss were derived from the partial photo ionization cross sections[20,21] and the photo ionization angular measurements[22,23]. The overall uncertainty on the absolute scale is about 28%, due to the uncertainty in the σ of 17% and 9% and in the β of 5% and 15% for Kr(4p) and Xe(5p), respectively.

From equation (7) it follows directly that the TDCS measured in the direction $\boldsymbol{k}_2 \parallel \boldsymbol{K}$ or $\boldsymbol{k}_2 \parallel -\boldsymbol{K}$ multiplied by K^2 tends to the same value

$$\sigma_{\text{lim}} := \lim_{K \to 0} K^2 \frac{d^3\bar{\sigma}}{d\Omega_1 d\Omega_2 dE} = \frac{1}{4\pi^3\alpha} \frac{k_1}{k_0} \frac{1}{(E_2 - E_b)} \frac{d\bar{\sigma}}{dE} [1 + \beta] \tag{21}$$

when K vanishes. Therefore the value of this limit is known from the kinematics of experiment and photo ionization data. In Table 1 we compare the value of this limit for experimental

Table 1. Comparison of the values for the limit determined experimentally and from the DWBA calculation using different forms of the local exchange potential for the slow outgoing electron.

	Kr(4p)	Xe(5p)
$\sigma_{\rm lim}$ exp.	1.01	0.387
$\sigma_{\rm lim}$ DWBA	0.371	0.348

and theoretical data. We find that the limit for Xenon is to within 25% of the experiment whilst we get a factor of 2 difference for Krypton. This discrepancy is clearly coming from the difference in the photoionization cross sections.

WHAT ELSE CAN GO WRONG?

Although there can be a large discrepancy between theory and experiment for the photo ionization cross section as encountered in the case of Krypton where theory and experiment disagree by a factor of 2, it does, however, not explain why the DWBA is a factor of 4-6 smaller for Krypton. This is particularly the case for Xenon where the disagreement between the photo ionization cross sections is a mere 11% and not a factor 3-4. In the following sections we will first discuss the normalisation of Helium before we discuss it for Krypton and Xenon.

THE NORMALISATION OF HELIUM

We have recalculated some of the Helium data taken at an impact energy of 600eV[5]. In Figure 7 we show a comparison between the absolute experimental data and our DWBA calculation. Figure 7 also shows a 1st Born calculation. Very good agreement is found. It is important to note that the difference between the DWBA and the 1st Born calculation is very small.

THE FITTING OF THE HELIUM DATA We have used a standard General Linear Least Squares fitting routine which can be found in the literature[24]. It has the advantage that one can not only choose the order of the polynomials but also by "freezing" the parameters of the fitting polynomial one can choose whether only even or odd or a combination of both should be used for the fitted polynomial. Klar et. al.[25] studied the structure of the form factor

$$\bar{\epsilon} := \sum_{m_b} \left| \left\langle \chi_b^-(\boldsymbol{r}_2) \left| e^{i\boldsymbol{K}\boldsymbol{r}_2} \right| \phi(\boldsymbol{r}_2) \right\rangle \right|^2 \qquad (22)$$

in equation (3) in more detail to give supplementary limitations to the extrapolation. In particular in the optical limit the expansion of the optical oscillator strength should only contain even powers of K.

Even without this limitation we found that the fitting of the theoretical curves especially the 1st Born curves were very stable and accurate indeed. In Figure 6 we show the 1st Born and DWBA calculation and an extrapolation to the zero momentum transfer. As can be seen the 1st Born and DWBA results are very close together over a wide range of K values. Especially close to the unphysical region the two theories are almost indistinguishable so that Lassettre's theorem which depends crucially on the applicability of the 1st Born approximation can indeed be used. We were able to reproduce the limit as given in equation (21) from the extrapolation of the 1st Born points to 4 digits (this limit is indicated by a short horizontal bar in Figure 6). The extrapolation of the DWBA was slightly less accurate and stable. We could however reproduce the limit for most polynomials to better than 5%.

Due to the experimental setup there was a rather large gap between the region of K which is accessible to experiment and $K = 0$ the extrapolation point. In order to obtain reliable results Jung et. al.[5] chose a narrow spacing of ΔK and performed the measurements with high statistical accuracy. They remark that the error of the extrapolation procedure is smaller than 10%.

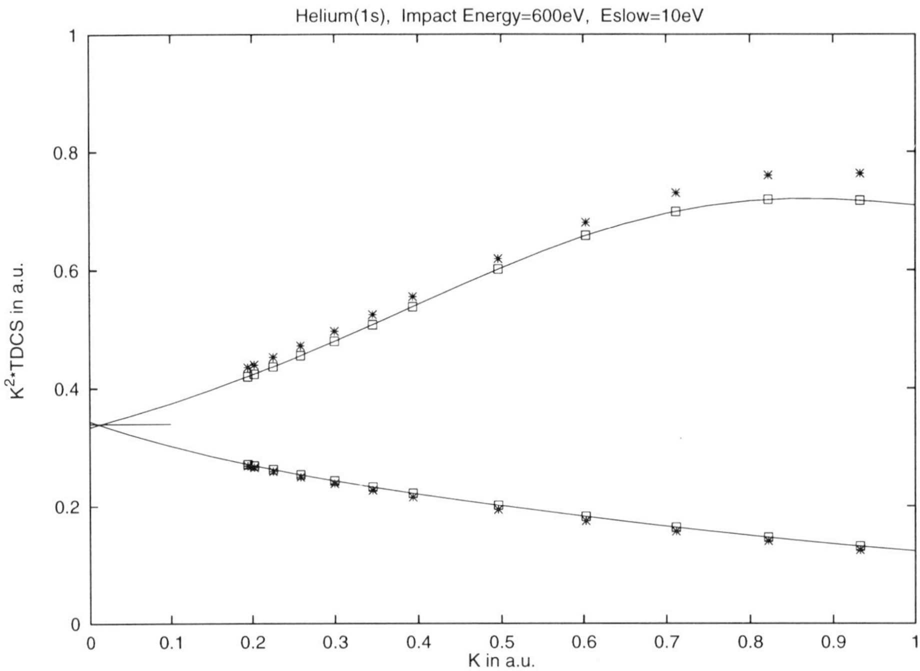

Figure 6. TDCS in coplanar asymmetric geometry at an impact energy of 600eV. Energy of slow outgoing electron is 10eV. Shown is standard DWBA (boxes) and 1st Born calculation (stars). Shown is also a numerical extrapolation through the DWBA calculation to the $\boldsymbol{K} = 0$ point. The horizontal bar denotes the analytic limit of the 1st Born.

THE NORMALISATION OF KRYPTON AND XENON

We have seen that the DWBA is very well capable of reproducing the absolute cross sections in the case of Helium. Furthermore the limiting process is justified as the DWBA has gone over to the first Born which validates Lassettre's theorem. In the following sections we will discuss the fitting procedure for the experimental and theoretical curves for Krypton and Xenon.

THE FITTING OF THE THEORETICAL DATA

We have performed a similar fitting for the theoretical curves to determine the limit. In Figure 8 we compare the limiting procedure for a 1st Born and the DWBA calculation with the experimental data for Xenon. We have scaled the experimental data to give the best agreement with the recoil curve of the DWBA calculation. We found again similar to the Helium case that the fitting procedure for the first Born curve is very stable and reproduces the limiting point as given in equation (21) well. However, the DWBA disagrees with the 1st Born over a wide range and converges to the 1st Born only very closely to the unphysical region if not inside. It is therefore probably not too surprising that the extrapolation of the DWBA does not converge to the same point at $K = 0$. One can only conclude from a theoretical point of view that the distortion effects are still too strong and a 1st Born treatment as required by Lassettre's theorem is not justified.

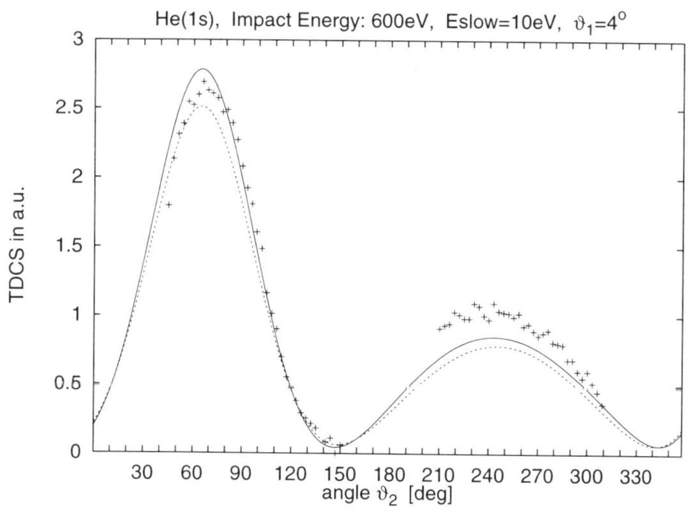

Figure 7. TDCS in coplanar asymmetric geometry at an impact energy of 600eV. Shown is the DWBA (solid curve) and 1st Born calculation (short dashed). Energy of slow outgoing electron is 10eV with angle of fast electron fixed at $\vartheta_1 = 10°$. Experimental data is on an absolute scale and given in a.u.

CAN THE DIFFICULTIES BE OVERCOME?

If we compare the Figures 6 with 8 we see that the most striking difference lies in the fact that there is almost no difference of the DWBA and the 1st Born calculation in the Helium case whereas for Krypton and Xenon both calculations are very different over a wide range and only start to merge very close to the unphysical region. Although any theory has to converge to the 1st Born eventually this can happen very deep inside the unphysical region. This is indeed the case for the DWBA. Even calculating the DWBA right up to the boundary of the unphysical region it is difficult to find a stable fit which correctly extrapolates to the

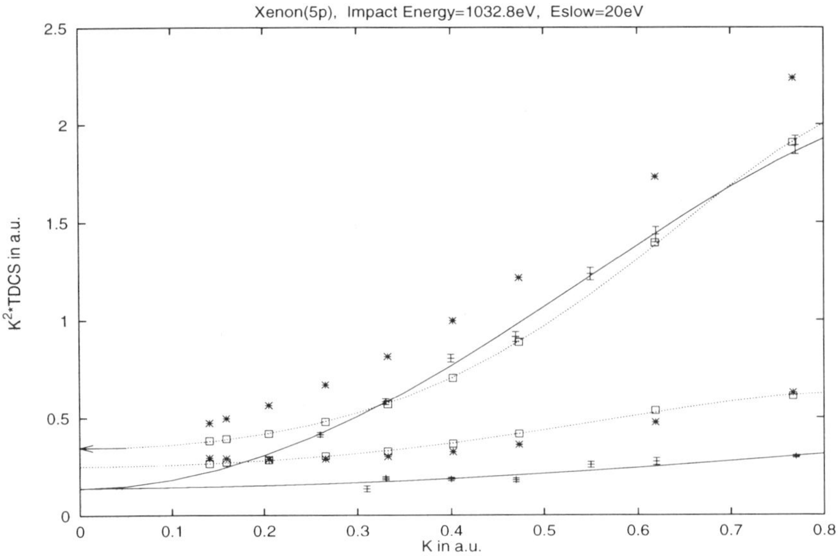

Figure 8. The limiting process of Xenon(5p) at an impact energy of 1032.8eV, energy of slow outgoing electron is 20eV. Shown is TDCS multiplied with the square of the momentum transfer $K^2 = |\boldsymbol{k}_0 - \boldsymbol{k}_1|^2$ against K. We compare the 1st Born (star sign) with the DWBA (box sign) and the experimental data which was scaled to give the best fit to the binary peak. Shown are furthermore a polynomial fit to the experiment (solid curve) and to the DWBA calculation (dotted). The DWBA uses the triplet potential for the slow outgoing electron. The horizontal bar at 0.3475 denotes the analytic limit of the 1st Born.

1st Born limit as can be seen in Figure 8.

Another important point related to the previous finding is that the curves bend upwards very strongly and unfortunately do so where no experimental data is given. Furthermore the fitting of the theoretical curves clearly indicates that the curves continue to bend upwards in the unphysical region, particularly the recoil curve. There is some indication in the fitting of the experimental data that such non-linear behaviour would be seen in any extrapolation of this data into the unphysical region. However, due to the instability of the fitting procedure for the experimental data the extrapolation is not very reliable so that one has to ask: to which extend do the experimental curves bend upwards?

It should be clear that a huge error can be introduced depending on how far the measurements would bend upwards in this region. We have rescaled the experimental data in Figures 8 to give the best agreement with the binary curve of the DWBA calculation since this is the region where we get the best agreement with experiment (see also Figures 1-4). The difference between the experimentally obtained limit and the 1st Born limit given by equation (21) disagree by more than a factor of 2.5. Unfortunately, the case of Krypton is more ambiguous. We can however conclude also for Krypton from a theoretical point of view that the DWBA converges to the 1st Born only very close to the unphysical region and that both curves tend to bend upwards rather strongly. Some indication of this is also found in the experimental data.

RESULTS

The overall shape agreement between experiment and theory is good. Especially in the binary region and for larger scattering angles of the fast outgoing electron ($\vartheta_1 = 8^\circ$ for

Krypton and Xenon) we get very good shape agreement with experiment. This is also the region where the 1st Born does rather poorly. Only in the recoil direction we get some deviation. This good shape agreement together with the very good absolute agreement found in the case of Helium can be seen as an additional indication that the DWBA is the appropriate theory to model these experiments and that it might give the right absolute value of the cross section.

Unfortunately due to the the instability coming from the scarcity of the experimental data points it is not possible to decide which effect is the dominant one: The instability of the fitting or the unjustified 1st Born treatment of the measurement. It would therefore be highly desirable to put these measurements onto an absolute scale by different means. This would shed light onto the question as to whether the DWBA reproduces correct absolute values particularly in the case of Xenon, but also as to how much the experimental data bends upwards in the unphysical region. This would in turn decide whether a stable fitting of the experimental data (more points) would allow for a normalisation or whether the distortion effects are so strong in the experimentally inaccessible and unphysical region so that Lassettre's theorem cannot be applied and any extrapolation in this energy regime is doomed to fail.

ACKNOWLEDGEMENTS

We are extremely grateful to Dr. Avaldi and Prof. Stefani for allowing us access to their experimental data and very many useful discussions. We gratefully acknowledge financial support for this work by the British Council and DAAD (ARC), NATO through grant CRG 950407 and the European Union (HCM CHRX-CT93-0350). J.R. holds a Junior Research Fellowship at Wolfson College Cambridge. We are grateful to Prof. J. H. Macek for some very helpful comments.

REFERENCES

1. G. Stefani, L. Avaldi, and R. Camilloni. *J. Physique*, C:6–1, 1993.
2. Colm T. Whelan, R. J. Allan, H. R. J. Walters, and X. Zhang. (e,2e), effective charges, distorted waves and all that! In Colm T. Whelan, H. R. J. Walters, A. Lahmam-Bennani, and H. Ehrhardt, editors, *(e,2e) and related processes*, page 1. Kluwer Academic Publisher, 1993.
3. Mitio Inokuti. inelastic collisions of fast charged particles with atoms and molecules - the bethe theory revisited. *Rev. Mod. Phys.*, 43:297, 1971.
4. A. Lahmam-Bennani, M. Cherid, and A. Duguet. A critical evaluation of the various methods for absolute scale determination in (e,2e) experiments. *J. Phys. B*, 20:2531, 1987.
5. K. Jung, R. Müller-Fiedler, P. Schlemmer, H. Ehrhardt, and H. Klar. Absolute triple differential cross sections of electron impact ionisation of helium at 600ev collision energy. *J. Phys. B*, 18:2955, 1985.
6. E. Fainelli, R. Camilloni, G. Petrocelli, and G. Stefani. *Nuovo Cimento D*, 9:33, 1987.
7. L. Avaldi, R. Camilloni, R. Multari, G. Stefani, X. Zhang, H. R. J. Walters, and Colm T. Whelan. *Phys. Rev. A*, 48:1195, 1993.
8. G. Stefani, R. Camilloni, and A. Giardini-Guidoni. *Phys. Lett.*, 64A:364, 1978.
9. A. Lahmam-Bennani, H. F. Wellestein, A. Duguet, and M. Lecas. *Rev. Sci. Instrum.*, 56:43, 1985.
10. Y.-K. Kim. *Phys. Rev. A*, 28:656, 1983.
11. N. Lassettre, Edwin, Ausma Skerbele, and Michael A. Dillon. Generalized Oscillator Strength for $1^1S \rightarrow 2^1P$ transition of helium. Theory of limiting oscillator strengths. *J. Chem. Phys.*, 50:1829, 1969.
12. B.H. Bransden and C. J. Joachain. *Physics of atoms and molecules*. Longman, London and New York, 1983.
13. J. Rasch. *(e,2e) processes with neutral atom targets*. PhD thesis, University of Cambridge, 1996.
14. S.J. Ward and J.H. Macek. Wave functions for contiuum states of charged fragments. *Phys. Rev. A*, 49, 1994.

15. Colm T. Whelan, R. J. Allan, J. Rasch, H. R. J. Walters, X. Zhang, J. Röder, K. Jung, and H. Ehrhardt. Coulomb three-body effects in (e,2e) collisions: The ionization of H in coplanar symmetric geometry. *Phys. Rev. A*, 50:4394, 1994.
16. J. Röder, J. Rasch, K. Jung, Colm T. Whelan, H. Ehrhardt, R. J. Allan, and H. R. J. Walters. Coulomb three-body effects in low energy impact ionisation of H(1s). *Phys. Rev. A*, 53:225, 1996.
17. J. Rasch and Colm T. Whelan. On the numerical evaluation of a class of integrals occuring in scattering problems. *Comp. Phys. Commun., in press*, 1996.
18. Colm T. Whelan. On the Bethe approximation to the reactance matrix. *J. Phys. B*, 26:2343, 1985.
19. Clementi and Roetti. Roothan-hartree-fock atomic wavefunctions. *Atomic Data and Nuclear Data Tables*, 14:177, 1974.
20. Anders Fahlman, Manfred O. Krause, Thomas A. Carlson, and Agneta Svensson. Xe 5s,5p correlation satellites in the region of strong interchannel interactions, 28-75 eV. *Phys. Rev. A*, 30:812, 1984.
21. S. Aksela, H. Aksela, M. Levasalmi, K. H. Tan, and G. M. Bancroft. partial photoionization cross sections of kr 3d, 4s, and 4p levels in the photon energy range 37-160 ev. *Phys. Rev. A*, 36:3449, 1987.
22. Manfred O. Krause, Thomas A. Carlson, and Pamela R. Woodruff. Angular distribution of photoelectrons of Xe 5p spin-orbit components between 20 and 105 eV. *Phys. Rev. A*, 24:1374, 1981.
23. S. H. Southworth, P. H. Kobrin, C. M. Truesdale, D. Lindle, S. Owaki, and D. A. Shirley. Photoelectron and Auger electron asymmetries: alignment of Xe+2 D5/2 by phototionization. *Phys. Rev. A*, 24:2257, 1981.
24. William H. Press, Saul A. Teukolsky, William T. Vetterling, and Brian P. Flannery. *Numerical Recipes in Fortran*. Cambridge University Press, 1992.
25. H. Klar, K. Jung, and H. Ehrhardt. Electron-impact ionization of helium by fast electrons at small momentum transfer: A quantum-defect analysis of experimental data. *Phys. Rev. A*, 29:405, 1984.

INDEX

Series Publications

Below is a chronological listing of all the published volumes in the *Physics of Atoms and Molecules* series.

ELECTRON AND PHOTON INTERACTIONS WITH ATOMS
Edited by H. Kleinpoppen and M. R. C. McDowell

ATOM–MOLECULE COLLISION THEORY: A Guide for the Experimentalist
Edited by Richard B. Bernstein

COHERENCE AND CORRELATION IN ATOMIC COLLISIONS
Edited by H. Kleinpoppen and J. F. Williams

VARIATIONAL METHODS IN ELECTRON–ATOM SCATTERING THEORY
R. K. Nesbet

DENSITY MATRIX THEORY AND APPLICATIONS
Karl Blum

INNER-SHELL AND X-RAY PHYSICS OF ATOMS AND SOLIDS
Edited by Derek J. Fabian, Hans Kleinpoppen, and Lewis M. Watson

INTRODUCTION TO THE THEORY OF LASER–ATOM INTERACTIONS
Marvin H. Mittleman

ATOMS IN ASTROPHYSICS
Edited by P. G. Burke, W. B. Eissner, D. G. Hummer, and I. C. Percival

ELECTRON–ATOM AND ELECTRON–MOLECULE COLLISIONS
Edited by Juergen Hinze

ELECTRON–MOLECULE COLLISIONS
Edited by Isao Shimamura and Kazuo Takayanagi

ISOTOPE SHIFTS IN ATOMIC SPECTRA
W. H. King

AUTOIONIZATION: Recent Developments and Applications
Edited by Aaron Temkin

ATOMIC INNER-SHELL PHYSICS
Edited by Bernd Crasemann

COLLISIONS OF ELECTRONS WITH ATOMS AND MOLECULES
G. F. Drukarev

THEORY OF MULTIPHOTON PROCESSES
Farhad H. M. Faisal

PROGRESS IN ATOMIC SPECTROSCOPY, Parts A, B, C, and D
Edited by W. Hanle, H. Kleinpoppen, and H. J. Beyer

RECENT STUDIES IN ATOMIC AND MOLECULAR PROCESSES
Edited by Arthur E. Kingston

QUANTUM MECHANICS VERSUS LOCAL REALISM: The Einstein-Podolsky-Rosen Paradox
Edited by Franco Selleri

ZERO-RANGE POTENTIALS AND THEIR APPLICATIONS IN ATOMIC PHYSICS
Yu. N. Demkov and V. N. Ostrovskii

COHERENCE IN ATOMIC COLLISION PHYSICS
Edited by H. J. Beyer, K. Blum, and R. Hippler

ELECTRON–MOLECULE SCATTERING AND PHOTOIONIZATION
Edited by P. G. Burke and J. B. West

ATOMIC SPECTRA AND COLLISIONS IN EXTERNAL FIELDS
Edited by K. T. Taylor, M. H. Nayfeh, and C. W. Clark

ATOMIC PHOTOEFFECT
M. Ya. Amusia

MOLECULAR PROCESSES IN SPACE
Edited by Tsutomu Watanabe, Isao Shimamura, Mikio Shimizu, and Yukikazu Itikawa

THE HANLE EFFECT AND LEVEL CROSSING SPECTROSCOPY
Edited by Giovanni Moruzzi and Franco Strumia

ATOMS AND LIGHT: INTERACTIONS
John N. Dodd

POLARIZATION BREMSSTRAHLUNG
Edited by V. N. Tsytovich and I. M. Ojringel

INTRODUCTION TO THE THEORY OF LASER–ATOM INTERACTIONS (Second Edition)
Marvin H. Mittleman

ELECTRON COLLISIONS WITH MOLECULES, CLUSTERS, AND SURFACES
Edited by H. Ehrhardt and L. A. Morgan

THEORY OF ELECTRON–ATOM COLLISIONS, Part 1: Potential Scattering
Philip G. Burke and Charles J. Joachain

POLARIZED ELECTRON/POLARIZED PHOTON PHYSICS
Edited by H. Kleinpoppen and W. R. Newell

INTRODUCTION TO THE THEORY OF X-RAY AND ELECTRONIC SPECTRA OF FREE ATOMS
Romas Karazija

VUV AND SOFT X-RAY PHOTOIONIZATION
Edited by Uwe Becker and David A. Shirley

DENSITY MATRIX THEORY AND APPLICATIONS (Second Edition)
Karl Blum

SELECTED TOPICS ON ELECTRON PHYSICS
Edited by D. Murray Campbell and Hans Kleinpoppen

PHOTON AND ELECTRON COLLISIONS WITH ATOMS AND MOLECULES
Edited by Philip G. Burke and Charles J. Joachain

COINCIDENCE STUDIES OF ELECTRON AND PHOTON IMPACT IONIZATION
Edited by Colm T. Whelan and H. R. J. Walters